Digital Design Fundamentals

Digital Design Fundamentals

KENNETH J. BREEDING

The Ohio State University

Prentice Hall, Englewood Cliffs, New Jersey 07632

Library of Congress Cataloging-in-Publication Data

Breeding, Kenneth J. (Kenneth James), (date)
    Digital design fundamentals.

    Bibliography: p.
    Includes index.
    1. Digital electronics.   2. Logic circuits.
3. Switching circuits.   I. Title.
TK7868.D5B72   1989   621.3815   88-9813
ISBN   0-13-212721-0

© 1989 by Prentice-Hall, Inc.
A Division of Simon & Schuster
Englewood Cliffs, New Jersey 07632

Printed in the United States of America

10   9   8   7   6   5   4   3

ISBN   0-13-212721-0

Prentice-Hall International (UK) Limited, *London*
Prentice-Hall of Australia Pty. Limited, *Sydney*
Prentice-Hall Canada Inc., *Toronto*
Prentice-Hall Hispanoamericana, S.A., *Mexico*
Prentice-Hall of India Private Limited, *New Delhi*
Prentice-Hall of Japan, Inc., *Tokyo*
Simon & Schuster Asia Pte. Ltd., *Singapore*
Editora Prentice-Hall do Brasil, Ltda., *Rio de Janeiro*

Contents

6   *ASYNCHRONOUS SEQUENTIAL CIRCUITS*   *154*

7   *PULSE-MODE OR MULTIPLY CLOCKED SEQUENTIAL CIRCUITS*   *206*

Preface

This book has been written to satisfy the need for a fundamentals text covering the complete range of foundation material in switching theory and digital design and using a mixed-logic symbology throughout. The book is intended as an introductory text dealing with the fundamental concepts used in designing general digital systems. Although many of the examples used come from computer applications of digital design, this is not intended to be a computer design text. The book is aimed at sophomore engineering students, typically those in electrical or computer engineering, who have the usual background in mathematics and physics. It is also assumed that the student has been introduced to the concept of a computer and programming, perhaps via a knowledge of some high-level programming language.

The reader is not expected to have ever programmed a computer at the assembler level and thus is not assumed to be familiar with the concepts of registers, memory addressing, coding, status, or interrupts. The principal purpose of this text, then, is to serve as an entry-level book for the study of general digital system design as it might apply in many areas, including communications, controls, and computers.

A major feature of this book is the use of a mixed-logic symbology in the design process throughout. Such usage is very common among practicing digital design engineers. This symbology can also be found in numerous catalogs of integrated circuits. The reason for the ever-increasing usage of a mixed-logic symbology is that, properly used, such symbols quickly and clearly convey the function of a design as well as the designer's intent. Because of this, it is important that the student understand the use of mixed logic, both in order to interpret the designs of others and to be able to communicate his or her own designs and ideas.

Although specific integrated circuits are not discussed (the 7474 and 74LS76 flip-flops are exceptions to this), the elements used throughout—the gates, flip-flops, and so on—are those that are readily available in TTL or CMOS technologies. Thus, with a standard small- to medium-scale integrated circuit catalog at hand, the reader can find real-world counterparts to the elements used. The ambitious reader, therefore, should be able to construct the circuits designed in the text using an IC catalog and parts obtained from a neighborhood electronic parts store.

The text is organized in such a way that Chapters 1 through 5 constitute the usual introductory material in digital design: number systems, Boolean and switching algebras, logic design, and clocked sequential circuits. This material is adequate for use in a one-quarter introductory digital design course. The remaining four chapters deal with some advanced topics in both combinational and sequential circuit design, as well as an introduction to large-scale system design. Thus, there is adequate material to make up a one-semester or a two-quarter course dealing with digital systems. Upon completion of this material the reader should be able to easily go on to courses dealing with computer architecture and design, systems programming, microprocessor system design and programming, and many application areas that require digital systems as part of the implementing hardware.

As just indicated, the book is organized into nine chapters. Each chapter, apart from the first, ends with an annotated bibliography which describes a number of books that cover material discussed in the chapter. The bibliography is followed, in all cases, by a fairly comprehensive set of exercise problems that illustrate the concepts presented. In many instances problems are given which introduce new and unusual concepts, as well. A brief description of each of the chapters follows.

Chapter 1 introduces the scope of the book and defines the concepts of a digital system and digital information.

Number systems, especially the binary system, are introduced in Chapter 2. This chapter also discusses binary arithmetic and introduces the idea of a signed 2's complement representation, showing how it is applied to the addition and subtraction process. Various commonly encountered codes are also discussed. The idea of a "table look-up" is introduced here to indicate how one might convert between these various codes. Finally, a computational method is shown for the conversion of BCD to binary and binary to BCD. This algorithm is not usually found in introductory texts.

Chapter 3 introduces the Huntington postulates that define a general Boolean algebra. These postulates are used to prove, algebraically, a few useful theorems. However, algebraic manipulation is not belabored. A two-element Boolean algebra, the switching algebra, is then introduced, and all subsequent results are proved via a process of complete enumeration. The various forms for switching expressions are examined. The next topics taken up are Karnaugh maps and the Quine-McCluskey algorithm for the simplification of switching expressions. In conjunction with the tabular (Quine-McCluskey) approach, cyclic covering tables and the Petrick algorithm are discussed. Finally, the Quine-McCluskey algorithm is extended to the problem of simplifying multiple functions. This latter material, too, is something not generally found in introductory texts.

The mixed-logic symbology, to be used throughout the remainder of the book, is defined in Chapter 4. Many design examples are carried out here to illustrate not only the design process, but the use of the symbology as well. This chapter also introduces the reader to the use of ROMs and PLAs for implementing complex switching functions. Some comments are also given concerning other symbology standards, one of which, IEEE Std. 91-1984, is further discussed in the Appendix.

The concept of a sequential circuit is introduced in Chapter 5. Various types of flip-flops are introduced here. Although transparent latches and master-slave flip-flops are discussed, edge-triggered flip-flops, such as the 7474 and 74LS76, are used throughout the remainder of the chapter and the book. Following the introduction of these devices, counters are discussed. Then there is a section on the analysis and design of more general clocked sequential circuits. This chapter also includes a discussion of simplification procedures for state tables and gives some rules of thumb concerning the assignment of state variables to the various states so as to reduce the complexity of the implementing equations. Finally, the conversion between Mealy and Moore models is discussed. This conversion process is not normally found in texts at this level, although the process is very straightforward. The timing differences between these models are also discussed.

Chapter 6 introduces the concept of asynchronous or fundamental-mode sequential circuits. Races and hazards are introduced and illustrated with various examples. An analysis procedure and an accompanying synthesis method are discussed. Procedures are described for avoiding races, and the essential hazard is introduced. Many analysis and design examples are given, including an extensive analysis of the 7474 edge-triggered D flip-flop. In order to illustrate some of the design concepts developed here, the design of the 7474 is then re-created. This examination of the 7474 highlights many of the difficulties associated with fundamental-mode circuits and is also not a topic found in the general texts in this area.

There are many problems in digital systems in which many unsynchronized signals are present. Chapter 7 deals with such multiply clocked, or pulse-mode, sequential circuits. The emphasis in this chapter is on showing how such circuits can always be designed in a reliable manner. This is done by using a unique double-ranked organization for the flip-flops used in the feedback paths. It is shown how such circuits can be designed by controlling either the flip-flop clock inputs or the flip-flop asynchronous set and clear inputs. Finally, a practical illustration of these procedures is given in the design of a single-step circuit for an Intel 8085 microprocessor.

Chapter 8 deals with special topics in combinational circuit design and analysis. The topics selected tend to be those that relate to current interests in VLSI design, artificial intelligence, and so on, and consist of bilateral networks, threshold logic, symmetric functions, functional decomposition, and iterative networks. In conjunction with functional decomposition, a discussion of the use of multiplexers for the implementation of general switching functions is given.

The last chapter, Chapter 9, introduces the concepts associated with the design of large-scale systems. The idea of a register is introduced, along with a register transfer notation which is useful for describing complex operations within large systems. These ideas are then used to show how large systems can be designed in a modular fashion by breaking them up into smaller, simpler subsystems. The use of flow charts for describing a given algorithm and their equivalence to state diagrams is also discussed in some detail.

Finally, a couple of examples are given. The first is a piece of hardware that can be used for carrying out a multiplication. The second, a bicycle speedometer, illustrates some of the problems associated with having incoming signals which are not synchronized with the system clock and shows how such signals can be synchronized to avoid any difficulties that might otherwise arise.

The Appendix introduces a current symbology standard, IEEE Std. 91-1984. This standard is very useful for describing the function of medium-scale devices such as counters and shift registers. The primitives of this symbology standard—the ones most commonly encountered—are the ones discussed. The use of this standard is illustrated with a large number of examples. Most of these examples show the IEEE standard symbol that would be used to represent many of the more complex circuits designed or discussed in earlier chapters.

It would be impossible to thank all of those people who made this book happen. However, I would most sincerely like to thank the many students who suffered through the use of earlier versions and in particular those students who found mistakes in the manuscript and brought them to my attention. I would especially like to thank my department chairman, H. C. Ko, without whose support I could never have finished this text. I would also like to thank Ms. Jackie Buckner for her efforts in reproducing and binding the manuscript each quarter it was used in the classroom and Ms. Jenny Curry who made sure that the students got copies. A very special thanks is in order for my very good friend and colleague, Professor Robert E. Fenton, who used the manuscript in his classes and made many valuable comments. Last, and perhaps most important, I would like to thank my wife, Julie, and kids, Dori and Greg, for their patience and their general support for this project.

Introduction to
Digital Systems

1

1.1 WHAT IS A DIGITAL SYSTEM?

Simply put, a digital system is a system that processes discrete information. The discrete entities making up this information may represent anything from simple arithmetic integers, letters of the alphabet, or other abstract symbols to values for a voltage, a pressure, or any other physical quantity. To a digital system, what these entities represent is not important in the processing of the information. What they represent is important, however, to the human observer who must interpret the results of the process. A digital system, then, is one that accepts as input digital information representing numbers, symbols, or physical quantities, processes this input information in some specific manner, and produces a digital output.

In a large number of computer applications, the computer is required to process information related to physical quantities, such as pressure or temperature. Since nature is not digital, however, (unless, of course, one wants to go to the quantum-mechanical level), the physical quantity of time or temperature or whatever must, somehow, be converted to a digital form before it can be processed by the computer. The usual way of doing this is to first take the physical quantity to be processed and convert it into a voltage or a current.[1] This is done by using a *transducer*—a device that converts energy coming into it in one form to energy in another form at its output. A thermocouple is a good example of a temperature transducer: it produces an output voltage proportional to its ambient temperature. This output voltage becomes an *analog* of the temperature of the device.

---

[1] At least, this is what might be expected of an electrical engineer. A mechanical engineer, on the other hand, might prefer to convert the physical quantity into a position of a lever or a gear.

We use analogs of physical quantities all the time. For example, the position of the mercury in a thermometer is an analog of the temperature, and the angular position of the hands of a clock is the analog of the time. The analog of a physical quantity is, like the quantity itself, usually a continuous variable. Since a computer operates only on discrete entities, which usually can be associated with numbers, the continuous variable representing the physical quantity must first be converted to a digital form. This conversion is carried out by an *analog-to-digital converter (ADC)*.[2] The digital output from the ADC, then, is a discrete approximation to the actual value of the continuous physical variable. The computer or other digital system can now process the information for whatever purpose is required.

Let us take a look at a typical digital system where these ideas are put together to perform a simple task. Suppose we have to maintain a given constant temperature in a liquid, such as the developer used in a photographic processing lab. To do this we must measure the temperature of the developer and then use the result of our measurement to turn on or turn off a heating element that surrounds the developer. To perform this task, a thermocouple might be used as the transducer that converts the temperature of the liquid to an analog voltage. This voltage would then be converted to a digital value of sufficient precision to ensure the accurate control of the temperature. The resulting digital value would then be used by some digital system, such as a microprocessor, to determine whether the heating element should be on or off. This digital system is also an example of a *feedback* control system, in which the result of an action taken by the system, in this case turning the heater element on or off, is "fed back" in order to determine whether a new and different action should be taken.

1.2 WHY ARE DIGITAL SYSTEMS SO PERVASIVE?

We might logically ask in the above example why we should use a digital system for this simple control function. After all, mechanical thermostats, which perform the given task, are readily available and inexpensive. To answer this question we need to look closer at why more and more of the everyday products that we encounter are becoming digital. There are three fundamental reasons why this is happening:

1. Flexibility
2. Reliability
3. Cost

Consider, for example, the temperature control system described above. It is obvious that the system described could easily be replaced by a mechanical thermostat. However, suppose we wish, at some later time, to add some features to the system, such as, for example, the ability to automatically change the temperature of the developer at different stages of

---

[2] A device which carries out the reverse process, converting a digital quantity back to an analog value, is called a *digital-to-analog-converter (DAC)*.

the development process. Such a *programmable* thermostat is easily achieved using digital systems. In fact, if we were to use a microprocessor as part of such a temperature control system, we could control not only the temperature of the developer but the entire film development process. This idea is precisely why ''same day'' film processing services are so readily available. Clearly, it would be difficult to obtain this degree of flexibility in any other way with the ease with which we can accomplish it using a digital system.

To get some idea of how reliable digital systems can be, we need only look at the way in which information is represented in these systems. A digital system processes information in a discrete form which is normally binary. The two values of a binary digit, or *bit*, are 1 and 0. These values are commonly represented in a digital system by two different voltages. In fact, the 1 is usually represented by a range of voltages and the 0 by another, nonoverlapping range of voltages. In one implementing technology, the TTL (transistor-transistor logic) technology, a 1 is represented by voltages in the range of 2 to 5 volts (V) and a 0 is represented by voltages in the range of 0 to 1 Volt. Because these values are represented by a range of voltage, any minor change in voltage level due to noise or other external events will not cause a 0 to be misinterpreted as a 1 or vice versa. As we shall see in the next chapter, arbitrary numbers and symbols can be represented by strings of 1s and 0s. It is possible to design these digital representations so that even if noise is so large as to change the voltage corresponding to a 1 to the range for a 0, for example, the original representation can be re-created. A good example of this is a compact disk (CD), in which digital information, representing sound, is encoded in such a way that a 1-mm hole could be drilled through the disk without the loss of a single note![3] Obviously, this degree of reliability makes digital systems extremely attractive for any application requiring highly reliable operation, and especially for applications where a human life depends on the outcome of this operation.

Digital systems from their very inception have been flexible and reliable. Their more recent use in every day items, such as watches, calculators, and household appliances, has come about because of their very low cost. The cost of digital devices has dropped dramatically over the past 30 years. This is illustrated by the cost of some of the 7400 series small-scale digital integrated circuits, which in the early 1960s was around $70 apiece. These devices, which are still available and extensively used, can be purchased for less than 15 cents today. A similar reduction in cost can be seen in one of the first microprocessors, the Intel 8080. This device, which appeared in production around 1972, originally cost about $300. Its price at one point in recent years dropped to around $2 or less. The cost of computer memory has followed similar trends. In the 1950s, memory costs were generally figured in the dollars per bit range, whereas today the cost is more likely to be in milli-cents per bit. These dramatic cost reductions have come about because of advances in integrated circuit technology, specifically, the ability to put hundreds of thousands of transistors on a piece of silicon roughly 6 mm (one-quarter of an inch) on a side. Clearly the trend is for increasingly complex functions to be integrated in silicon at increasingly reduced prices.

---

[3] An excellent discussion of this error-correction ability can be found in the article entitled ''The Digital Reproduction of Sound,'' by John Monforte, which appeared in the December 1984 issue of *Scientific American*.

1.3 ORGANIZATION OF THE BOOK

The purpose of this book, then, is to introduce the student to the basic concepts required to design a digital system. For this purpose the book is organized into nine chapters, each dealing with a subject either essential or just very helpful to the design of digital systems. A number of examples are given throughout the text in order to illustrate the various concepts. Each chapter ends with an annotated bibliography giving sources for further information on topics discussed within the chapter and a set of exercise problems which further illustrate these principles. Chapters 2 through 5 cover the essential material required for the design of any digital system, whether it be a computer or a simple controller, such as the temperature controller described in this chapter. Chapters 6 and 7 describe concepts which can make large-scale systems easier to design and more efficient in implementation. Chapter 8 discusses a number of special issues that are becoming, for one reason or another, such as VLSI (very large-scale integration) design and artificial intelligence, of increasing interest. Finally, Chapter 9, which does not heavily depend on the material in Chapters 6, 7, and 8, describes in some detail how these ideas can be put together to construct a large-scale digital system—for example, in the design of a computer or a controller for an industrial process. A very brief description of the subjects covered in each of these chapters follows.

Chapter 2 discusses number representations and methods of information coding. This chapter also discusses binary arithmetic in some detail.

Chapter 3 defines and details the algebra required for digital system design—Boolean algebra and its subset switching algebra.

Chapter 4 introduces the fundamental building block of digital system design, the logic gate. A symbology standard that helps to clarify the operation of circuits designed using these gates is also discussed. Together with the switching algebra presented in Chapter 3, this chapter serves as an introduction to combinational circuit design. Combinational circuits are those in which the output is a function only of the circuit inputs at any given instant of time.

Chapter 5 introduces a class of circuits called *sequential circuits*, in which circuit outputs are fed back to the input. This causes the output to become a function of not only the current input but also some past sequence of inputs. This chapter also introduces the flip-flop circuit element and shows how this device can be used in the sequential circuit to control the time at which the outputs change. Since this time of change is controlled by a single system clock, circuits of this type are generally referred to as synchronous or clocked sequential circuits. This is the class of sequential circuits that is generally used to control the operations within a computer.

In Chapter 6, sequential circuits that are not controlled by a clock are investigated. Since no clock is present in such a system to synchronize the circuit outputs, such circuits are referred to as asynchronous sequential circuits, or sequential circuits operating in the fundamental mode. Flip-flops themselves are analyzed and designed in this chapter, along with many other very useful fundamental-mode devices.

Chapter 7 deals with sequential circuits in which more than one clock signal is present. We will refer to such systems as multiply clocked sequential circuits. This chapter also briefly discusses a particular subclass called pulse-mode circuits, in which the input clock signals are considered to be very short pulses.

In Chapter 8 a number of special topics are introduced that are important in various application areas of digital systems; for example, VLSI design and artificial intelligence.

Finally, Chapter 9 applies the ideas developed in preceding chapters to the design of large-scale digital systems. This chapter gives a model for such systems and presents methods for organizing their design.

Number Systems

2

2.1 INTRODUCTION

It may be obvious that a digital computer operates only on numbers. The way in which the machine operates on these numbers, however, is a function of what the numbers represent (do they represent themselves, other numbers, or alphanumeric characters?) and in what form they are represented. Clearly, the design of the *central processing unit*, the portion of the computer that handles all arithmetic and logic operations, cannot be carried out without a complete knowledge of the form in which the numbers are represented in the machine. Furthermore, this form is generally quite different from the way numbers must be represented to the human operator, and so there has to be some type of conversion in the computer input/output system.

The purpose of this chapter, then, is to discuss the various ways in which numbers and other quantities are represented and manipulated in a computer. In addition, various forms of data encoding, as well as binary arithmetic, will be examined.

2.2 BASE CONVERSION

The number system we most often use is the decimal system. For various reasons, which will be examined later, the decimal number system is not a convenient one for a computer to use. Computers work most efficiently on information that is binary. Since computers are not good with decimal numbers and people are generally not very proficient with the use of binary numbers, some type of conversion between these sytems must occur at the interface between people and computers. In this section we will examine the various issues involved in the conversion.

2.2.1 Radix r to Decimal Conversion

A positional notation has long been used for writing numbers. In such a representation the position of each digit indicates the weight associated with the digit.[1] In particular, the number 276.5 would be interpreted as

$$2 \times 10^2 + 7 \times 10^1 + 6 \times 10^0 + 5 \times 10^{-1} = 276.5 \qquad (2.2.1)$$

The various powers of 10 used in this representation, which are the respective weights, are indicative of the assumption that the number 276.5 was written as a decimal number, or a number written in *base 10*. The base of a number system is also referred to as the *radix* of the system.

In general, the radix of a system can be anything; 5 or 12 or -3, or even an irrational number, such as π or e. Usually, however, the radix of number systems is taken as a positive integer. When a number is written in a base other than 10, the radix used must somehow be noted so that the number can be properly interpreted. Usually this is indicated by placing the number in parentheses and attaching a subscript at the end to indicate the base. Thus $(1321)_4$ indicates that the number 1321 has a radix of 4 and would be interpreted as follows:

$$(1321)_4 = 1 \times 4^3 + 3 \times 4^2 + 2 \times 4^1 + 1 \times 4^0 \qquad (2.2.2)$$

Note that if the arithmetic in Equation (2.2.2) is carried out in the decimal system, the number $(1321)_4$ must be the same as the number 121 in base 10!

In general, a number of radix r, A_r, can be written as

$$A_r = (a_n a_{n-1} \ldots a_0 . a_{-1} \ldots a_{-m})_r = \sum_{i=-m}^{n} a_i r^i \qquad (2.2.3)$$

where the a_i are digits in the radix r system and where the point (.) is termed the *radix point*, which, as is customary, separates the integral and fractional parts of the number. Carrying out the arithmetic of Equation (2.2.3) in the decimal number system results in the decimal equivalent of A_r. For example, consider the problem of finding the decimal number equivalent to $(364.213)_7$. The value is found by using the notation of Equation (2.2.3) as follows:

$$(364.213)_7 = 3 \times 7^2 + 6 \times 7^1 + 4 \times 7^0 + 2 \times 7^{-1} + 1 \times 7^{-2} + 3 \times 7^{-3}$$

$$= (193.314868 \ldots)_{10}$$

where the trailing points indicate that additional fractional digits occur.

---

[1] The Roman numeral system is an example of a system which uses a nonweighted notation for representing numbers.

2.2.2 Decimal to Radix r Conversions

Conversion from some radix r to decimal is quite straightforward, as just indicated. The question that naturally arises next is how to convert from decimal to an arbitrary radix equivalent. To see how this process may be carried out, let B_{10} be a given decimal number which is to be converted to a number A_r radix r. That is,

$$B_{10} = A_r = (a_n a_{n-1} \ldots a_0)_r \tag{2.2.4}$$

or, expanding A_r,

$$B_{10} = a_n r^n + a_{n-1} r^{n-1} + \cdots + a_1 r^1 + a_0 \tag{2.2.5}$$

Now, if B_{10} is divided by r, Equation (2.2.5) becomes

$$\frac{B_{10}}{r} = (a_n r^{n-1} + \cdots + a_2 r + a_1) + \frac{a_0}{r} \tag{2.2.6}$$

$$= \text{Int}\left(\frac{B_{10}}{r}\right) + \text{Frac}\left(\frac{B_{10}}{r}\right)$$

where Int and Frac indicate the integral and fractional parts of B_{10}/r. From Equation (2.2.6), we see that

$$a_0 = \text{Rem}\left(\frac{B_{10}}{r}\right) \tag{2.2.7}$$

where Rem means the remainder of B_{10}/r. If this process is now repeated starting with Int (B_{10}/r), the next remainder will be a_1 and the next integral part will be $a_n r^{n-2} + a_{n-1} r^{n-3} + \cdots + a_2$. Continuing this process until no integral part remains will produce the digits of A_r.

Consider as an example the problem of finding the base 3 equivalent of $(278)_{10}$. The work may be carried out as follows:

Quotient	Remainder
3)278	
3)92	$2 = a_0$
3)30	$2 = a_1$
3)10	$0 = a_2$
3)3	$1 = a_3$
3)1	$0 = a_4$
0	$1 = a_5$
Stop	

Thus

$$(278)_{10} = (101022)_3.$$

As a check, convert $(101022)_3$ back to decimal:

$$(101022)_3 = 1 \times 3^5 + 1 \times 3^3 + 2 \times 3 + 2 = (278)_{10}$$

Numbers, in general, have fractional parts as well as integral parts. Conversion of these fractional parts to an equivalent radix r representation may be carried out in a manner similar to the conversion of the integral parts. Let B_{10} now represent a fractional decimal number equivalent to a fractional number A_r in radix r. Thus

$$B_{10} = A_r = (0.a_{-1}a_{-2} \cdots a_{-m})_r \qquad (2.2.8)$$
$$= a_{-1}r^{-1} + a_{-2}r^{-2} + \cdots + a_{-m}r^{-m}$$

Multiplying the result of Equation (2.2.8) by r yields

$$rB_{10} = a_{-1} + (a_{-2}r^{-1} + \cdots + a_{-m}r^{-m+1}) \qquad (2.2.9)$$

from which the integral part becomes a_{-1}. The fractional part, $(a_{-2}r^{-1} + \cdots + a_{-m}r^{-m+1})$, when multiplied by r yields a_{-2}, and so on. Thus repeated multiplication by r yields the successive digits of the radix r representation of the fractional number B_{10}.

As an example, consider the conversion of $(0.27)_{10} = (?)_4$. The process goes as follows:

Integer	Fraction
	.27
	$\times 4$
$a_{-1} = 1$	.08
	$\times 4$
$a_{-2} = 0$	.32
	$\times 4$
$a_{-3} = 1$	.28
	$\times 4$
$a_{-4} = 1$	.12
	.
	.
	.

Thus $(0.27)_{10} = (0.1011 \ldots)_4$, and as a check,

$$(0.1011 \ldots)_4 = 1 \times 4^{-1} + 1 \times 4^{-3} + 1 \times 4^{-4} + \cdots$$
$$= (0.2695 \ldots)_{10}$$

As is generally the case, this conversion process yields a nonexact equivalent. This fact must be taken into account when computation is done with a computer not using the decimal system.

The conversion of general decimal numbers with both integral and fractional parts can now easily be handled by simply converting each part separately and combining the results. For example, solve the equation $(123.56)_{10} = (?)_7$. First, convert the integral part:

$$
\begin{array}{r}
7)\underline{123} \\
7)\underline{\ 17} \quad 4 \\
7)\underline{\ \ 2} \quad 3 \\
0 \quad 2
\end{array}
$$

Next, convert the fractional part:

$$
\begin{array}{r}
.56 \\
\underline{\times 7} \\
3 \quad .92 \\
\underline{\times 7} \\
6 \quad .44 \\
\underline{\times 7} \\
3 \quad .08 \\
\underline{\times 7} \\
0 \quad .56 \\
. \\
. \\
.
\end{array}
$$

The result then becomes

$$(123.56)_{10} = (234.3630\ldots)_7$$

where, as usual, the trailing points mean the result is not exact.

Conversion between two nondecimal systems can be handled most easily by using the decimal system as an intermediate step. For example, the problem of solving $(1354.24)_6 = (?)_4$ would be accomplished by first converting from base 6 to base 10 and then converting this base 10 number to base 4. Thus

$$(1354.24)_6 = (358.4444\ldots)_{10}$$

$$= (11212.1301\ldots)_4$$

2.2.3 Counting in a Radix r System

In the conversion process just described, it is interesting to note that the only numerical values the digits may take fall in the range of 0 to $r - 1$. Furthermore, note that

$$10_r = 1 \times r^1 + 0 \times r^0 = r_{10} \tag{2.2.10}$$

Because of these two observations, counting in radix r always produces the sequence of numbers $0, 1, 2, \ldots, (r - 1), 10, 11, 12, \ldots, 1(r - 1), \ldots$. Figure 2.2.1 shows the counting sequence for various radices.

Decimal	$r = 2$	$r = 3$	$r = 8$
0	0	0	0
1	1	1	1
2	10	2	2
3	11	10	3
4	100	11	4
5	101	12	5
6	110	20	6
7	111	21	7
8	1000	22	10
9	1001	100	11
10	1010	101	12
11	1011	102	13
12	1100	110	14

Figure 2.2.1   Counting in various systems of different radix r.

When $r > 10$, a problem arises in the representation of those digits x in the range $9 < x < r$, since no standard symbols exist for these numbers. By convention, capital letters are used to represent these digits. Thus, for $r = 16$ (the hexadecimal system), the counting sequence would be 0, 1, 2, 3, 4, 5, 6, 7, 8, 9, A, B, C, D, E, F, 10, . . . , where $(A)_{16} = (10)_{10}$, $(B)_{16} = (11)_{10}$, and so on.

2.2.4 Binary, Octal, and Hexadecimal Conversions

Normally, computations within a computer are carried out in the binary, or base 2, system. This is principally because digital circuits are usually two-state devices. Circuit elements having more than two states do exist, but these generally suffer from low reliability and other difficulties, some of which will be alluded to later.

Conversion from binary to decimal and vice versa is carried out as described above but is generally much easier than conversions between decimal and a radix larger than 2. An example will help illustrate: Convert $(132)_{10}$ to $(x)_2$. The conversion goes as follows:

$$
\begin{array}{ll}
2)\underline{132} & \\
2)\underline{66} & 0 \\
2)\underline{33} & 0 \\
2)\underline{16} & 1 \\
2)\underline{8} & 0 \\
2)\underline{4} & 0 \\
2)\underline{2} & 0 \\
2)\underline{1} & 0 \\
0 & 1
\end{array}
$$

and thus $(132)_{10} = (10000100)_2$, which, as a check, yields

$$2^7 + 2^2 = 128 + 4 = (132)_{10}$$

In a binary number, each binary digit, or *bit*, is weighted as a power of 2. Thus, as this example illustrates, conversion from binary to decimal requires only the addition of the powers of 2 corresponding to the 1s in the number.

Generally, working with binary numbers is somewhat cumbersome, because of the large number of bits required to make up even small decimal equivalents. For this reason, the *octal* and *hexadecimal*, or just *hex*, systems are commonly used to represent these numbers. To see the relationship between binary, octal, and hex, consider the binary number 110101011:[2]

$$110101011 = 1 \times 2^8 + 1 \times 2^7 + 0 \times 2^6 + 1 \times 2^5 + 0 \times 2^4$$
$$+ 1 \times 2^3 + 0 \times 2^2 + 1 \times 2^1 + 1 \times 2^0$$

$$= (1 \times 2^2 + 1 \times 2^1 + 0 \times 2^0)2^6 + (1 \times 2^2 + 0 \times 2^1 + 1 \times 2^0)2^3$$
$$+ (0 \times 2^2 + 1 \times 2^1 + 1 \times 2^0)2^0$$

$$= 6 \times (2^3)^2 + 5 \times (2^3)^1 + 3 \times (2^3)^0$$

$$= 6 \times 8^2 + 5 \times 8^1 + 3 \times 8^0$$

$$= (653)_8.$$

This example illustrates the extreme ease of conversion from binary to octal. The conversion simply involves grouping the bits in threes and writing the decimal value of each group. Thus

$$(\underline{101}\ \ \underline{111}\ \ \underline{100})_2$$
$$= (\ 5 \qquad 7 \qquad 4\ )_8.$$

In an exactly analogous fashion, the conversion from binary to hex can be simply carried out by grouping the bits in fours. Consider, for example, the following conversion:

$$(\underline{0001}\ \ \underline{0111}\ \ \underline{1100})_2$$
$$= (\ 1 \qquad 7 \qquad C\ )_{16}$$

where C represents the twelfth hex digit.

If it is necessary to convert a number from hex to octal or vice versa, it is generally easier to use binary rather than decimal as the intermediate step. For example,

$$(1A8E)_{16} = (?)_8 = (0001\ 1010\ 1000\ 1110)_2$$

$$= (001\ 101\ 010\ 001\ 110)_2$$

$$= (1\ \ 5\ \ 2\ \ 1\ \ 6)_8.$$

The result here is obtained by doing nothing more than writing the hex number in binary and then regrouping the bits to form the octal result.

---

[2] The subscript 2 is omitted here because the number was described as a *binary* number.

2.3 BINARY ARITHMETIC

Carrying out arithmetic operations in binary may take a bit of getting used to, but it is generally simpler than it is in decimal, since the addition and multiplication tables are so simple. These tables are given in Figure 2.3.1.

$$0 + 0 = 0 \qquad 0 \times 0 = 0$$
$$0 + 1 = 1 \qquad 0 \times 1 = 0$$
$$1 + 1 = 10 \qquad 1 \times 1 = 1$$
$$\text{(a)} \qquad\qquad \text{(b)}$$

Figure 2.3.1   (a) Addition and (b) multiplication tables for binary numbers.

Consider as an example of the addition process the sum of the two binary numbers $A = 10111010$ and $B = 110111$. The addition is carried out as follows:

```
      11111        (carries from preceding bit position)
      10111010
  +     110111
      11110001
```

As a check, we note that $A = (186)_{10}$ and $B = (55)_{10}$ and thus the decimal value of $A + B$ is 241, which is equal to binary 11110001.

In carrying out the addition in the above example, a number of incidents occurred where more than two bits had to be added. This, of course, was caused in each case by the carry generated by the addition of the previous bits. An alternative representation for the addition table of Figure 2.3.1 which includes the carry to be added as well as the carry generated is given in Figure 2.3.2. It will be shown later that this table can be used, in the form given, to generate hardware that performs binary addition in a computer's central processing unit. In using this table it should be noted that the number in the "carry in" column of the table is the carry that has been generated by addition of the numbers in the previous column, $i - 1$, of bits and the carry out is the carry in of the next column of bits, $i + 1$.

As with decimal arithmetic, multiplication uses both the multiplication table and the addition table. The process is carried out by first multiplying the multiplicand by each digit

Carry in	A_i	B_i	Sum$_i$	Carry out
0	0	0	0	0
0	0	1	1	0
0	1	0	1	0
0	1	1	0	1
1	0	0	1	0
1	0	1	0	1
1	1	0	0	1
1	1	1	1	1

Figure 2.3.2   Addition table including carries.

of the multiplier to form a set of partial products. These partial products are then added to form the final product. For example,

$$\begin{array}{rl}
101100 & \text{multiplicand} \\
\times \quad \underline{1011} & \text{multiplier} \\
101100 & \\
101100 & \text{partial products} \\
000000 & \\
\underline{101100} & \\
111100100 & \text{product}
\end{array}$$

This result is easily checked by multiplying the decimal equivalents of the binary multiplicand and mutiplier as follows:

$$(101100)_2 \times (1011)_2 = 44 \times 11 = 484 = (111100100)_2.$$

Subtraction and division introduce the same extra complexities in binary as they do in decimal arithmetic: borrowing, and estimating quotient digits. Consider first the problem of subtraction. A subtraction table may be set up in the manner shown in Figure 2.3.3; the table is similar in form to the addition table given in Figure 2.3.1. Using this table, the difference between the two binary numbers 10000 and 101 is computed as follows:

$$\begin{array}{rl}
0\ 1\ 1\ 1 & \text{resulting bit after borrow} \\
\cancel{1}\ {}^1\cancel{0}\ {}^1\cancel{0}\ {}^1\cancel{0}\ {}^1 0 & \\
-\quad \underline{1\ 0\ 1} & \\
1\ 0\ 1\ 1 &
\end{array}$$

Some other examples are as follows:

$$1010 - 1 = 1001$$

$$110010 - 101 = 101101$$

$$1101 - 100101 = -11000$$

It will be shown in the next section that subtraction can actually be performed by first "coding" the subtrahend and then using addition, thus avoiding the various complications arising because of the borrows.

Long division may be carried out in binary in a manner equivalent to decimal division, but it is generally much easier, since there is virtually no need for estimation of quotient digits. An example will best illustrate the process. Consider the problem of determining

$$\begin{array}{ll}
0 - 0 = 0 & \\
1 - 0 = 1 & \\
1 - 1 = 0 & \\
0 - 1 = 1 & \text{with a borrow from the next} \\
 & \text{higher bit position}
\end{array}$$

Figure 2.3.3. Binary subtraction table.

100101/101. There are many ways of organizing the work. However, a classic approach is as follows:

$$
\begin{array}{r}
111 = \text{quotient} \\
101\overline{)100101} \\
\underline{101} \\
1000 \\
\underline{101} \\
111 \\
\underline{101} \\
10 = \text{remainder}
\end{array}
$$

Thus $100101/101 = 111$ with a remainder of 10; or, as a check, in decimal the division problem becomes $37/5 = 7$ with a remainder of 2. As a second example, consider the problem $11010111/110$:

$$
\begin{array}{r}
100011 \\
110\overline{)11010111} \\
\underline{110} \\
1011 \\
\underline{110} \\
1011 \\
\underline{110} \\
101
\end{array}
$$

Here the result is 100011 with a remainder of 101, or, in decimal, as a check, $215/6 = 35$ with a remainder of 5.

Notice that in both examples, estimating whether or not the divisor will go into a partial dividend requires only the step of determining whether or not the partial dividend is greater than or equal to the divisor. If it is, the value that is put into the quotient has to be a 1; the divisor is then subtracted from the partial dividend. If it is not, then a 0 is placed in the quotient and the next bit of the dividend is brought down; the process is repeated until a 1 can be placed in the quotient and the divisor can be subtracted from the resulting partial dividend.

2.4 COMPLEMENT ARITHMETIC

It was mentioned in Section 2.3 that subtraction can be carried out by using addition if the subtrahend is "coded" properly. The implication of this, with regard to the design of a computer, is that a single piece of hardware, an adder, can be used to perform all arithmetic operations. This happens because binary multiplication involves addition only and binary division involves subtraction only. This clearly simplifies the design process as well as the designed hardware. The purpose of this section, then, is to describe this coding and show how it can be used for number representation and arithmetic operations.

2.4.1 Radix and Diminished Radix Complements

Let A be an n-digit integer number in radix r representation. Then the *radix complement* of A is defined as

$$A^* = \text{radix complement of } A = r^n - A \qquad (2.4.1)$$

and the *diminished radix complement* is defined as

$$A^+ = \text{diminished radix complement of } A = r^n - A - 1 \qquad (2.4.2)$$

To see how the complement representation of a number can be used in the subtraction process, let A and B be two n-digit numbers[3] and suppose that $B - A$ is to be determined. The claim is that the difference can be found by adding the radix complement of A to B, or

$$A^* + B = r^n - A + B \qquad (2.4.3)$$
$$= r^n + (B - A)$$

Recall from the preceding section that $r_{10} = 10_r$. Thus r^n in radix r arithmetic is just 1 followed by n zeros. Two possible cases occur here. First, assume that $B > A$. Then the result is positive and Equation (2.4.3) yields the correct n-digit difference preceded by a 1. An example may help. Let $A = 0592$ and $B = 3456$ be two 4-digit decimal numbers. From Equation (2.4.1),

$$A^* = 10000 - 0592 = 9408 \qquad (2.4.4)$$

and adding this complement to B we obtain

$$B + A^* = 3456 + 9408 = 1\ 2864.$$

Ignoring the 1, 2864 is, of course, the correct answer.

The second case occurs when $A > B$. In this case, Equation (2.4.3) may be written as $r^n - (A - B)$, and since $A - B$ is now positive, the result is, by definition of the radix complement in Equation (2.4.1), the radix complement of $A - B$! For example, let $A = 6734$ and $B = 523$; then, as before,

$$A^* = 10000 - 6734 = 3266 \qquad (2.4.5)$$

and adding this to B, we have

$$B + A^* = 523 + 3266 = 3789$$

$$= 10000 - 6211$$

In the first case, where $B - A$ was positive, $n + 1$ digits appeared in the complement sum, with the leftmost being ignored and the remaining digits being the magnitude of the correct answer. In the second case, where $B - A$ was negative, the complement sum had only n digits and was the radix complement of the answer. In this case the magnitude of

---

[3] If the two numbers do not have the same number n of significant digits, then the smaller may have zeros appended on the left to make up the necessary n digits.

the answer can be found by taking the radix complement. In both cases, however, the difference of A and B was found by the use of addition (ignoring for the moment that the complement was found by subtraction).

This process is particularly simple when applied to binary numbers. Consider, for example, the subtraction of $A = 110101$. from $B = 111001$, i.e. the problem, $B - A = (?)$. The first step is to find the 2's complement of A. In this case both numbers have 6 bits, making $n = 6$ in Equation (2.4.1), and so

$$A^* = 1000000 - 110101$$

Before carrying out this subtraction, note that from Equation (2.4.2)

$$A^* = A^+ + 1 \qquad\qquad (2.4.6)$$

so that

$$\begin{aligned} A^* &= (1000000 - 1 - A) + 1 \\ &= (111111 - 110101) + 1 \\ &= (001010) + 1 \\ &= 001011 \end{aligned}$$

The important thing to observe from this is the extreme ease with which the 1's complement is found: simply interchange 1s and 0s, which requires no subtraction at all. The 2's complement is then obtained by adding 1. Continuing with the problem now requires that A^* be added to B to obtain the answer:

$$A^* + B = 001011 + 111001 = 1 \quad 000100$$

Since the result has a carry out of the sixth bit position, the result is positive and has a value of 100. This, of course, can be checked by simply subtracting the original two arguments.

In this example, absolutely no subtraction was used to obtain the difference between two binary numbers, since the 2's complement of A was found by interchanging 1s and 0s and then adding 1 to the result. It is important to remember that both numbers must contain the same number of bits at the start. Some further examples will illustrate this procedure.

$$1011011 - 0010110 = 1011011 + 1101001 + 1 \qquad\qquad \text{(positive)}$$
$$= 1 \quad 1000101$$
$$10011 - 10111 = 10011 + 01000 + 1 = 11100 \qquad\qquad \text{(negative 00100)}$$
$$110100110 - 11001 = 110100110 + 111100110 + 1$$
$$= 1 \quad 110001101 \qquad \text{(positive)}$$
$$1 - 100000 = 000001 + 011111 + 1 = 100001 \qquad\qquad \text{(negative 011111)}$$

In each of these examples the 2's complement was generated by taking the 1's complement and adding 1. A very simple, and mechanical, alternative to this is the following. Starting on the right and moving to the left, copy the rightmost zeros until reaching the first 1. Copy this 1. From this point on copy the complements of the remaining bits. For example, to convert $A = 10110100$ to its 2's complement form, we proceed as follows:

	Complement	Copy
A	10110	100
A*	01001	100

This simple procedure works because in taking the 1's complement of A, the rightmost three bits would become 011. After adding 1 to obtain the 2's complement, these bits become 100, the original right three bits.

Before proceeding to examine how the sign of a number can be made part of the number, let us go back for a moment to the decimal system. Equation (2.4.6) may be used to compute the 10's complement of A^* in Equation (2.4.4) in a particularly simple manner. Specifically,

$$A^* = (10000 - 1 - 0592) + 1$$

$$= (9999 - 0592) + 1$$

$$= 9407 + 1$$

$$= 9408$$

Notice that although subtraction was required to get the 9's complement, it was a particulary simple subtraction requiring *no borrows*. This occurred because each digit of A was subtracted from 9 to obtain the corresponding digit of A^+. Thus, converting A to A^+ requires at each digit position only a knowledge of that digit and not the whole number. This can be done by a "table look-up" procedure (as will be described in Section 2.5.4), which requires no subtraction at all.

Subtraction can also be carried out using the diminished radix complement. In this case, however, the carries cannot be ignored. Using the diminished radix complement, if a carry is generated it must be added to the result. This addition is referred to as an *end-around carry*. For example, consider the subtraction of 101101 from 111001. Taking the 1's complement of 101101 and adding, we obtain the result

$$\begin{array}{r} 111001 \\ + \; 010010 \\ \hline 001011 \\ \longrightarrow 1 \qquad \text{add end-around carry} \\ \hline 001100 \end{array}$$

As was the case for the radix complement, the generation of a carry implies that the result of the addition is positive. The absence of a carry implies a negative result. Problem 2.11 at the end of this chapter explores the reason for the end-around carry.

One problem with the diminished radix complement is that the representation for the number 0 is not unique. To see this, note that the 1's complement of 000000 is 111111. Thus both of these numbers must represent the number 0. The nonuniqueness of the number

0 is one of the reasons that the diminished radix complement is seldom used in actual designs.

2.4.2 Binary Signed Representations

In the examples just worked, the sign of the result was inferred by whether or not a carry was generated out of the high-order bit position. It would be extremely useful if the sign of the number could be carried as part of the number itself. In the decimal system that most of us have grown up with, this is handled in a *sign magnitude* representation in which each number is preceded by a sign, such as $+149$ or -3765. When the sign is missing, the number is usually considered to be positive. Such a representation can work in a computer as well. However, the representation almost always used by the computer hardware is a *signed 2's complement* representation. In this representation, the leftmost (or most significant) bit is taken as the sign. The sign of the number is minus if this bit is 1 and plus if it is 0. The bits following the sign are either the magnitude of the number, if the sign is plus, or the 2's complement of the magnitude of the number, if the sign is minus. This representation has many advantages, not the least of which is that a string of computations may be carried out without regard to the resulting sign at each step. The sign of the answer will be found as the sign bit of the final result.

In a *signed 2's complement* representation it is always assumed that the number of bits in the operands is the same. In a large number of microprocessors this number is 8 bits, which is defined as a *byte*. Some examples of numbers represented in this manner are the following:

$$00111010 = +58$$
$$11100101 = -0011011 = -27$$
$$10000001 = -1111111 = -127$$
$$01111111 = +127$$

By convention, the number 10000000 is taken as -128. This makes a certain degree of sense, because $-127 - 1 = -128$, which, when carried out in signed 2's complement arithmetic, yields $10000001 + 11111111 = 10000000$. Thus numbers represented in this form using 8 bits, or a byte, can take on values ranging from -128 to $+127$.

To illustrate how this representation carries the sign through a string of computations, consider a couple of examples using a 4-bit signed 2's complement form (4 bits is often referred to as a *nibble*). Let $A = 0011$ $(+3)$ and $B = 0100$ $(+4)$. Then

$$A + B = 0011 + 0100 = 0111 \qquad (+7)$$
$$A - B = 0011 + 1100 = 1111 \qquad (-1)$$
$$-A + B = 1101 + 0100 = 1\ \ 0001 \qquad (+1)$$

where we ignore the carry, as before; and

$$-A - B = 1101 + 1100 = 1\ \ 1001 \qquad (-7)$$

where again we ignore the carry. Notice in these examples that the sign bit is treated in exactly the same way as any other bit and that the carries out of the sign bit position are ignored.

The addition of two n-bit numbers can result in a number whose value requires more then n bits to represent. Such a situation is referred to as *overflow* if the result is positive and *underflow* if the result is negative. It is important that the occurrence of overflow or underflow be detected so that decisions are not based on incorrect results. In a signed 2's complement representation, overflow or underflow occurs whenever the sign of the two arguments is the same but different from the sign of the result. For example, let $A = 01101000$ and $B = 01011000$ be two 8-bit signed 2's complement numbers ($A = +104$ and $B = +88$). Since the sum of these two numbers (192) is greater than $+127$, the largest number possible in an 8-bit signed 2's complement representation, an overflow will occur when we add the two numbers. In particular, $A + B = 11000000$, which indicates a negative result. Problem 2.15 gives another indication of the occurrence of overflow or underflow.

2.5 CODING

It is usually the case that we interact with a computer via a keyboard in which each key represents some piece of information such as an alphabetic or numeric character or a control character (e.g., a tab, a space, or a line feed). The key inputs must be converted to some binary form before the computer can process them. This is usually done by assigning a specific pattern of bits to a byte so that there is a byte stored somewhere in the computer's memory to correspond to each keyboard input. One such code is the ASCII code, which will be discussed in Section 2.5.4.

Information may also enter the computer from external sensors (such as thermometers or strain gauges), from switches, from shaft position indicators, and from many other devices. All of this information must be converted in some way to binary for proper handling by the computer. Furthermore, it may be convenient, in some applications, to handle the numbers internally as decimal digits which have been suitably encoded in some binary form. This type of representation is very common in hand-held calculators and other devices where nuemric information must be constantly entered by a human, processed, and finally returned to the human in numerical form.

Many other reasons exist for coding information; among them are encryption and error detection and correction. The purpose of this section, then, is to describe a few of the commonly used codes and discuss how they are internally handled and how we can convert from one code to another.

2.5.1 Binary-Coded Decimal (BCD) and Excess-3 Codes

One of the most common internal representations for decimal numbers is the *binary-coded decimal*, or *BCD*, representation. In this form, the ten decimal digits are represented by a 4-bit binary number whose value is the decimal digit. For example, the digit 9 is coded

Decimal	BCD	Excess 3
0	0000	0011
1	0001	0100
2	0010	0101
3	0011	0110
4	0100	0111
5	0101	1000
6	0110	1001
7	0111	1010
8	1000	1011
9	1001	1100

Figure 2.5.1   BCD and excess-3 codes.

as 1001. Figure 2.5.1 gives the code for each of the ten digits. In this form a number such as 1853 would be represented internally as 0001 1000 0101 0011. Addition can be carried out in BCD by adding the two numbers as if they were binary but with some slight modification to the computational process. Consider, for example, the addition problem 253 + 314. In BCD this becomes

$$
\begin{array}{c}
\ 0010 \quad 0101 \quad 0011 \\
+\ 0011 \quad 0001 \quad 0100 \\
\hline
\ 0101 \quad 0110 \quad 0111 = 567
\end{array}
$$

which is, of course, the correct answer. The addition in this example was carried out by simply adding the two numbers in binary. Note that no carries between digits were generated, because the sum in each digit column never exceeded 9. If the sum of two digits is a number greater than 9, then one of two things can happen: either the resulting 4 bits is not a legal BCD code (i.e., it is not one of the ten in Figure 2.5.1), or a carry occurs out of the 4-bit group. An example of the first situation would be the addition of, say, 6 + 8, which in BCD becomes

$$
\begin{array}{c}
\ 0110 \\
+\ 1000 \\
\hline
\ 1110
\end{array}
$$

which is not a legal BCD number. Adding 6 to this result will yield the correct answer (why?). Thus the answer is

$$
\begin{array}{c}
\ 1110 \\
+\ 0110 \\
\hline
1\ \ 0100 = 14 \qquad \text{in BCD}
\end{array}
$$

The second case will occur for additions such as 8 + 9:

$$
\begin{array}{c}
\ 1000 \\
+\ 1001 \\
\hline
1\ \ 0001
\end{array}
$$

In this case we note that although the low-order 4 bits represents a legitimate BCD number, the result of the addition yields a number greater than 9, as indicated by the carry, and so the correct answer may be obtained once again by adding 6. Thus

$$
\begin{array}{r}
1 \quad 0001 \\
+ \quad \underline{0110} \\
1 \quad 0111 \;=\; 17 \qquad \text{in BCD}
\end{array}
$$

Consider, as a somewhat more complex example, the problem of finding the sum of 769 and 358, which in BCD becomes

$$
\begin{array}{lllll}
& 0111 & 0110 & & 1001 \\
+ & \underline{0011} & \underline{0101} & & \underline{1000} \\
& 1010 & 1011 & 1 & 0001 \\
+ & \underline{0110} & \underline{0110} & \downarrow & \underline{0110} & \text{add in the 6s} \\
1\;0000 & 1\;\,0001 & 1 & 0111 \\
+ & 1^{\swarrow} & \;\;1^{\swarrow} & & & \text{add in the carries} \\
\hline
1\;0001 & 0010 & & 0111 & = 1127 \text{ in BCD}
\end{array}
$$

It may happen in carrying out the BCD addition that the result after adding in the 6s and the carries is still not a correct BCD number. If this occurs, we simply apply the correction procedure once more. For example, consider the sum of 37 and 64, which is found as follows:

$$
\begin{array}{llll}
& 0011 & 0111 \\
+ & \underline{0110} & \underline{0100} \\
& 1001 & 1011 \\
+ & & \underline{0110} & \text{add in the 6s} \\
& 1001 & 1\;\,0001 \\
+ & \;1^{\swarrow} & & \text{add in the carry} \\
\hline
& 1010 & 0001 \\
+ & \underline{0110} & & \text{add in the 6} \\
1\;0000 & & 0001 & = 101 \text{ in BCD}
\end{array}
$$

A modified version of BCD, which has some attractive features when subtraction is required, is the *excess-3* code. This is basically the same coding as BCD except that each digit has 3 added to it. Figure 2.5.1 gives the specific code values. The attractive characteristic of the excess-3 code is that it is self-complementing, i.e., the 1's complement of the coded number yields the 9's complement of the number itself. For example, 3 has a code of 0110, whose 1's complement is 1001, which is the excess-3 code for 6, the 9's complement of 3. Thus subtraction in this binary-coded decimal form can be easily carried out using the diminished radix complement scheme described earlier.

To see how the self-complementing feature of the excess-3 code can be used for subtraction, consider first the addition of two excess-3 numbers, A and B. Adding these two excess-3–encoded numbers is perhaps most easily carried out by first converting each to its BCD equivalent, then adding the results, as described above, and, finally, converting the result back to excess-3. To convert an excess-3–encoded number to BCD is a very simple process. Let $X' = X + 3$ be an excess-3 digit, where the X is the decimal equivalent of X'. To obtain X we need only add 13 to X' and take the result modulo (16).[4] For example, let $X' = 0111$, the excess-3 code for 4. Then $X' + 1011 = 0111 + 1101 = 1\,0100$, where we obtain the result modulo (16), 0100, by ignoring the carry. Converting all of the digits of each of the numbers A and B in this way, we obtain the respective BCD representations. For example, to convert the excess-3–encoded representation of the number 97, 1100 1010, to its equivalent BCD representation, we simply add 1101 (13 decimal) to each digit, as follows:

$$
\begin{array}{cccc}
 & 1100 & & 1010 \\
+ & 1101 & & 1101 \\
\hline
1 & 1001 & 1 & 0111
\end{array}
$$

Ignoring the carries generated at each digit position, we obtain the result, 1001 0111, which is, of course, the BCD representation for the decimal number 97.

Now, to perform the subtraction of two excess-3 numbers, say $A' - B'$, we first take the 9's complement of B' by interchanging 0s and 1s in the coded digits. This produces the number B'^+. Next we convert the numbers A' and B'^+ to their BCD equivalents, A and B^+. Once this is done, we can add the results in accordance with diminished radix complement arithmetic to produce the difference. The final excess-3–encoded result is found by adding 3 to each digit of the difference. Problems 2.19 and 2.20, at the end of the chapter, explore this process a bit more.

2.5.2 BCD to Binary and Binary to BCD Conversions

When numbers are entered into a computer from, say, a keyboard, they are encoded in some way. A string of encoded digits then needs to be converted to a binary number so that the computer can process the numeric information thus presented. Suppose that the encoding is in BCD.[5] The problem then becomes one of converting these digits to binary. Recall from Section 2.2.2 that converting from decimal to binary requires repeatedly dividing the decimal number by 2 and using the remainder digits as the successive bits of the equivalent binary number. The same can be done for BCD in a very simple way if we make a few observations first.

In the decimal system (or any radix r, system, for that matter) division by 10_r requires only that the radix point be moved one digit position to the left. The remainder is the digit that moves to the right of the radix point. Another way of thinking about this is to assume that the radix point stays fixed and that the number *shifts to the right one digit position*. Thus, in binary, the number 1001, which is the BCD code for 9, when divided by 2 by shifting right, becomes (maintaining 4 bits in the answer) 0100, with 1 being shifted out of the low-order position. This results in the correct answer of 4 with a remainder of 1.

---

[4] $y = x$ modulo (n) means that y is the remainder obtained upon dividing x by n.

[5] It will be shown in Section 2.5.4 that whatever code is used, it can be converted to BCD. Thus this statement is made without loss of generality.

This idea can be used to divide a string of BCD digits by 2 in a very simple manner. Take, for example, 3609/2. In BCD this becomes (0011 0110 0000 1001)/2. Shifting each digit to the right one position will divide that digit by 2, but for the division to be correct for the entire number, a correction must be made as follows. If a 1 is shifted out of some digit position, then 5 that is, (0101) must be added to the next lower digit position (why?). Thus, for this example the division may be carried out as follows:

$$
\begin{array}{lllll}
(& 3 & 6 & 0 & 9 \ )/2 \\
= (0011 & 0110 & 0000 & 1001)/2 \\
= \ 0001 & 1 \ 0011 & 0000 & 0100 & \text{remainder of 1} \\
& + \ 0101 & & & \text{add 5 as necessary} \\
0001 & \overline{1000} & 0000 & 0100 \\
1 & 8 & 0 & 4 & \text{remainder of 1}
\end{array}
$$

Using this simple method of division by 2, we can carry out the conversion from BCD to binary. Consider the conversion of 0101 0011 to binary. The work can be organized as follows:

	BCD value		/ Binary result (remainder)
	0101	0011 /	
shift	0010	1 0001 / 1	
add 5		+ 101	
	0010	0110 / 1	
shift	0001	0011 / 01	
shift	0000	1 0001 / 101	
add 5		+ 101	
	0000	0110 / 101	
shift	0000	0011 / 0101	
shift	0000	0001 / 10101	
shift	0000	0000 / 110101 = 53 decimal	

As in Section 2.2.2, this process stops as soon as the dividend goes to zero.

The process of going from binary to BCD is exactly the reverse of the above conversion except that 0101 is subtracted from any BCD digit greater than or equal to 5 before the shift is made and a 1 is set up as a carry into the next higher-order digit position (why?). An example will illustrate the process. Consider the conversion of 1101101 to BCD. The work may be organized as follows:

	BCD result / Binary value	
	0000	0000 / 1101101
shift left	0000	0001 / 101101
shift left	0000	0011 / 01101
shift left	0000	0110 / 1101
subtract 5		− 0101
	0000	1 0001 / 1101
shift left	0001	0011 / 101

```
shift left              0010    0111 / 01
subtract 5                     −0101
                        0010  1 0010 / 01
shift left              0101    0100 / 1
subtract 5            − 0101
                0000  1 0000    0100 / 1
shift left      0001    0000    1001          and stop
```

The answer, 109 in BCD, is easily seen to be the correct decimal equivalent of 1101101 binary.

2.5.3 Other Codes for Representing Numbers

The BCD code is an example of a *weighted code* in which each bit position has a corresponding weight associated with it. The number represented by the code character is found by adding the weights corresponding to each 1 in the code. The weights for the BCD code are 8421, and in fact, the BCD code is sometimes referred to as an "8421 code." Other weighted codes exist and have been used in various computer systems over the years. A weighted code, in order to represent the decimal digits, must have weights which can sum to each of the 9 digits. These weights need not, however, be positive. Figure 2.5.2 gives some examples of other weighted codes.

As mentioned earlier, one reason for coding a number might be to permit error detection. One simple error-detection code is the 2-out-of-5 code, in which each digit is represented by a character having 5 bits with two of them always 1 and the remaining three bits always 0. Since there are 10 such combinations, each decimal digit will correspond to one such combination. In this representation, if an error occurs, say one that causes a bit to be set to a 1, then the error is readily detected by the fact that the received code does not have exactly two 1s and three 0s. Another simple error-detection mechanism is the addition of one extra bit whose value is determined so that the number of 1s in the representation is even (or odd, if one prefers). Such a bit is called a *parity* bit. If an error occurs in the handling of a number with the result that a single bit is changed, then the total number of 1s will now be odd and it will be evident that an error has occurred. Other

Decimal	2421	8421	32211
0	0000	0000	00000
1	0001	0111	00001
2	0010	0110	00100
3	0011	0101	00101
4	0100	0100	00111
5	0101	1011	01101
6	0110	1010	10101
7	0111	1001	10111
8	1110	1000	11101
9	1111	1111	11111

Figure 2.5.2 Some examples of weighted codes.

codes exist which are capable of not only detecting errors but correcting them as well. Problems 2.22 and 2.23 at the end of the chapter explore some of these coding techniques.

Another very useful and commonly encountered code is the *Gray code*. In this code, successive digits differ in only one bit position. For example, a Gray code sequence for 3 bits would be 000, 001, 011, 010, 110, 111, 101, 100. The generation of this Gray code sequence is very simple. The pattern of changing values of the least significant bit for the first four digits is 01 followed by its reflection 10; then the sequence 01 followed by 10 is repeated as many times as necessary. The next bit from the right has a pattern over *eight* digits of 0011 (twice the number of 0s and 1s) followed by its reflected value 1100. The next bit has the pattern 00001111 (again, twice the number of 0s and 1s) followed by *its* reflected value 11110000. This process continues for as many bits as are in the code. For example, a Gray code for 4 bits is generated as illustrated in Figure 2.5.3. The Gray code is used extensively for shaft encoders and other applications requiring a single bit change between characters.

2.5.4 Alphanumeric Codes

Numeric information is not the only information that computers process. Alphabetic characters, punctuation marks, special characters such as mathematical symbols, and many other nonnumeric items must be encoded into a binary form before the computer can properly handle them. One such code is ASCII (American Standard Code for Information Interchange), which is used extensively for representing characters that come from a keyboard. In this code, 7 bits are used to represent all upper- and lowercase alphabetic and numeric

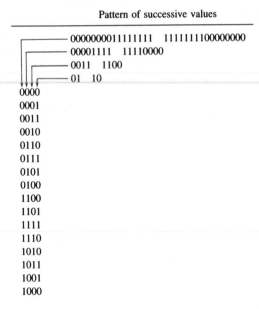

Figure 2.5.3 Generation of a 4-bit Gray code.

Character	ASCII code	EBCDIC code	Character	ASCII code	EBCDIC code
A	41	C1	0	30	F0
B	42	C2	1	31	F1
C	43	C3	2	32	F2
D	44	C4	3	33	F3
E	45	C5	4	34	F4
F	46	C6	5	35	F5
G	47	C7	6	36	F6
H	48	C8	7	37	F7
I	49	C9	8	38	F8
J	4A	D1	9	39	F9
K	4B	D2	blank	20	40
L	4C	D3	!	21	5A
M	4D	D4	"	22	7F
N	4E	D5	#	23	7B
O	4F	D6	$	24	5B
P	50	D7	%	25	6C
Q	51	D8	&	26	50
R	52	D9	'	27	70
S	53	E2	(	28	4D
T	54	E3	)	29	5D
U	55	E4	*	2A	5C
V	56	E5	+	2B	4E
W	57	E6	,	2C	6B
X	58	E7	−	2D	60
Y	59	E8	.	2E	4B
Z	5A	E9	/	2F	61

Figure 2.5.4 A partial listing of ASCII and EBCDIC codes.

characters, as well as all of the usual punctuation marks and typewriter control information, such as line feeds, tabs, and carriage returns. In many computers, this scheme is extended by using an eighth bit to obtain 128 more characters; usually, these 128 are graphic symbols which can be displayed on the terminal screen. Figure 2.5.4 gives an abbreviated list of ASCII codes and the characters represented by each. This table also lists another code used for alphanumerics called EBCDIC (Extended BCD Interchange Code), which uses all eight bits of a byte to represent the information. In ASCII, the word HELLO would be stored internally as the five bytes (written in hex form) 48, 45, 4C, 4C, 4F.

It quite often happens in computer systems that the codes used for input and the codes used for output are different. Suppose, for example, that a keyboard used for data input produces EBCDIC and a display terminal used for output requires ASCII. Obviously, the computer must make a conversion between these two forms if information is to be displayed properly. This is easily done by referring to the table shown in Figure 2.5.4, which could be stored in the computer's memory somewhere. If, for example, the EBCDIC code D5 (the code for the letter N) is entered, the corresponding ASCII code can be found by first locating D5 in the table and observing that the corresponding ASCII code is 4E. This process is called a *table look-up* and is a very important technique in the design and use of computers. We will see much more of this process later.

2.6 AN ANNOTATED BIBLIOGRAPHY

There are many excellent references that discuss much of the material covered in this chapter. The following five are good examples of these.

DIETMEYER, D. L., *Logic Design of Digital Systems*, Allyn & Bacon, Boston, 1978.

FLETCHER, W. I., *An Engineering Approach to Digital Design*, Prentice-Hall, Englewood Cliffs, N.J., 1980.

HILL, J. F., and G. R. PETERSON, *Introduction to Switching Theory and Logical Design*, 2nd ed., Wiley, New York, 1974.

KOSTOPOULOS, G. K., *Digital Engineering*. Wiley-Interscience, New York, 1975.

MANO, M. M., *Digital Logic and Computer Design*, Prentice-Hall, Englewood Cliffs, N.J., 1979.

A very thorough discussion of binary arithmetic can be found in the classical text by Flores. Flores discusses algorithms for both signed and unsigned addition, subtraction, multiplication, and division, in Chapters 2 and 3. This book also covers many advanced topics dealing with computer computational algorithms. A good discussion of signed arithmetic, as handled in a typical modern microprocessor, can be found in Appendix C of the text by Camp et al.

CAMP, R. C., T. A. SMAY, and C. J. TRISKA, *Microprocessor Systems Engineering*, Matrix Publishers, Inc., Chesterland, Ohio, 1979.

FLORES, I., *The Logic of Computer Arithmetic*, Prentice-Hall, Englewood Cliffs, N.J., 1963.

The subject of BCD arithmetic is covered by a number of authors. Chapter 2 of Givone and Roesser gives some good introductory examples. A very extensive discussion of BCD arithmetic operations, including multiplication and division, can be found in Chu's book. This discussion is somewhat advanced, however.

CHU, Y., *Computer Organization and Microprogramming*, Prentice-Hall, Englewood Cliffs, N.J., 1972.

GIVONE, D. D., and R. P. ROESSER, *Microprocessors/Microcomputers: An Introduction*, McGraw-Hill, New York, 1980.

The problem of converting between BCD and binary was addressed many years ago in the paper by Couleur. A design based on the ideas presented there is given in Chapter 9 of the text by Rhyne. Rhyne also discusses excess-3 arithmetic. The book by Short gives a simple microprocessor program for converting BCD to binary in Section 7.3–4. This method is quite different from the one presented here.

COULEUR, J. F., "BIDEC—A Binary-to-Decimal or Decimal-to-Binary Converter," *IEEE Trans. Electronic Computers*, Vol. EC-7, No. 6, December 1958, pp. 313–316.

RHYNE, V. T., *Fundamentals of Digital System Design*, Prentice-Hall, Englewood Cliffs, N.J., 1973.

SHORT, K. L., *Microprocessors and Programmed Logic*, Prentice-Hall, Englewood Cliffs, N.J., 1981.

The book by Floyd gives a good discussion of excess-3 addition in Chapter 2. Floyd shows an alternative to the method presented here. He also, briefly, discusses the self-complementing properties of this code.

FLOYD, T. L., *Digital Fundamentals*, 2nd ed., Chas. E. Merrill, Columbus, Ohio, 1982.

There are many classic texts that deal with the coding of information for purposes of error detection and correction. Three recent volumes, however, discuss this process in a fairly clear and elementary manner. Schwartz has included, in Chapter 6 of the third edition of his classic text on information theory, a basic discussion of coding in general. Wilkinson presents an excellent introductory discussion of the Hamming codes. Bertsekar and Gallager present a very nice introduction to various applications of parity for the detection and correction of errors.

BERTSEKAS, D., and R. GALLAGER, *Data Networks*, Prentice-Hall, Englewood Cliffs, N.J., 1987.

SCHWARTZ, M., *Information, Transmission, Modulation, and Noise*, 3rd ed., McGraw-Hill, New York, 1980.

WILKINSON, B., *Digital System Design*, Prentice-Hall International, 1987.

2.7 PROBLEMS

2.1. Write the decimal equivalent of the following numbers:
   (a) $(375)_9$
   (b) $(12211)_3$
   (c) $(0.225)_6$
   (d) $(1A3.5A)_{11}$

2.2. Convert each of the following decimal numbers to the equivalent number in the indicated base:
   (a) $1375 = (?)_8$
   (b) $995 = (?)_{13}$
   (c) $0.3378 = (?)_2$
   (d) $1075 = (?)_4$
   (e) $735.263 = (?)_{-8}$

2.3. Convert the following numbers:
   (a) $(1076)_8 = (?)_7$
   (b) $(0.225)_6 = (?)_9$
   (c) $(122.13)_{-4} = (?)_3$
   (d) $(1BA6.2B)_{13} = (?)_2$

2.4. Convert the following decimal numbers to binary:
   (a) 365
   (b) 12
   (c) 3709
   (d) 123.662
   (e) $\pi = 3.14159. . .$

2.5. Convert the following binary numbers to decimal:
    (a) 1101011
    (b) 1011
    (c) 1101.1110
    (d) 1111.101
    (e) 11010000.0001

2.6. Convert as indicated:
    (a) $(11101.110)_2 = (?)_8 = (?)_{16}$
    (b) $(A1EF)_{16} = (?)_2 = (?)_8$
    (c) $(1375)_8 = (?)_2 = (?)_{16}$

2.7. On an examination, a student wrote $(2756)_6$ as the answer to a question. Since 7 and 6 are greater than 5, the largest digit permitted in the radix 6 system, the answer must be wrong. What would you guess is the most likely decimal equivalent of this number, and why?

2.8. Perform the indicated arithmetic, maintaining your answer in sign magnitude form.
    (a) $(135C)_{16} + (22A54)_{16} = ?$
    (b) $(110101)_2 + (1011)_2 = ?$
    (c) $(23)_4 \times (31)_4 = ?$
    (d) $(1766)_8 - (23)_8 = ?$
    (e) $(1101101)_2/(101)_2 = ?$

2.9. Perform the indicated arithmetic on the following binary numbers given in sign magnitude form:
    (a) 101101 − 1101 = ?
    (b) 1110001 − 1110100 = ?
    (c) 10111.101 + 1001.011 = ?
    (d) (−1101) × (110) = ?
    (e) 10101/11 = ?
    (f) 10001.101 × 111.001 = ?
    (g) 1101 − 110110 = ?

2.10. Using the radix complement representation, perform the following subtractions. Assume that the numbers are all positive.
    (a) $(1765)_{10} - (351)_{10} = ?$
    (b) $(576)_{10} - (901)_{10} = ?$
    (c) $(1101011)_2 - (10111)_2 = ?$ ones complement
    (d) $(101101)_2 - (110101)_2 = ?$
    (e) $(100011)_2 - (100100)_2 = ?$
    (f) $(1111111)_2 - (1)_2 = ?$

2.11. When two numbers are subtracted using the radix complement, a carry generated in the high-order digit position is ignored. Show that this carry must be added to the result if the subtraction is carried out using the diminished radix complement. This carry is termed an *end-around carry*. (*Hint*: Recall that the definition of the diminished radix complement is just the radix complement minus 1.)

2.12. Perform the following binary subtractions, using the diminshed radix representation. Assume that the numbers are unsigned positive binary numbers.
    (a) 11010 − 1011
    (b) 101110001 − 1110011
    (c) 110111 − 1000011
    (d) 10100 − 11001

(e) $1101 - 1111$

(f) $1001 - 1001$

2.13. Convert the following decimal numbers to 8-bit signed 2's complement form:

(a) 165

(b) 23

(c) -23

(d) 100

(e) -97

2.14. Assuming that the following binary numbers are in signed 2's complement form, what is the decimal value, in sign magnitude form, of the indicated arithmetic?

(a) $00101101 + 00011110$

(b) $11100101 + 11110110$

(c) $11011011 + 00101101$

(d) $11100101 + 01011011$

(e) $00101101 - 01110111$

(f) $11010111 - 11110100$

(g) $11110101 - 11100011$

2.15. *Prove*: In an *n*-bit signed 2's complement representation, overflow or underflow in the addition of two numbers is indicated either if a carry comes into the sign bit and no carry goes out or if no carry enters the sign position but a carry goes out. (*Hint*: Remember that overflow occurs if the two arguments have the same sign but produce a result having a different sign.)

2.16. Encode the following decimal numbers in BCD and excess-3:

(a) 137

(b) 2345

(c) 12.36

(d) 5.9556

(e) 1941

2.17. Convert each decimal number to BCD and carry out the indicated arithmetic, leaving the result in sign magnitude BCD form.

(a) $375.2 + 26$

(b) $193 + 488$

(c) $275 - 366$

(d) $1234 + 999$

(e) $378 - 149$

2.18. Why is an illegal BCD digit converted to a legal BCD digit by adding 6?

2.19. Using the 9's complement, perform the following decimal arithmetic. Leave your answers in complemented form and indicate which results are negative.

(a) $3789 - 145$

(b) $1234 - 678$

(c) $375 - 421$

(d) $137.225 - 49.117$

2.20. Repeat Problem 2.19 after encoding the decimal numbers in excess-3 code. Use the fact that the 9's complement of a decimal number is the 1's complement of its excess-3–coded form.

2.21. Devise a method for representing the sign in a signed 10's complement representation.

2.22. We say that the *distance* between two *n*-bit numbers is the number of bit positions in which the two numbers differ. A code is said to be of "minimum distance *k*" if the minimum distance between any two coded numbers is k. Devise a minimum distance 2 coding for the decimal digits, 0 to 9. (*Hint*: The BCD code with one extra bit will do the trick.)

2.23. The most likely error that can occur in the transmission of data is a change in one bit due to noise in the transmission path. Show that if information is encoded in some minimal distance 3 code, a single-bit error not only can be detected but also can be corrected. Devise such a code for the decimal digits and show an example of how a single-bit error can be corrected. (*Hint*: Add three parity bits so they check parity over three unique subsets of the now 7 bits, four data and three parity.)

2.24. Write your name in ASCII and EBCDIC.

Boolean and Switching Algebra

3

3.1 INTRODUCTION

All engineering disciplines have a mathematical base on which the development of concepts depends. The design of digital systems, including computers, is no different. Here the mathematical base is called *Boolean algebra*.[1] As one might guess, this mathematical system is named after someone named Boole, in fact, George Boole, who was one of the first people to develop a rigorous mathematical structure for investigating the way we reason. Boole's treatise, published in 1854, was entitled *An Investigation of the Laws of Thought*.[2] No practical application was made of Boolean algebra until the late 1930s. A. Nakashima, in Japan, in 1937, and, in the following year, C. E. Shannon, at the Massachusetts Institute of Technology, each independently applied the algebra of Boole to the analysis of networks of relays. This was a very important application, since the telephone system at this time was growing very rapidly and required very large relay networks for switching and otherwise handling calls. If such a system was to grow in a controlled way, it was essential that a rigorous mathematical base be developed to describe the general interconnections. Obviously, the application of Boolean algebra has expanded dramatically over the intervening years as digital systems have grown and become increasingly more pervasive in our world.

Because of the importance of switching algebra to the design not only of computers but of communications systems, control systems, and any other system that requires or

---

[1] It is unfortunately true that many people tend to use the terms ''Boolean algebra'' and ''switching algebra'' interchangeably. As we will see shortly, a switching algebra is, strictly speaking, a subset of Boolean algebra.

[2] This book was reprinted by Dover Publications in 1954.

uses digital technology, it is important that we understand the intricacies of the algebra. Thus, in this chapter, Boolean algebra and its subset, *switching algebra*, will be defined. We will also investigate some of the implications of these definitions and examine the various methods that can be used for handling and simplifying equations.

3.2 THE HUNTINGTON POSTULATES

Algebras are defined by listing a set of statements which are taken to be fact. These statements are termed the *axioms* or the *postulates* of the algebra. One of the goals of the mathematician is to reduce the number of postulates required to define an algebra to a minimum consistent set. In 1904, E. V. Huntington set himself the task of reducing the definition of Boolean algebra to this minimal set of postulates. He found that all of the results and implications of the algebra described by Boole could be derived from only six basic postulates. Using these six, Huntington defined a Boolean algebra as follows:

The Huntington postulates (*E. V. Huntington, 1904*).   The set $\langle B, +, \cdot, ^{-} \rangle$, where B is the set of elements or constants of the algebra, the symbols $+$ and $\cdot$ are two binary operators, and the overbar $^{-}$ is a unary operator, is a Boolean algebra if the following hold true:[3]

1. *Closure*. For all elements a and b in the set B,
    (i) $a + b$ is an element of B and
    (ii) $a \cdot b$ is an element of B.
2. (i) There exists a 0 element in B such that for every element a in B, $0 + a = a + 0 = a$ and
    (ii) there exists a 1 element in B such that for every element a in B, $1 \cdot a = a \cdot 1 = a$.
3. *Commutativity*. For all elements a and b in the set B,
    (i) $a + b = b + a$ and
    (ii) $a \cdot b = b \cdot a$
4. *Distributivity*. For all elements a, b, and c in the set B,
    (i) $a \cdot (b + c) = a \cdot b + a \cdot c$ and
    (ii) $a + (b \cdot c) = (a + b) \cdot (a + c)$
5. For every element a in the set B, there exists an element $\bar{a}$ in the set B such that
    (i) $a + \bar{a} = 1$ and
    (ii) $a \cdot \bar{a} = 0$.
6. There exist at least two distinct elements in B.

A *switching algebra* is a Boolean algebra in which the number of elements in the set B is precisely 2.

---

[3] The terms "binary operator" and "unary operator" refer to the number of arguments involved in the operation: two or one, respectively.

In this definition, the two binary operators, represented by the signs $+$ and $\cdot$, are called the OR and the AND, respectively, and the unary operator, represented by the overbar $\overline{}$, is called the NOT or the *complement* operator. The specific behavior of these operators can be deduced from the postulates, as we will show in a moment. Before we take a close look at switching algebra, which is really the main subject for the remainder of this book, let us consider some of the algebraic implications of these postulates by stating and proving some theorems which will be useful later.

Theorem 3.2.1 (Idempotence).   For all elements a in the set B,

(i) $a + a = a$ and

(ii) $a \cdot a = a$.

Proof.   Consider first $a + a$:

$$
\begin{aligned}
a + a &= (a + a) \cdot 1 && \text{(Postulate 2(ii))}\\
&= (a + a) \cdot (a + \overline{a}) && \text{(Postulate 5(i))}\\
&= a + a \cdot \overline{a} && \text{(Postulate 4(ii))}\\
&= a + 0 && \text{(Postulate 5(ii))}\\
&= a && \text{(Postulate 2(i))}
\end{aligned}
$$

The proof of the second part follows similarly:

$$
\begin{aligned}
a \cdot a &= a \cdot a + 0 && \text{(Postulate 2(i))}\\
&= a \cdot a + a \cdot \overline{a} && \text{(Postulate 5(ii))}\\
&= a \cdot (a + \overline{a}) && \text{(Postulate 4(i))}\\
&= a \cdot 1 && \text{(Postulate 5(i))}\\
&= a && \text{(Postulate 2(ii))}\\
& && \text{QED}
\end{aligned}
$$

An interesting observation should be made here, and that is that both the postulates and Theorem 3.2.1 are stated in two parts. The difference between the two parts is that all ANDs and ORs and all 1s and 0s are interchanged. This, in fact, is the definition of the *dual of a Boolean expression*. Thus, part (ii) of Theorem 3.2.1 is the dual of part (i). Furthermore, note that the proof of part (ii) uses, at each step, the dual of the postulate used in proving the corresponding step of part (i). This results in the *principle of duality*.

The Principle of Duality.   If a Boolean statement is proved true, then the dual of the statement is also true.

Using this principle, we need only prove the first half of a statement, since the dual portion is provable by using the dual postulates. Consider as an example the next theorem.

Theorem 3.2.2. For all elements a in the set B,

(i) $a \cdot 0 = 0 \cdot a = 0$ and

(ii) $a + 1 = 1 + a = 1$.

Proof. Consider part (i):

$$a \cdot 0 = 0 + a \cdot 0 \qquad \text{(Postulate 2(i))}$$

$$= a \cdot \bar{a} + a \cdot 0 \qquad \text{(Postulate 5(ii))}$$

$$= a \cdot (\bar{a} + 0) \qquad \text{(Postulate 4(i))}$$

$$= a \cdot (\bar{a}) \qquad \text{(Postulate 2(i))}$$

$$= 0 \qquad \text{(Postulate 5(ii))}$$

Also, by postulate 3(ii), $a \cdot 0 = 0 \cdot a$. Since the result is true for $a \cdot 0 = 0$, by the principle of duality it is also true for $a + 1 = 1$.   QED

The reader should fill in the proof of the second part of Theorem 3.2.2.

Postulate 5(i) states that the complement of an element is in the set B but says nothing about the possibility that an element might have another complement. In fact, as the next theorem demonstrates, the complement of an element is unique, a very important fact to remember.

Theorem 3.2.3. Let a be an element of B. Then $\bar{a}$ is unique.

Proof. We will prove this by assuming that $\bar{a}$ is not unique and show that this results in a contradiction. Assume that a has two distinct complements (not equal), $\bar{a}$ and b. Then, by Postulate 5, we must have that

$$a + b = 1 \qquad \text{and} \qquad a + \bar{a} = 1$$

and

$$a \cdot b = 0 \qquad \text{and} \qquad a \cdot \bar{a} = 0$$

Then

$$\begin{aligned}
\bar{a} &= \bar{a} \cdot 1 \qquad &\text{(Postulate 2(ii))} \\
&= \bar{a} \cdot (a + b) \\
&= \bar{a} \cdot a + \bar{a} \cdot b \qquad &\text{(Postulate 4(i))} \\
&= 0 + \bar{a} \cdot b \\
&= \bar{a} \cdot b \qquad &\text{(Postulate 2(i))}
\end{aligned}$$

Next, in a similar way, consider b:

$$\begin{aligned}
b &= b \cdot 1 \\
&= b \cdot (a + \bar{a}) \\
&= b \cdot a + b \cdot \bar{a} \\
&= 0 + b \cdot \bar{a} \\
&= \bar{a} \cdot b
\end{aligned}$$

From these two cases we observe that

$$b = \overline{a} \cdot b = \overline{a}$$

which contradicts the assumption that the two complements of a were distinct.   QED

Nothing has been said to this point about the number of elements in the set B other than that it must be at least 2. It turns out that a general Boolean algebra has 2^n elements.[4] We have already mentioned that a switching algebra is basically a two-element Boolean algebra which, obviously, has the two elements 0 and 1. From this point on, we will restrict our attention to switching algebras only. A few of the problems given at the end of the chapter will examine some simple aspects of general Boolean algebras.

The AND, OR, and NOT operators have not yet been formally defined. However, the way in which they operate on 0 and 1 may be deduced from the postulates and the theorems just presented. Obviously, for the binary operators, AND and OR, there are four possibilities for values of the two switching variables operated on. Let x and y be two such switching variables, where a switching variable is taken to mean a variable that can take on only the value 0 or 1. Now consider the AND operation, $x \cdot y$. All of the possibilities for x and y, along with the resulting value of the AND, $x \cdot y$, are given in the following table:

AND

x	y	$x \cdot y$	
0	0	0	(Theorem 3.2.1(ii))
0	1	0	(Theorem 3.2.2(i))
1	0	0	(Postulate 3(ii))
1	1	1	(Theorem 3.2.1(ii)

In a similar manner, or by using the principle of duality, the defining table of values for the OR operator becomes

OR

x	y	$x + y$
0	0	0
0	1	1
1	0	1
1	1	1

By simply observing from Theorem 3.2.3 that the complement of a value is unique, we find, since there are only two possible values that a switching variable can take on, that the NOT operator must be defined as follows:

NOT

x	$\overline{x}$
0	1
1	0

[4] The proof of this is beyond the scope of this book and will not be given here, but a readable proof can be found in Elliott Mendelson, *Boolean Algebra and Switching Circuits*, Schaum's Outline Series, by McGraw-Hill, New York, 1970, beginning on p. 135, Sec. 5.2.

Restricting our attention to a switching algebra and then using the definitions for the three operators just given, we can easily deduce further results for the algebra by *completely enumerating* all possible values for the switching expression.[5]

Theorem 3.2.4 (Involution).   Let x be a switching variable. Then

$$\overline{(\overline{x})} = x$$

Proof.   We will prove this by complete enumeration:

x	$\overline{x}$	$\overline{(\overline{x})}$
0	1	0
1	0	1

Since the left column is identical to the right column and since we have listed all possibilities, we have proved the result.   QED

Another example will further illustrate the process of enumeration.

Theorem 3.2.5.   Let x and y be two switching variables. Then

(i) $x + x \cdot y = x$ and
(ii) $x \cdot (x + y) = x$

Proof.   Again, the proof will be by complete enumeration:

x	y	$x \cdot y$	x	$+$	$x \cdot y$	
0	0	0	0	+	0	= 0
0	1	0	0	+	0	= 0
1	0	0	1	+	0	= 1
1	1	1	1	+	1	= 1

Since the column labeled x and the column labeled $x + x \cdot y$ are exactly the same, we have proved the result. Part (ii) is, of course, true by the principle of duality.   QED

A number of useful identities may be proved using the idea of complete enumeration. The following theorem lists a few of these identities. The proof is left as an exercise for the reader.

Theorem 3.2.6.   Let x, y, and z be switching variables. Then

1. *Associativity*
   (i) $x \cdot (y \cdot z) = (x \cdot y) \cdot z$
   (ii) $x + (y + z) = (x + y) + z$
2. (i) $x + \overline{x} \cdot y = x + y$
   (ii) $x \cdot (\overline{x} + y) = x \cdot y$

[5] These results also apply to the more general Boolean algebra.

3. *Consensus*

   (i) $x \cdot y + \bar{x} \cdot z + y \cdot z = x \cdot y + \bar{x} \cdot z$

   (ii) $(x + y) \cdot (\bar{x} + z) \cdot (y + z) = (x + y) \cdot (\bar{x} + z)$

3.3 DE MORGAN'S THEOREM

The complement of a variable in a switching algebra was defined by Postulate 5 of the Huntington postulates. We know that since the algebra is closed, by Postulate 1, that $x + y$ results in an element in the algebra and thus has a complement, $\overline{(x + y)}$. What we would like to know, however, is what this complement is in terms of the variables and their complements. De Morgan's theorem addresses this question.

    Theorem 3.3.1 De Morgan's Theorem.　Let x and y be two switching variables. Then

   (i) $\overline{(x + y)} = \bar{x} \cdot \bar{y}$

   (ii) $\overline{(x \cdot y)} = \bar{x} + \bar{y}$

    Proof.　This may easily be verified by complete enumeration, as follows:

x	y	$x + y$	$\overline{(x + y)}$	$\bar{x}$	$\bar{y}$	$\bar{x} \cdot \bar{y}$
0	0	0	1	1	1	1
0	1	1	0	1	0	0
1	0	1	0	0	1	0
1	1	1	0	0	0	0

    Since column $\overline{(x + y)}$ is identical to column $\bar{x} \cdot \bar{y}$ and all possibilities are listed, the result is proved. Part (ii) is true by the principle of duality.　　　　　QED

    This result is especially useful for the evaluation of complements of switching expressions. For example, suppose we are given the expression $\overline{[\bar{x} + y \cdot (\bar{z} + w)]}$ involving the variables w, x, y, and z and are asked to put this in a form where the complements are associated only with individual variables and not with groups of variables. Using De Morgan's theorem and some of the results from Section 3.2, this can easily be done:

$$\overline{[\bar{x} + y \cdot (\bar{z} + w)]} = \overline{(\bar{x})} \cdot \overline{[y \cdot (\bar{z} + w)]} \qquad \text{(Theorem 3.3.1(i))}$$

$$= \overline{(\bar{x})} \cdot [\bar{y} + \overline{(\bar{z} + w)}] \qquad \text{(Theorem 3.3.1(ii))}$$

$$= \overline{(\bar{x})} \cdot [\bar{y} + \overline{(\bar{z})} \cdot \bar{w}] \qquad \text{(Theorem 3.3.1(i))}$$

$$= x \cdot (\bar{y} + z \cdot \bar{w}) \qquad \text{(Theorem 3.2.4)}$$

$$= x \cdot \bar{y} + x \cdot z \cdot \bar{w} \qquad \text{(Postulate 4(i))}$$

Note, in this example, that application of De Morgan's theorem requires that the original expression first be partitioned into two pieces separated by either a $+$, as was the case here, or a center point. Continuing this on each of the resulting pieces allows successive application of these laws. This result may be extended to the complement of the AND or the OR of more than two variables by the following corollary to theorem 3.3.1

Corollary 3.3.2. Let $x_1, x_2, \ldots, x_n$ be n switching variables. Then

(i) $\overline{(x_1 \cdot x_2 \cdots x_n)} = \bar{x}_1 \cdot \bar{x}_2 + \cdots + \bar{x}_n$ and

(ii) $\overline{(x_1 + x_2 + \cdots + x_n)} = \bar{x}_1 + \bar{x}_2 \cdots \bar{x}_n.$

A second example illustrates this extension.

$$\overline{\{[\bar{x} \cdot \overline{(y + z)}] \cdot (y + w \cdot \bar{z}) \cdot (x + z)\}}$$
$$= \overline{[\bar{x} \cdot \overline{(y + z)}]} + \overline{(y + w \cdot \bar{z})} + \overline{(x + z)}$$
$$= \overline{(\bar{x})} + \overline{[\overline{(y + z)}]} + \bar{y} \cdot \overline{(w \cdot \bar{z})} + \bar{x} \cdot \bar{z} \qquad (3.3.1)$$
$$= x + y + z + \bar{y} \cdot [\overline{w} + \overline{(\bar{z})}] + \bar{x} \cdot \bar{z}$$
$$= x + y + z + \bar{y} \cdot (\overline{w} + z) + \bar{x} \cdot \bar{z}$$

As we shall see later, De Morgan's theorem plays a very important part in the design of the hardware of a computer.

3.4 SWITCHING FUNCTIONS

A *switching function* may be defined simply as a mapping from the set of binary n-tuples[6] into the set $\{0, 1\}$ and may be denoted, in the usual way, as, for example, $f(x_1, x_2, \ldots, x_n)$, where the x_i are switching variables. Since there are n variables, each of which can take on one of two values, 0 and 1, there must be a total of 2^n possible assignments for these n variables. For each of the possible assignments, the function f will, of course, take on a value of either 0 or 1.

There are quite a number of different ways in which a switching function may be represented. The expressions given in Section 3.3 are examples of switching functions; for instance, Equation (3.3.1) is a switching function on four variables. The purpose of this section, then, is to describe some of the more commonly used methods for denoting switching functions and to show how the functions can be derived and how they can be converted from one form to another.

3.4.1 Truth Tables

As defined above, a switching function is just an association of 0 and 1 with each of the possible assignments of the variables of a function. Because of this, a simple way of representing a switching function is to make a list of the possible variable assignments and note the value the function takes on for each assignment. Such a list is called a *truth table*.

[6] An *n*-tuple is an ordered set of *n* numbers, such as the 6-tuple (101101), which has 6 digits.

As an example, some function $f(x, y, z)$ might have the truth table shown in Figure 3.4.1. From this table we can determine what value $f(x, y, z)$ will take on for any possible assignment of the three variables. Thus, we can observe that if $x = 1$, and $y = 0$, and $z = 1$, then $f(x, y, z) = 1$.

x	y	z	$f(x, y, z)$
0	0	0	0
0	0	1	1
0	1	0	1
0	1	1	0
1	0	0	0
1	0	1	1
1	1	0	0
1	1	1	0

Figure 3.4.1   A truth table for a function $f(x, y, z)$.

To see how a truth table might be created, suppose we would like to describe a function $g(w, x, y, z)$ whose value is 1 whenever the decimal equivalent of the four variables, taken as a 4-bit number, is greater than 9. Such a function would be useful for checking whether or not a 4-bit number represents a legitimate BCD code. The truth table for this function would be as shown in Figure 3.4.2. Note that whenever w, x, y, and z, taken as a 4-bit number with w the high-order bit, takes on a value greater than 9, g takes on a value of 1.

Suppose now we are given the function[7]

$$h(x, y, z) = \bar{x} + y\bar{z} \qquad (3.4.1)$$

w	x	y	z	$q(w, x, y, z)$
0	0	0	0	0
0	0	0	1	0
0	0	1	0	0
0	0	1	1	0
0	1	0	0	0
0	1	0	1	0
0	1	1	0	0
0	1	1	1	0
1	0	0	0	0
1	0	0	1	0
1	0	1	0	1
1	0	1	1	1
1	1	0	0	1
1	1	0	1	1
1	1	1	0	1
1	1	1	1	1

Figure 3.4.2   The truth table for a BCD code checker.

[7] In what follows, the AND symbol, ($\cdot$), will be omitted if no confusion can occur. Thus $y \cdot \bar{x}$ will be written as $y\bar{x}$.

and are asked to construct the corresponding truth table. To do this, we simply note that $h = 1$ whenever $x = 0$, without regard to the values of y and z, and that $h = 1$ whenever $y = 1$ and $z = 0$, without regard to the value of x. The truth table for $h(x, y, z)$ thus becomes as shown in Figure 3.4.3.

x	y	z	$h(x, y, z)$
0	0	0	1
0	0	1	1
0	1	0	1
0	1	1	1
1	0	0	0
1	0	1	0
1	1	0	1
1	1	1	0

Figure 3.4.3 The truth table for the function given in Equation (3.4.1).

From the way in which we represent switching functions by truth tables, it is easy to count the number of possible switching functions on n variables. For each possible assignment of the n variables, we can define a function whose value is 0 and we can define another whose value is 1. Since there are 2^n possible assignments on the n variables, there must be $2^{(2^n)}$ possible switching functions on those n variables. For $n = 2$, then, there must be 16 possible functions, and for $n = 4$, there are 65,536 possible functions. The table in Figure 3.4.4 lists all of the functions on two variables and lists names given to some of these functions.

$xy = 00$	01	10	11	Function	Name
0	0	0	0	0	
0	0	0	1	xy	AND
0	0	1	0	$x\bar{y}$	Implication
0	0	1	1	x	
0	1	0	0	$\bar{x}y$	
0	1	0	1	y	
0	1	1	0	$\bar{x}y + \bar{y}x$	Exclusive OR
0	1	1	1	$x + y$	OR
1	0	0	0	$\overline{(x + y)}$	NOR
1	0	0	1	$\bar{x}\bar{y} + xy$	Equivalence
1	0	1	0	$\bar{y}$	
1	0	1	1	$x + \bar{y}$	
1	1	0	0	$\bar{x}$	
1	1	0	1	$\bar{x} + y$	
1	1	1	0	$\overline{(xy)}$	NAND
1	1	1	1	1	

Figure 3.4.4 A listing of the switching functions on two variables and the names given to some of these functions.

3.4.2 Canonical Forms

The truth table representation for a switching function has its uses, but it is certainly not very compact, especially for functions of a large number of variables. There are several alternatives to this tabular representation. One of the simplest is to list only the assignments for which a function is 1 or, alternatively, list those for which the function is 0. Such a list is, of course, unique for any given function, and is referred to as a canonical representation.[8]

One way of writing a canonical representation is as an equation or an expression in terms of the variables. Consider, as an example, the function $f(x, y, z)$ given by the table of Figure 3.4.1. The function f is 1 whenever $x = 0$, $y = 0$, and $z = 1$ *or* whenever $x = 0$, $y = 1$, and $z = 0$ *or* whenever $x = 1$, $y = 0$, and $z = 1$. It is easily verified, by simply substituting these values for the variables, that the expression $\bar{x}\bar{y}z + \bar{x}y\bar{z} + x\bar{y}z$ takes on the value 1 only when these particular variable assignments are made, and so f can be written as

$$f(x, y, z) = \bar{x}\bar{y}z + \bar{x}y\bar{z} + x\bar{y}z \qquad (3.4.2)$$

Equation (3.4.2) is made up of the "sum" of three "product" terms, where each "product" term is the AND of a set of *literals*. A literal is defined here as a variable or the complement of a variable. Thus, this equation is referred to as a *sum of products* (SOP) expression. In this case the expression consists of three product terms and nine literals. If a product term involves all the variables of a function, it is referred to as a *minterm*. Equation (3.4.2) is made of minterms only and is therefore called a *canonical minterm expression or expansion* of the function $f(x, y, z)$.

Consider now Equation (3.4.1). Although this equation is an SOP expression, it is not a canonical minterm expression for $h(x, y, z)$, since the product terms are not minterms. However, using the truth table shown in Figure 3.4.3 and proceeding as was done to derive Equation (3.4.2), we can easily find a canonical minterm expansion to be

$$h(x, y, z) = \bar{x}\bar{y}\bar{z} + \bar{x}\bar{y}z + \bar{x}y\bar{z} + \bar{x}yz + xy\bar{z} \qquad (3.4.3)$$

It was mentioned earlier that a canonical representation can also be made up of a list of the variable assignments which make the function 0. To write a canonical expression for a function based on the 0 values, all we need do is change our point of view with regard to the function. Consider, as an example, the function $h(x, y, z)$ given in Equation (3.4.3). If an assignment on the three variables makes the function h equal to 0, then this assignment must make $\bar{h}$ equal to 1. Therefore, first write the canonical minterm expansion for $\bar{h}$. From Figure 3.4.3, this becomes

$$\bar{h}(x, y, z) = x\bar{y}\bar{z} + x\bar{y}z + xyz \qquad (3.4.4)$$

---

[8] The term "canonical," as used here, refers to a list of items that defines a function with which other functions can be compared to determine equivalence.

What we are after is the canonical representation for h, not $\bar{h}$. From Theorem 3.2.4, we know that $h = \overline{(\bar{h})}$, and so all that needs to be done to obtain h is to complement Equation (3.4.4) using De Morgan's theorem. This yields

$$h(x, y, z) = (\bar{x} + y + z)(\bar{x} + y + \bar{z})(\bar{x} + \bar{y} + \bar{z}) \qquad (3.4.5)$$

This expression is in quite a different form from that of Equation (3.4.4). Here we have the "product" of three "sum" terms, and so we will describe this form as a *product of sums* (POS) expression. Each sum term involves all of the variables of the function and is called a *maxterm*. Thus, Equation (3.4.5) is referred to as a *canonical maxterm expression or expansion* of the function.

Since a canonic representation for a switching function is nothing but a list, it need not be given in literal form; other possibilities exist. One very common canonic representation for functions is found by treating the variable assignments as binary numbers and then listing the decimal equivalents of those assignments which cause the function to be 1 or, alternatively, 0. Using this representation, $h(x, y, z)$ of Figure 3.4.3 would be written as

$$h(x, y, z) = \Sigma\, m(0, 1, 2, 3, 6) \qquad (3.4.6)$$

where the Σ implies "sum" or OR and the lowercase m implies minterms. Using the assignments which make h zero, the representation in a similar form would be

$$h(x, y, z) = \Pi\, M(4, 5, 7) \qquad (3.4.7)$$

where the Π implies "product" or AND and the uppercase M implies maxterms. We will refer to these canonical forms as the *canonical minterm and maxterm index list representations*. Note that conversion from one of these forms to the other is simply a matter of listing the elements in one that do not appear in the other.

As a final example, consider the BCD checker of Figure 3.4.2. the min- and maxterm list representations for $g(w, x, y, z)$ become

$$g(w, x, y, z) = \Sigma\, m(10, 11, 12, 13, 14, 15)$$
$$= \Pi\, M(0, 1, 2, 3, 4, 5, 6, 7, 8, 9) \qquad (3.4.8)$$

which are found by simply listing the assignments for which $g = 1$, in the first case, and for which $g = 0$, in the second.

3.4.3 Conversion of SOP and POS Expressions to Canonic Forms

The equation

$$f(A, B, C) = \bar{A}B + AC \qquad (3.4.9)$$

is in an SOP form, but it is not a canonical minterm expansion of f, because the product terms are not minterms. It was shown above that the canonic expansion of f can be found from the truth table for the function. It is not necessary, however, to generate the truth table to get this expansion. In Equation (3.4.9), we note that the first product term becomes a minterm if variable C is included and the second becomes a minterm if variable B is included. This is easily done by ANDing each product term with 1 in the following form:

$$f(A, B, C) = \overline{A}B \cdot 1 + A \cdot 1 \cdot C$$
$$= \overline{A}B(C + \overline{C}) + A(B + \overline{B})C \qquad (3.4.10)$$
$$= \overline{A}BC + \overline{A}B\overline{C} + ABC + A\overline{B}C$$

This expression is now a canonical minterm expansion of f with alternate index list forms of

$$f(A, B, C) = \Sigma\, m(3, 2, 7, 5)$$
$$= \Pi\, M(0, 1, 4, 6) \qquad (3.4.11)$$

It sometimes occurs that an equation is simpler in the SOP form than in the POS form, and so it is useful to be able to convert between these two representations. Again, consider Equation (3.4.9). To get this expression into a POS form, we may apply Postulate 4(ii) as necessary to break up the product terms. Thus,

$$f(A, B, C) = \overline{A}B + AC$$
$$= (\overline{A}B + A)(\overline{A}B + C)$$
$$= (A + \overline{A})(A + B)(C + \overline{A})(C + B)$$
$$= 1 \cdot (A + B)(\overline{A} + C)(B + C) \qquad (3.4.12)$$
$$= (A + B)(\overline{A} + C)(B + C)$$
$$= (A + B)(\overline{A} + C) \qquad \text{(by consensus)}$$

In this case, the complexity of the expressions is the same: two sum terms and four literals.

The sum terms in Equation (3.4.12) can be converted to maxterms by using the dual process that was used to obtain the minterms in Equation (3.4.10). The sum term $A + B$, for example, becomes

$$A + B = A + B + 0$$
$$= A + B + C\overline{C}$$
$$= (A + B + C)(A + B + \overline{C})$$

In a like manner, the remaining sum terms can be converted to maxterms to produce the canonical maxterm expansion for f of

$$f(A, B, C) = (A + B + C)(A + B + \overline{C})(\overline{A} + B + C)(\overline{A} + \overline{B} + C) \qquad (3.4.13)$$

Postulate 4(i) can be used to carry out the reverse process of going from POS to SOP expressions. The reader should expand Equation (3.4.13) using this postulate to verify that $f(A, B, C)$ of this equation is equal to $f(A, B, C)$ of Equation (3.4.10).

3.5 SIMPLIFICATION OF SWITCHING FUNCTIONS

In general, the cost of implementing an equation in hardware is related directly to the number of terms and the number of literals in each term of the expression. It is therefore important that we be able to reduce the complexity of an equation before it is cast in hardware. The purpose of this section is to examine some of the various ways by which we can simplify switching expressions before we implement them.

There are three fundamental approaches we will consider here which can be used to simplify switching expressions. The first approach uses the postulates and other results to reduce the form of an expression algebraically. This approach generally requires a good deal of experience to accomplish a reduction with any degree of facility and is therefore used sparingly. It is important, however, that we develop some feeling for this process if we are going to understand the other approaches. The second approach is a pictorial or diagrammatic approach that uses a map, called a *Karnaugh map*, on which the function is plotted. From this plot, groups of minterms that can combine to form a single product term are easily identified. This approach, however, is limited in practice to functions of six variables or fewer. The final technique is one that can be implemented on a computer and can, therefore, handle functions of an arbitrarily large number of variables. Typical of procedures of this type is one called the *Quine-McCluskey algorithm*. Each of these reduction methods will be described in what follows.

3.5.1 Algebraic Manipulation

Since the objective here is to simplify switching functions, we need to define just what is meant by simplification. Basically, an expression will be considered simplified whenever it contains a minimal number of literals and terms, either product or sum terms. By minimal, we mean that any other expression having fewer terms and literals will not represent the original function, i.e., will not produce the truth table of the original function. If the minimal expression is in SOP form, it will be called, naturally enough, a *minimal sum of products* (SOP) expression, and if it is in POS form, then it is a *minimal product of sums* (POS) expression.

Since we are dealing with a switching algebra, we may use the theorems and postulates of the algebra to find this minimal form. There are three basic results on which the reduction procedures heavily depend. For the minimizing of SOP expressions, these are

Result 1. $xy + \bar{x}y = y$ (easily verified using distributivity)

Result 2. $x + \bar{x}y = x + y$ (Theorem 3.2.6, part 2(i))

Result 3. $\bar{x}z + xy + yz = \bar{x}z + xy$ (Theorem 3.2.6, part 3(i)(consensus))

Of course, the dual of these results would be used for minimizing functions given in POS form. Other results are used on occasion, but these three apply most often. A simple algebraic reduction procedure consists of applying result 1 to the function until it cannot be applied further, and then doing the same with result 2. When result 2 can no longer be applied, we go back to result 1. We continue until neither result 1 nor result 2 applies, and then we go to result 3. When none of results 1, 2, and 3 can be applied, we may assume that the minimal form has been found. It turns out that this assumption is not always correct. However, the resulting form is usually close to minimal. An example will help illustrate this processs. Let

$$f(w, x, y, z) = \overline{w}\overline{x}z + \overline{w}xz + xyz + wxy \qquad (3.5.1)$$
$$= [\overline{x}(\overline{w}z) + x(\overline{w}z)] + xyz + wxy$$
$$= \overline{w}z + xyz + wxy \qquad (\text{result } 1)$$
$$= \overline{w}z + w(xy) + z(xy)$$
$$= \overline{w}z + wxy \qquad (\text{result } 3)$$

As can be seen from this example, it is not always obvious how to factor the expression at each step so as to apply one of the three results listed above. This, of course, is the part that takes experience. Consider another example. Let

$$g(A, B, C, D) = AB\overline{C} + ABC + BCD + \overline{A}CD + A\overline{B}\,\overline{C}D \qquad (3.5.2)$$
$$= [(AB)\overline{C} + (AB)C] + BCD + \overline{A}CD + A\overline{B}\,\overline{C}D$$
$$= AB + BCD + \overline{A}CD + A\overline{B}\,\overline{C}D \qquad (\text{result } 1)$$
$$= A[B + \overline{B}(\overline{C}D)] + BCD + \overline{A}CD$$
$$= A[B + \overline{C}D] + BCD + \overline{A}CD \qquad (\text{result } 2)$$
$$= AB + A\overline{C}D + BCD + \overline{A}CD$$
$$= [A(B) + \overline{A}(CD) + BCD)] + A\overline{C}D$$
$$= (AB + \overline{A}CD) + A\overline{C}D \qquad (\text{result } 3)$$
$$= AB + \overline{A}CD + A\overline{C}D$$

As a final example, consider the simplification of expression (3.3.1), which is repeated here:

$$x + y + z + \overline{y}(\overline{w} + z) + \overline{x}\overline{z} = (x + \overline{x}\overline{z}) + [y + \overline{y}(\overline{w} + z)] + z$$
$$= x + \overline{z} + y + \overline{w} + z + z \qquad (3.5.3)$$

Although none of the three results listed above can be applied at this stage, this expression clearly simplifies to the constant value 1, because it contains the term $z + \overline{z} = 1$ and, in the switching algebra, $1 +$ "anything" $= 1$.

These examples illustrate some of the difficulties involved in simplifying switching expressions algebraically. Although algebraic simplification is not totally straightforward, it can often result in a simplified form much more rapidly than use of the other two methods to be described shortly. It is therefore important that the reader develop some degree of facility with this process. A number of problems are given at the end of the chapter to help in this regard.

3.5.2 Prime Implicants

In an SOP expression, each of the product terms is called an *implicant* of the function, because it "implies" the function in the sense that if the product term is 1 then the function is also 1. Suppose some function $h(w, x, y, z)$ has, among others, the four minterms $\overline{w}\overline{x}yz$, $\overline{w}xyz$, $w\overline{x}yz$, and $wxyz$. Each of these product terms is, of course, an implicant of h. In simplifying h, we note that the first two minterms can combine by observing that the sum is equal to $\overline{w}(\overline{x}yz + xyz)$, which, upon applying result 1 to the term in parentheses, reduces to $\overline{w}yz$. In a similar way, the second two minterms combine to yield wyz. These two product terms are also implicants of h, since $h = 1$ if either is 1. Furthermore, they are smaller, in terms of the number of literals, than the original minterms. However, these two product terms will also combine to give the term yz, which is yet a smaller implicant of h. Note here that each of $\overline{w}yz$ and wyz, both implicants of h, also implies yz, because any assignment that makes either of them 1 also makes $yz = 1$. Continuing this line of thought, it would seem that the simplification process involves finding the set of "smallest" implicants of the given function. Specifically, we will define a *prime implicant* as an implicant of a function which does not imply any other implicant of the function. Thus yz is a prime implicant of h.

For any given switching function, it should be fairly obvious that the set of prime implicants is unique, since they are derived from a unique set of minterms. The question is whether we need to use all of the prime implicants to represent the function in minimal form. To answer this question, note that the prime implicant yz *covers* or is "made up" of the four minterms $\overline{w}\overline{x}yz$, $\overline{w}xyz$, $w\overline{x}yz$, and $wxyz$. Now, if all of the minterms of a function are covered by some proper subset of the set of prime implicants, including in the final expression those not in this subset would yield an expression for the function which is larger than necessary. As an example, let $f(x, y, z) = \overline{x}y + xz + yz$, which has as minterms $\overline{x}yz$, $\overline{x}y\overline{z}$, xyz, and $x\overline{y}z$. Since none of the terms $\overline{x}y$, xz, and yz implies any of the others, they must be the prime implicants of f. However, $\overline{x}y$ covers the minterms $\overline{x}yz$ and $\overline{x}y\overline{z}$, and xz covers xyz and $x\overline{y}z$. Since all four of the minterms of f are covered by these two prime implicants, f can be written as $f(x, y, z) = \overline{x}y + xz$, which we already knew because of the consensus theorem, Theorem 3.2.6, part 3(i).

From these observations, we may conclude that the determination of a minimal SOP expression involves, *first*, finding all of the prime implicants of the function and then, *second*, finding a minimal subset of these prime implicants which covers all of the minterms of the function. Such a subset is called a *minimal cover for the function*. Similar observations may be made to find the minimal POS expression of the function. The two commonly used methods for finding a minimal closed cover are discussed in the next two sections.

3.5.3 Karnaugh Maps

In 1953 M. Karnaugh published an article describing a geometrical method for finding a minimal closed cover. This approach has been designated, naturally enough, the *Karnaugh map* method and is based on mapping minterms onto a surface in such a way that minterms that differ in one literal are adjacent to each other on the surface. The reason for this mapping is that when two minterms differ in one literal, they can be combined to form a product term which has this literal missing. For example, the two minterms $\overline{A}BC$ and ABC differ in only one literal, and therefore the sum reduces to $\overline{A}BC + ABC = BC$. Figure 3.5.1 shows a two-variable Karnaugh map. Each square in the map corresponds to a minterm; these minterms are indicated in the figure. Observe that every pair of adjacent squares corresponds to two minterms which differ in exactly one literal. Notice that all of the minterms in the column labeled B contain the literal B and that all of those in the other column contain $\overline{B}$. Similarly for the rows. This figure also gives the mapping of some

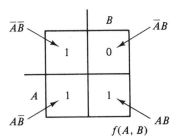

Figure 3.5.1 A two-variable Karnaugh map.

function $f(A, B)$, with a 1 in each square corresponding to a minterm of f. The other squares are automatically, at least for the moment, set to 0. The prime implicants are easily found by grouping the 1 cells into as large a block of adjacent cells as possible. For example, the pair of cells $A\overline{B}$ and $\overline{A}\,\overline{B}$ group together to give $\overline{B}$. A single square in this map is termed a *1-cube*. When two adjacent squares are taken together, the result is a *2-cube*. Two 2-cubes that are adjacent, or have a long edge in common, can be grouped to give a *4-cube*, and so on. The largest cubes of 1s, then, represent the prime implicants (why?). Thus, in Figure 3.5.2, the prime implicants are A and $\overline{B}$, and so $f(A, B) = A + \overline{B}$, with the coverings explicitly indicated by the circled regions in the figure.

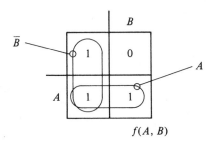

Figure 3.5.2 The coverings for the prime implicants of $f(A, B)$.

By taking two 2-variable Karnaugh maps and placing them side by side after reflecting one of the two variable maps, we obtain a three-variable map. Taking two three-variable maps and placing them side by side, again after reflecting one of the maps, we get a four-variable map. This process can be continued indefinitely, although the practical limit is for maps of six variables. Figure 3.5.3 shows maps for three and four variables.

Consider the three-variable map for a moment. This map consists of overlapping regions, three of which are indicated in the figure, each corresponding to an uncomplemented variable. Each region contains all of the possible minterms of three variables in which the variable that names the region appears (in this case) uncomplemented. Thus the region, or the 4-cube, marked y covers the four minterms $\overline{x}y\overline{z}$, $\overline{x}yz$, $xy\overline{z}$, and xyz. The intersection of two 4-cubes, such as y and z, forms a 2-cube—in this case, one covering the two minterms $\overline{x}yz$ and xyz, which combine to yield the product term yz, shown shaded in Figure 3.5.3(a). The intersection of three 4-cubes forms a 1-cube, which contains exactly one minterm. The portion of the map not covered by a variable corresponds to that covered by the complement of the variable. Notice that the leftmost and the rightmost columns of the

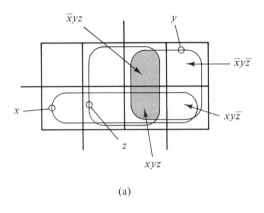

(a)

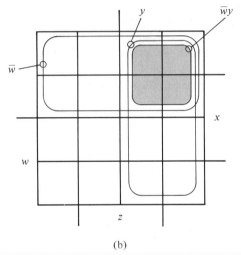

(b)

Figure 3.5.3 Karnaugh maps. (a) A three-variable map with some regions formed by the intersection of 4-cubes. (b) A four-variable map and the intersection of two 8-cubes to form a 4-cube.

three-variable map are adjacent, since they have the literal $\overline{z}$ in common. Thus, we may think of this map as being wrapped around a cylinder.

The four-variable map is similar, except that a region corresponding to a literal, such as $\overline{w}$ or y, as shown in Figure 3.5.3(b), is an 8-cube. Observe that the intersection of the 8-cubes for $\overline{w}$ and y forms the 4-cube corresponding to the product term $\overline{w}y$. Note also that the left and right columns of the four-variable map are adjacent, as are the top and bottom rows. Thus, in this case, we can think of the map as being located on a torus, or "doughnut."

Consider the mapping of some function $g(x, y, z)$ shown in Figure 3.5.4. The problem is to list all of the prime implicants and to find a minimal cover from this set. First, we must find the prime implicants from the map by finding the largest possible cubes that cover subsets of minterms of the function g. For example, the two adjacent 1s in the upper left-hand corner of the map, shown circled in the figure, form a 2-cube not adjacent to any other 2-cube. Thus this 2-cube must correspond to a prime implicant. The prime implicant

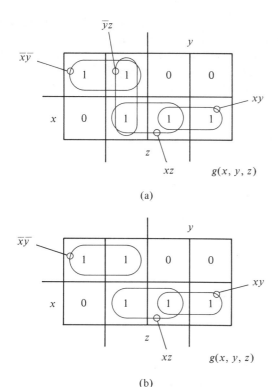

(a)

(b)

Figure 3.5.4 Mapping of a function $g(x, y, z)$. (a) The four prime implicants of $g(x, y, z)$. (b) A minimal closed cover for $g(x, y, z)$.

can be found as the intersection of the two 4-cubes which cover this 2-cube. In this case, the two 4-cubes in question are $\overline{y}$ and $\overline{x}$, and so the prime implicant formed by this 2-cube is $\overline{x}\overline{y}$. In a similar manner, three other prime implicants are found to be $\overline{y}z$, xz, and xy, as shown in Figure 3.5.4(a).

Next we must find a minimal subset of these prime implicants that covers all of the minterms of $g(x, y, z)$. To do this, first note the following. Minterm $\overline{x}\overline{y}\overline{z}$ is covered only by the prime implicant $\overline{x}\overline{y}$, and minterm $xy\overline{z}$ is covered only by prime implicant xy. Thus, these two prime implicants *must* be included in the minimal SOP form for $g(x, y, z)$. Such a prime implicant—one that covers a minterm not covered by any other prime implicant— is called an *essential prime implicant*. The remaining three minterms can be covered in two possible ways. However, since minterms $\overline{x}\overline{y}z$ and xyz are both covered by the two essential prime implicants, all we need worry about is the one minterm remaining uncovered, $x\overline{y}z$, which can be covered by either of the remaining prime implicants. For no particular reason, we will choose xz. Thus, a minimal sum of products representation for g is $g(x, y, z) = \overline{x}\overline{y} + xy + xz$; this is indicated by the circled terms in Figure 3.5.4(b).

Suppose, now, we are given the four-variable function

$$f(w, x, y, z) = \overline{w}\overline{y} + \overline{w}y\overline{z} + wxy + xy\overline{z} \tag{3.5.4}$$

and are asked to find a minimal sum of products representation. We will begin by plotting the function in a Karnaugh map as shown in Figure 3.5.5(a). To do this we simply place

1s in the squares covered by each product term. For example, $\overline{w}\overline{y}$ represents the intersection of the two 8-cubes $\overline{w}$ and $\overline{y}$ and is shown as the 4-cube in the upper left-hand corner of the figure. The remaining covers are also shown in the figure. Now that the function is plotted, we can find a minimal set of prime implicants that can be used to represent the function. Figure 3.5.5(b) shows the required cover, which results in the reduced expression for f of

$$f(w, x, y, z) = \overline{w}\overline{y} + \overline{w}\overline{z} + wxy \qquad (3.5.5)$$

Note that all three prime implicants are essential and cover all of the minterms.

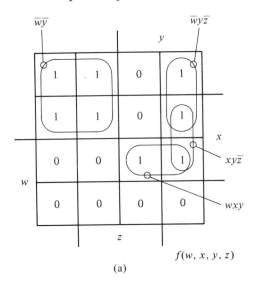

(a)

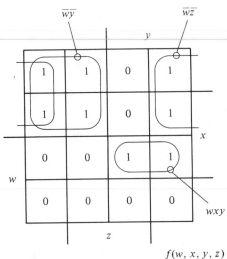

(b)

$f(w, x, y, z)$

Figure 3.5.5 Plots of $f(w, x, y, z)$: (a) plot of Equation (3.5.4); (b) plot of the reduced equation, Equation (3.5.5).

The form of the maps that we have just used is not always convenient. Consider, for example, the function $h(x, y, z)$ given in Figure 3.4.3. Since this is a truth table given in terms of 1s and 0s, it would be easier to plot the function if the Karnaugh map were labeled in terms of the variable values. Figure 3.5.6(a) shows a map that is so labeled and gives the plot of h. It can be seen that this mapping results in the minimal SOP representation for h of $h(x, y, z) = \bar{x} + y\bar{z}$, as shown in Figure 3.5.6(b).

Algebraic equations and truth tables are only two ways of representing switching functions. A minterm or maxterm index list is yet another way to present a function. Suppose we are to minimize the function

$$g(a, b, c, d) = \Sigma\, m(0, 4, 6, 7, 12, 13, 14, 15)$$

Figure 3.5.7 shows a labeling that makes plotting of g easy. In this form, each square is labeled with the corresponding minterm index value. For example, the square labeled 6 corresponds to the assignment on (a, b, c, d) of (0110) and the minterm $\bar{a}bc\bar{d}$, which is the intersection of the four 8-cubes $\bar{a}$, b, c, and $\bar{d}$. Thus, to plot g, all we need do is place a 1 in the square representing each of the minterms of g and place 0s elsewhere. Using this plot, we easily determine that the minimal SOP representation for g, as shown by the cover in the figure, is

$$g(a, b, c, d) = ab + bc + \bar{a}\bar{c}\bar{d}$$

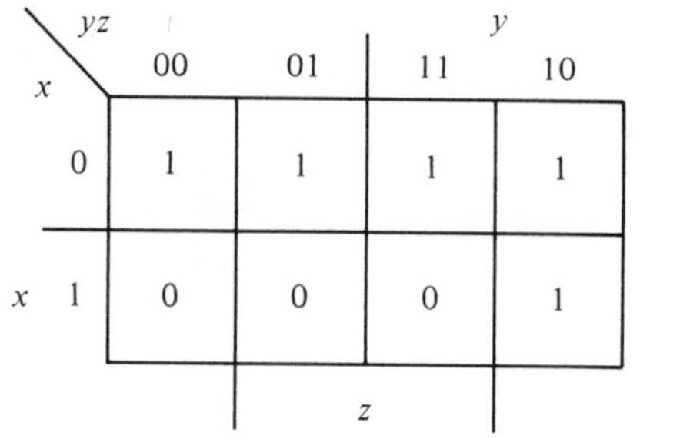

(a)

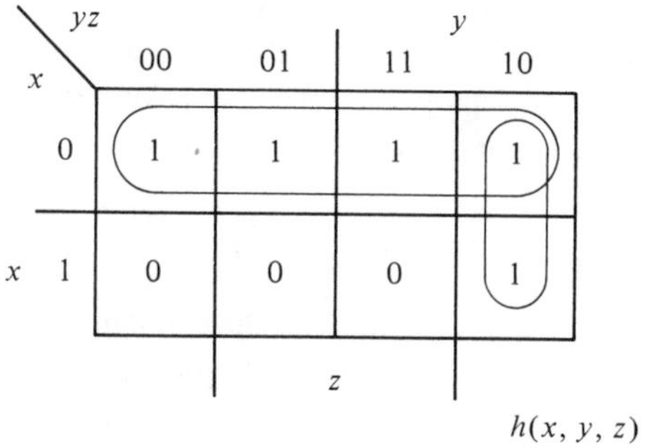

$h(x, y, z)$

(b)

$h(x, y, z)$

Figure 3.5.6 Karnaugh map labeling used with truth tables. (a) Plot of $h(x, y, z)$ from the truth table of Figure 3.4.3. (b) A minimal cover for $h(x, y, z)$.

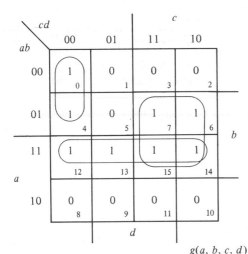

Figure 3.5.7 Karnaugh map labeling used with minterm index lists.

It should be apparent at this point that the three forms for the Karnaugh map given here each have their own value. Thus, for example, if we are given an expression as an index list and asked to write a minimal SOP expression for the function, we would plot the function on a Karnaugh map having each square identified by its index value and then replot the function in a map showing the regions associated with each variable. This was precisely what was done in the last example, shown in Figure 3.5.7. In general, therefore, which form of the map we use depends on how the function to be plotted is given and in what form we are required to express the function.

We have said very little, to this point, about how we would simplify expressions which are given in product of sums form. There is nothing difficult about handling such representations if we think of each sum term as the complement of a product term of the complement of the function. Then, instead of plotting a 1 on a map in the respective position, we plot a 0. An example will illustrate this approach. Let

$$F(a, b, c, d) = (a + \overline{b} + \overline{c})(\overline{a} + c + d)(\overline{b} + d) \tag{3.5.6}$$

The term $(a + \overline{b} + \overline{c}) = \overline{(\overline{a}bc)}$ and so we will plot 0s in the 2-cube corresponding to $\overline{a}bc$. Doing the same with the other two terms results in the plot shown in Figure 3.5.8 after 1s are placed in the remaining squares. From this plot we observe that Equation (3.5.6) is a minimal product of sums expression, since we have a minimal cover for the 0s of $F(a, b, c, d)$. The equivalent minimal sum of products expression can be found by covering the 1s and is given as

$$F(a, b, c, d) = \overline{a}\overline{b} + \overline{c}d + ad + \overline{b}c \tag{3.5.7}$$

This process is easily reversed to obtain a minimal product of sums expression from any given Karnaugh map by simply covering the 0s and writing the sum terms corresponding to each grouping.

It was mentioned earlier that the map method is practical for functions of six variables

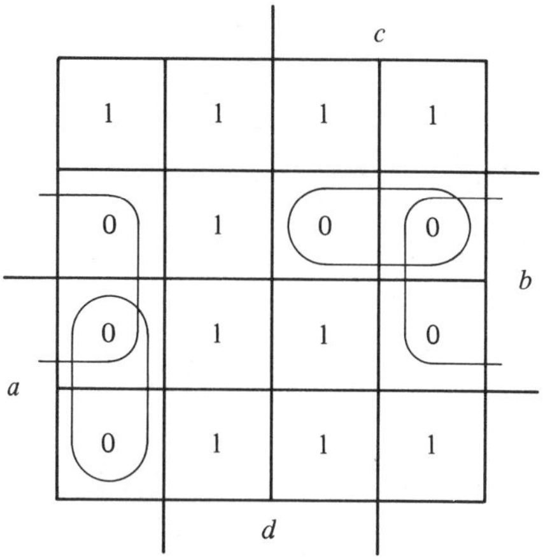

Figure 3.5.8 Mapping of the function
$F(a, b, c, d)$ $F(a, b, c, d)$ of Equation (3.5.7).

or fewer. There are two forms usually used for maps of five and six variables. In one form, a five-variable map is made up of two four-variable maps laid one on top of the other, with the one on top corresponding to $\bar{a}$ and the one on the bottom corresponding to a. This form is shown in Figure 3.5.9(a). In the other representation, a reversed image of a four-variable map, corresponding to a, is place beside a normal four-variable map, corresponding to $\bar{a}$. This form is shown in Figure 3.5.9(b). Figure 3.5.9 shows the two forms used for mapping the five-variable function

$$G(a, b, c, d, e) = \bar{b}\bar{c}\bar{d} + ab\bar{c} + \bar{b}\bar{c}e + \bar{a}\bar{d}e \tag{3.5.8}$$

The coverings for two of the prime implicants are shown in this figure. It appears that the adjacencies are a little easier to visualize in the form given in Figure 3.5.9(a). In particular, note the cover for the term $\bar{b}\bar{c}e$ given in Figure 3.5.9(b) and compare it with the same cover given in part (a). These ideas can be extended in the obvious way to produce the corresponding maps for six variables.

3.5.4 Don't Care Conditions

It occasionally happens that a switching function is defined in such a way that not all possible assignments of the variables occur. Such functions are said to be *incompletely specified*. For example, let the variables w, x, y, and z be used to encode BCD numbers and then define the function $f(w, x, y, z)$ as being 1 whenever the variables represent a BCD number divisible by 3. Otherwise, $f(w, x, y, z)$ is 0. To obtain an algebraic representation for f, we will plot the function on a Karnaugh map and then determine a minimal SOP expression. Assuming that only legitimate BCD numbers can occur, the question becomes, What do we plot as values in the map positions corresponding to the assignments of the variables that will not occur? Obviously, since they don't occur, we really *don't care* what values are plotted. However, a judicious choice may help in reducing the complexity of the realizing expression.

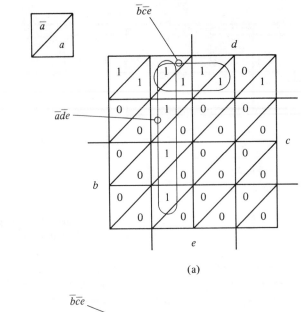

(a)

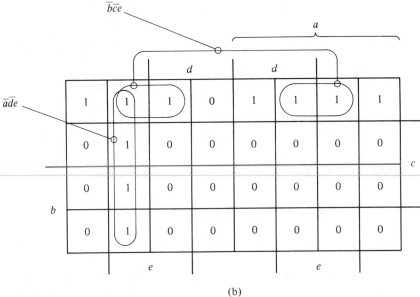

(b)

Figure 3.5.9 Two versions of five-variable Karnaugh map plottings of Equation (3.5.8). (a) A five-variable map made from two overlaid four-variable maps. (b) A five-variable map made of side-by-side four-variable maps.

Since we don't care what value f takes on for variable assignments that won't occur, we will plot a dash (–) in the map position corresponding to these assignments. In deriving a minimal expression for the function, we may consider the dash as either a 1 or a 0, as we wish. Thus, in finding the largest covers for the map entries containing a 1, we may use the dashed entries as 1s if this will make our cover larger. Using the don't cares in this way, f is plotted as in Figure 3.5.10, from which the expression is written as

$$f(w, x, y, z) = wz + \bar{x}yz + xy\bar{z} + \bar{w}\bar{x}\bar{y}\bar{z} \qquad (3.5.9)$$

Note that had the don't cares been assigned, a priori, the value 1, then two more terms would have to be added to Equation (3.5.9), and had they been assigned the value 0, each term in this equation would contain at least three literals. Thus the use of the don't cares has produced a simpler expression than otherwise possible.

It is important to observe that although six of the possible variable assignments in the above problem were assumed to be don't cares, the function $f(w, x, y, z)$ of Equation (3.5.9) does take on a value if any of these assignments is made. For example, if $(w, x, y, z) = (1100)$, a don't care in the map, Equation (3.5.9) reduces to 0. On the other hand, if $(w, x, y, z) = (1101)$, also a don't care, $f = 1$. As we will see in the next chapter, physical realizations are based on switching expressions, such as given in Equation (3.5.9). Thus a physical output will be produced for all possible physical inputs regardless of whether the problem statement includes don't cares or not.

When a switching function is defined using a minterm or maxterm index list, some method must be found to indicate the terms which are to be considered don't cares. This is usually done by writing them as an index list preceded by the letter d. Thus, in the above example,

$$\begin{aligned} f(w, x, y, z) &= \Sigma\, m(0, 3, 6, 9) + d(10, 11, 12, 13, 14, 15) \\ &= \Pi\, M(1, 2, 4, 5, 7, 8) + d(10, 11, 12, 13, 14, 15) \end{aligned} \qquad (3.5.10)$$

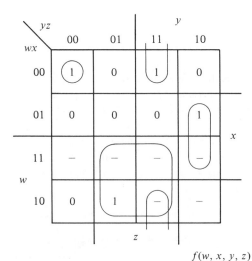

Figure 3.5.10   Using don't cares to simplify the function $f(w, x, y, z)$ of Equation (3.5.9).

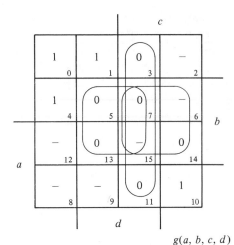

Figure 3.5.11 Plot of $g(a, b, c, d)$ of Equation (3.5.11) and a minimal POS cover.

Note that the don't cares are the same regardless of whether we are given a minterm or a maxterm list.

Before proceeding to a discussion of the Quine-McCluskey algorithm, consider the simplification of the following function of $g(a, b, c, d)$. Let

$$g(a, b, c, d) = \Pi M(3, 5, 7, 11, 13, 14) + d(2, 6, 8, 9, 12, 15) \qquad (3.5.11)$$

This function is plotted in Figure 3.5.11, from which the reader should verify that one of the possible minimal SOP expressions and one of the possible POS expressions are, respectively,

$$g(a, b, c, d) = \bar{b}\bar{c} + \bar{c}\bar{d} + \bar{b}\bar{d}$$
$$= (\bar{b} + \bar{c})(\bar{c} + \bar{d})(\bar{b} + \bar{d}) \qquad (3.5.12)[9]$$

3.5.5 The Quine-McCluskey Algorithm

When dealing with functions of more than five variables, the Karnaugh map method for finding a minimal sum of products representation becomes extremely unwieldy. Furthermore, the map method, which is easy for a *person* to use because people are good at recognizing visual patterns, is not good for computer implementation, since computers, at the moment, are not good at recognizing such patterns. A tabular method, which is easily implemented on a computer, is thus desirable for handling functions of large numbers of variables. One such method is the Quine-McCluskey algorithm. As with all methods of simplification, the tabular method consists of two parts: finding the prime implicants, and then finding a minimal cover.[10]

[9] Notice, in this particular case, that the POS form is the dual of the SOP form. Functions having this property are called *self-dual functions*.

[10] Recent algorithms have been developed which are more efficient than the Quine-McCluskey algorithm for computer computation. Some of these are referenced in the bibliography at the end of the chapter.

3.5.5.1 Finding the prime implicants.   In this method, determination of the prime implicants is based solely on the fact that $xp + \bar{x}p = p$, where p is some product of the remaining variables. The process begins with the minterm list. Each term is compared with each of the others to see whether the two differ in exactly one position. Whenever two terms do differ in exactly one position, the two are combined to form a product term having one fewer literal than the original two. After all terms are compared and all possible combinations are made, this comparison is repeated on the resulting new product terms. When a product term cannot be compared on this basis with any other term, it is a prime implicant. In carrying out these steps, we can reduce our effort by grouping the minterms on the basis of the number of 1s and then making comparisons only between groups that differ by one 1. The reason for this is that minterms can differ in only one position only if one has exactly one more 1 than the other. An example will help illustrate the process. Let

$$f(w, x, y, z) = \Sigma\, m(0, 2, 4, 5, 8, 10, 11, 12, 13, 15) \qquad (3.5.13)$$

For the moment, we will use the assignments corresponding to these minterm indices in finding the prime implicants (PIs). Thus minterm 5 corresponds to assignment (0101) and to the product term $\bar{w}x\bar{y}z$. Figure 3.5.12 shows the successive steps in the search process.

List 1 is found by listing the minterm assignments in groups according to the number of 1s in each assignment.

List 2 is derived from list 1 by combining those terms in list 1 that differ in exactly one position. This position, corresponding to the variable that is removed, is indicated with a dash (–) in list 2. For example, the two terms 0000 and 0010, corresponding to the minterms $\bar{w}\bar{x}\bar{y}\bar{z}$ and $\bar{w}\bar{x}y\bar{z}$, combine to give the term 00–0, corresponding to the product term $\bar{w}\bar{x}\bar{z}$. Since the assignments 0000 and 0010 have combined and therefore cannot be prime implicants, we place a check (✔) beside each. We continue, however, comparing 0000 with the rest of the elements in the second group, namely, 0100 and 1000. The resulting list 2 entries are 0–00 and –000. We then repeat the process of comparing each entry in

List 1		List 2		List 3	
0000	✔	00–0	✔	–0–0	*
0010	✔	0–00	✔	––00	*
0100	✔	–000	✔	–10–	*
1000	✔	–010	✔		
0101	✔	010–	✔		
1010	✔	–100	✔		
1100	✔	10–0	✔		
1011	✔	1–00	✔		
1101	✔	–101	✔		
1111	✔	101–	*		
		110–	✔		
		1–11	*		
		11–1	*		

Figure 3.5.12   Determination of the prime implicants for the function given in Equation (3.5.13).

the second group of list 1 with all entries in the third group and checking off those that combine. The results form the second group of list 2. We continue the process until all possible comparisons have been made.

List 3 is derived in a similar manner to that of list 2. Note, however, that in making the various comparisons in list 2, the two terms being compared must have the dash in the same variable position or else they cannot be products of the same variables and therefore cannot combine. For example, 00–0, in the first group of list 2, can be compared only with 10–0 in the second group. In this case, these two will combine to form –0–0, shown in list 3.

As we progress through the various comparisons, generating the sequence of lists, eventually we will find terms that will not combine with any other term in the list. Such terms are the prime implicants and are marked with an asterisk (∗). The term 101– in list 2, for example, which corresponds to $w\bar{x}y$, will not combine with anything in the next group and so is a prime implicant. This process yields the prime implicants indicated by the asterisks in Figure 3.5.12, namely, $w\bar{x}y$, wyz, wxz, $\bar{x}\bar{z}$, $\bar{y}\bar{z}$, and $x\bar{y}$.

An alternative representation is to use the minterm indices. Figure 3.5.13 gives the equivalent reduction procedure using these numbers. This table is organized exactly as before. Making the minterm comparisons is a bit different, however. A number in a group is compared with a number in a group below it by subtracting the former from the latter. If the difference is a power of 2, and is positive, then the two numbers combine to form a reduced product term in the next list. For example, comparing minterm 4 with 12, we get $12 - 4 = 8$, which is a power of 2, and positive. Therefore we combine the two to give the reduced product term 4, 12(8), where the (8) indicates the bit position that is missing in the second list. This corresponds to the comparison of 0100 (the 4) and 1100 (the 12), which produces the term –100, where the fourth bit position, corresponding to $2^3 = 8$, is the one missing. On the other hand, comparing 12, in the third group, with 11 in the fourth produces a difference of -1. These two minterms cannot combine, as is easily verified by comparing the corresponding assignments, namely, 1100 and 1011, which differ in three positions.

List 1		List 2		List 3	
0	✔	0, 2(2)	✔	0, 2, 8, 10(2, 8)	∗
2	✔	0, 4(4)	✔	0, 4, 8, 12(4, 8)	∗
4	✔	0, 8(8)	✔	4, 5, 12, 13(1, 8)	∗
8	✔	2 10(8)	✔		
5	✔	4, 5(1)	✔		
10	✔	4, 12(8)	✔		
12	✔	8, 10(2)	✔		
11	✔	8, 12(4)	✔		
13	✔	5, 13(8)	✔		
15	✔	10, 11(1)	∗		
		12, 13(1)	✔		
		11, 15(4)	∗		
		13, 15(2)	∗		

Figure 3.5.13 Using the minterm indices to find the prime implicants of $f(w, x, y, z)$ given in Equation (3.5.13).

Minterms

Prime implicants	0	2	4	5	8	10	11	12	13	15
10, 11(1)						X	X			
11, 15(4)							X			X
13, 15(2)									X	X
* 0, 2, 8, 10(2, 8)	X	X			X	X				
0, 4, 8, 12(4, 8)	X		X		X			X		
* 4, 5, 12, 13(1, 8)			X	X				X	X	

Figure 3.5.14   Covering table for $f(w, x, y, z)$ of Equation (3.5.13).

It is important, in forming the index list for the reduced product terms, that the lists be kept in lexicographical order. Doing this makes forming the third list from the second easy. If two terms have the same set of numbers in parentheses, then they may be compared by subtracting the first numbers in each list. The two terms then combine to form a term in list 3 if the difference is a power of 2 and positive. For example, 0, 2(2) and 8, 10(2) both have 2 in parentheses, and so we subtract the first number in each to get $8 - 0 = 8$. Thus, these two product terms combine to give 0, 2, 8, 10(2, 8) in list 3, which corresponds to –0–0 or $\bar{x}\bar{z}$.

3.5.5.2 Finding the minimal cover.   Once all of the prime implicants have been determined, a minimal subset must be found which covers the given function. This is done by setting up a covering table that shows all of the prime implicants and the minterms covered by each. The first step in determining a cover is to find all of the essential prime implicants. Figure 3.5.14 shows the table for the function $f(w, x, y, z)$ given in Equation (3.5.13). This table is set up in terms of the prime implicants given in index list form from Figure 3.5.13.

The rows of this table list the prime implicants (PIs) and identify, by an X, the minterms covered by each prime implicant. The essential prime implicants are readily found from this table by counting the number of prime implicants which cover the minterms. If a minterm is covered by only one PI, i.e., only one X appears in that minterm's column, then that prime implicant is essential. In this example we find two essential prime implicants, which are the two marked by asterisks (∗) in Figure 3.5.14.

Since minterms 0, 2, 4, 5, 8, 10, 12, and 13 are covered by the two essential prime implicants, we may reduce the size of the table by removing these columns and removing the rows corresponding to the essential PIs. Figure 3.5.15 shows the resulting table. Note

Minterms

Prime implicants	11	15
10, 11(1)	X	
* 11, 15(4)	X	X
13, 15(2)		X
0, 4, 8, 12(4, 8)		

Figure 3.5.15   Reduced covering table for $f(w, x, y, z)$ of Equation (3.5.13).

List form	Assignment	Product term
0, 2, 8, 10(2, 8)	–0–0	$\bar{x}\bar{z}$
4, 5, 12, 13(1, 8)	–10–	$x\bar{y}$
11, 15(4)	1–11	wyz

Figure 3.5.16 The prime implicants used to cover the function of equation (3.5.13).

that prime implicant 0, 4, 8, 12(4, 8) covers neither of the minterms 11 and 15 and so could be removed from the table. From this figure it is easily seen that the remaining two minterms, 11 and 15, are covered by the prime implicant 11, 15(4). Thus a minimal sum of products representation for f is found using the three prime implicants shown in their various forms in Figure 3.5.16. The resulting expression for f is

$$f(w, x, y, z) = \bar{x}\bar{z} + x\bar{y} + wyz \qquad (3.5.14)$$

The reader should verify this result by plotting f on a Karnaugh map.

3.5.5.3 Incorporation of don't care conditions.

Although the basic procedure just outlined also applies when don't care conditions enter the problem, some modifications are required in setting up the covering table. Because of these modifications, some complications may arise in finding a minimal cover. To illustrate, let us consider the function

$$g(a, b, c, d) = \Sigma\, m(0, 1, 3, 5, 13, 15) + d(2, 6, 10, 11, 12) \qquad (3.5.15)$$

The first step, as before, is to find the prime implicants. Since we wish to use the don't cares to maximize the number of minterms covered by each PI, we will include the don't cares in the minterm list used to find the prime implicants. Figure 3.5.17 shows the resulting prime implicant generation. Note that prime implicant 2, 6(4) is made up of don't cares only and so it will be ignored.

Since we don't care whether the terms 2, 6, 10, 11, and 12 are covered (they are, after all, don't care terms), we will not include these in the covering table. Using these

List 1	List 2	List 3
0 ✔	0, 1(1) ✔	0, 1, 2, 3(1, 2) *
1 ✔	0, 2(2) ✔	2, 3, 10, 11(1, 8) *
2 ✔	1, 3(2) ✔	
3 ✔	1, 5(4) *	
5 ✔	2, 3(1) ✔	
6 ✔	2, 6(4) *	
10 ✔	2, 10(8) ✔	
12 ✔	3, 11(8) ✔	
11 ✔	5, 13(8) *	
13 ✔	10, 11(1) ✔	
15 ✔	12, 13(1) *	
	11, 15(4) *	
	13, 15(2) *	

Figure 3.5.17 Derivation of the prime implicants of Equation (3.5.15).

Minterms

Prime implicants	0	1	3	5	13	15
1, 5(4)		X		X		
5, 13(8)				X	X	
12, 13(1)					X	
11, 15(4)						X
13, 15(2)					X	X
* 0, 1, 2, 3(1, 2)	X	X	X			
2, 3, 10, 11(1, 8)			X			

Figure 3.5.18 Covering table for $g(a, b, c, d)$ of Equation (3.5.15).

prime implicants, and ignoring the don't cares, the covering table becomes as shown in Figure 3.5.18. This table shows only one essential prime implicant, namely, 0, 1, 2, 3(1, 2), and so we need not consider the first three columns further. Each of the remaining minterms is covered by more than one prime implicant. Our job now is to select a minimal subset of these prime implicants which covers all of the required minterms. In general, there may be many ways in which this can be done. Usually, however, we will be interested in only a single solution and not all such possible covers. Section 3.5.5.4 describes a method that can be used to determine all of the covers that require a minimal number of product terms and literals. For now, however, let us concentrate on finding a single, minimal cover. Proceeding, then, we note in Figure 3.5.18 that prime implicant 11, 15(4) covers only minterm 15, whereas prime implicant 13, 15(2) covers both minterms 13 and 15. We say that a row of the table is *dominated* by another row if all of the minterms covered by the dominated row are also covered by the other row. A dominated row may, therefore, be removed from the table if the number of literals associated with the prime implicant of the dominated row is not less than the corresponding number for the dominating row. Thus, since 11, 15(4) is dominated by 13, 15(2) and both have the same number of literals, we may ignore the prime implicant 11, 15(4) and be assured that the resulting expression is no more complicated than any other possible expression for the function. Note, also, that after the minterms covered by the essential prime implicant are removed, prime implicant 1, 5(4) becomes dominated by 5, 13(8). Since both have the same number of literals, we can also ignore prime implicant 1, 5(4). The resulting, reduced covering table is shown in Figure 3.5.19.

After the table has been reduced, the remaining two prime implicants become essential. These are generally referred to as *secondary essential prime implicants*, since they become essential only after all other essential and dominated prime implicants are eliminated. Using

Minterms

Prime implicants	5	13	15	
* 5, 13(8)	X	X		secondary essential
* 13, 15(2)		X	X	secondary essential

Figure 3.5.19 Reduced covering table for $g(a, b, c, d)$ of Equation (3.5.15).

the one essential and the two secondary essential prime implicants, the function $g(a, b, c, d)$ reduces to

$$g(a, b, c, d) = \bar{a}\bar{b} + b\bar{c}d + abd \qquad (3.5.16)$$

where $\bar{a}\bar{b}$ corresponds to the essential prime implicant and $b\bar{c}d$ and abd correspond to the secondary essential prime implicants given in Figure 3.5.19. As we shall see in Section 3.5.5.4, this is not the only possible minimal expression for $g(a, b, c, d)$.

3.5.5.4 The Petrick algorithm. As indicated above, it quite often happens that there is more than one possible cover for a given function. In fact, it may happen that after all essential and secondary essential prime implicants are found, the remaining minterms can be covered in many ways. This would be the case if each column in the reduced covering table were to contain at least two X's. A table in which this is the case is said to be *cyclic*. As an example, let $h(a, b, c, d)$ be given by

$$h(a, b, c, d) = \Sigma\, m\, (1, 5, 7, 8, 10, 14) + d(0, 6, 9, 11, 13, 15) \qquad (3.5.17)$$

After the prime implicants are determined, the resulting covering table is as shown in Figure 3.5.20 and is seen to be cyclic, since every minterm is covered by at least two prime impicants.

To find a minimal cover, we can reason as follows. Minterm 1 is covered if we use prime implicant A or B; minterm 5 is covered if we use prime implicant B or D; and so on, for each of the minterms. The function will be covered if minterm 1 is covered *and* minterm 5 is covered *and* minterm 7 is covered *and* the other minterms through minterm 14 are covered. Now, if we use A to mean "use prime implicant A," then we can write a logical equation which expresses the requirements for the cover, as follows:

THE FUNCTION IS COVERED

$$= (A + B)(B + D)(D + E)(A + C)(C + F)(E + F) \qquad (3.5.18)$$

Reducing this expression by using the laws of Boolean algebra, we get

THE FUNCTION IS COVERED

$$= BCE + ABEF + BCDF + ABDF + ACDE + ADEF + ACDF + ADF \qquad (3.5.19)$$

Equation (3.5.19) can be interpreted as follows: The function is covered if we use prime implicants B and C and E or we use the prime implicants A and B and E and F or   . . .

Prime implicants		Minterms					
		1	5	7	8	10	14
A	0, 1, 8, 9(1, 8)	X			X		
B	1, 5, 9, 13(4, 8)	X	X				
C	8, 9, 10, 11(1, 2)				X	X	
D	5, 7, 13, 15(2, 8)		X	X			
E	6, 7, 14, 15(1, 8)			X			X
F	10, 11, 14, 15(1, 4)					X	X

Figure 3.5.20   A cyclic covering table for $h(a, b, c, d)$ of Equation (3.5.17).

Thus we have found *all* covers for the function. We need only select one from among this set which requires the smallest number of prime implicants and literals. In this case, there are two that require three prime implicants: *BCE* and *ADF*; all of the others require four prime implicants. Since both of these sets of prime implicants require the same number of literals, we can select either. Let us select prime implicants *B*, *C*, and *E*. These are as follows:

PI	Index list	Assignment	Literals
B	1, 5, 9, 13(4, 8)	—01	$\bar{c}d$
C	8, 9, 10, 11(1, 2)	10—	$a\bar{b}$
E	6, 7, 14, 15(1, 8)	–11–	bc

from which $h(a, b, c, d)$ becomes

$$h(a, b, c, d) = \bar{c}d + a\bar{b} + bc \qquad (3.5.20)$$

If we had used prime implicants *A*, *D*, and *F*, we would have obtained the second minimal SOP expression,

$$h(a, b, c, d) = \bar{b}\bar{c} + bd + ac \qquad (3.5.21)$$

Let us now go back to the covering table for the function $g(a, b, c, d)$ given in Figure 3.5.18. After we remove the columns associated with the essential prime implicant 0, 1, 2, 3(1, 2), the reduced table becomes as shown in Figure 3.5.21. We can find all of the possible covers, including the one found in Section 3.5.5.3, using the Petrick algorithm, as follows:

$$\begin{aligned}
\text{ALL MINTERMS ARE COVERED} &= (A + B)(B + C + E)(D + E) \\
&= (AC + AE + B)(D + E) \qquad (3.5.22) \\
&= ACD + AE + BD + BE
\end{aligned}$$

From Equation (3.5.22), we see that there are actually three ways in which the minterms 5, 13, and 15 can be covered using only two prime implicants. Thus, the function $g(a, b, c, d)$ can be expressed in a minimal SOP form in three ways, namely,

$$\begin{aligned}
g(a, b, c, d) &= \bar{a}\bar{b} + b\bar{c}d + abd \\
&= \bar{a}\bar{b} + b\bar{c}d + acd \qquad (3.5.23) \\
&= \bar{a}\bar{b} + \bar{a}\bar{c}d + abd
\end{aligned}$$

	Prime implicants	Minterms		
		5	13	15
A	1, 5(4)	X		
B	5, 13(8)	X	X	
C	12, 13(1)		X	
D	11, 15(4)			X
E	13, 15(2)		X	X

Figure 3.5.21 Covering table for $g(a, b, c, d)$ of Equation (3.5.15) after the essential prime implicant is removed.

where the product term $\bar{a}b$ is the essential prime implicant. Note here that the first of the three expressions in Equation (3.5.23) is the one found in Section 3.5.5.3.

3.5.5.5 Summary of the Quine-McCluskey algorithm.
In summary, the Quine-McCluskey algorithm for finding a minimal sum of products expression for a given function follows the steps given below. If it is necessary to find all possible minimal covers, then steps 4 and 5 should be ignored.

Step 1. Using the don't cares, if any, find the set of all prime implicants of the function by the procedure outlined in Section 3.5.5.1.

Step 2. Construct a covering table as described in Section 3.5.5.2.

Step 3. Identify all of the essential prime implicants and form a reduced covering table.

Step 4. Reduce the table further by removing the dominated rows whose corresponding prime implicants are no simpler than the rows that dominate them.

Step 5. Identify the secondary essential prime implicants and reduce the covering table again.

Step 6. Use the Petrick algorithm to select a minimal cover for the remaining minterms, if any.

3.5.6 Using the Quine-McCluskey Algorithm to Simplify Multiple Functions

In the design of large digital systems, it very often happens that many functions must be generated all of which are functions on the same set of variables. As we shall see in the next chapter, a physical piece of hardware is required to implement each product term and each literal. It therefore behooves us to reduce the total number of these terms if we wish to obtain functions which can be physically implemented with the least amount of hardware. We could, of course, find minimal SOP expressions for the functions by applying the Quine-McCluskey algorithm to each. However, the total number of product terms and literals required to implement all of the functions may be reduced if we recognize that proper selection of product terms may make it possible to share terms among functions. For example, the two functions

$$f_1(a, b, c, d) = \bar{a}bc + a\bar{d}$$

and

$$f_2(a, b, c, d) = \bar{a}\bar{b} + b\bar{c} + a\bar{d}$$

have the term $a\bar{d}$ in common. Thus, these two functions require the generation of only four product terms, using a total of nine literals, to implement.

The Quine-McCluskey algorithm can easily be modified to generate minimal covers for several functions which maximize the number of terms that are common among the functions. The basic idea of this modification is to find all of the prime implicants for each function and then find all of the prime implicants that are shared among all possible combinations of functions. For example, suppose we are to implement the functions g_1, g_2, and g_3. We would first find all of the PIs for each of these functions. Next we would find all of the PIs that are common to pairs of functions, namely, the PIs of the functions g_1g_2, g_1g_3, and g_2g_3. Finally, we would find all of the PIs common to all of the functions, namely, $g_1g_2g_3$. A covering table can then be set up using these prime implicants and from it a minimal cover can be found using the general procedures described in Section 3.5.5.4.

Let us illustrate this process with a simple example. Suppose we are required to implement the following two functions:

$$f_1(W, X, Y, Z) = \Sigma\, m(3, 4, 5, 6, 7, 11, 12, 13, 14) \qquad (3.5.24)$$

$$f_2(W, X, Y, Z) = \Sigma\, m(3, 5, 11, 13, 15) \qquad (3.5.25)$$

The minterms that are shared between these functions are found by taking the product of f_1 and f_2. This yields

$$f_1(W, X, Y, Z)f_2(W, X, Y, Z) = \Sigma\, m(3, 5, 11, 13) \qquad (3.5.26)$$

Using the tabular procedure described in Section 3.5.5.1, we can easily derive the prime implicants for each of these functions. Using these prime implicants, we can then set up the covering table as described in Section 3.5.5.2. The resulting table is shown in Figure 3.5.22. Notice that this table is organized vertically as three tables, one for each of the functions f_1, f_2, and f_1f_2. The minterms to be covered are those associated with the individual functions, f_1 and f_2, only. For reference purposes, we have labeled each row with a letter along the right side of the table.

The next step in the simplification procedure is to find the essential prime implicants. In this case there is only one: prime implicant c, corresponding to 4, 6, 12, 14(2, 8). This is marked by the asterisk ($*$) in the table. After we remove this row and the columns corresponding to the minterms covered by this prime implicant, indicted by X's at the bottom of the table, the table reduces to that shown in Figure 3.5.23.

The table will next be further reduced by finding and eliminating all of the dominated rows (we are interested here in only one solution, not all solutions). It can be seen that row j dominates rows a and i and row k dominates rows e and f. We will therefore remove rows a, e, f, and i. The resulting table is shown in Figure 3.5.24, from which we see that prime implicants j and k, 5, 13(8) and 3, 11(8), respectively, become secondary essential PIs. Again, the minterms covered by these prime implicants are indicated by the X's at the bottom of the table.

The only minterms not yet covered are minterm 7 of function f_1 and minterm 15 of f_2. Using the Petrick algorithm, we find that a cover occurs if we use b and g, or b and h, or d and g, or d and h. Since prime implicant b has two literals and prime implicant d has three, we will select from either d and g, or d and h. Let us arbitrarily pick d, 3, 7(4), to cover minterm 7 and g, 13, 15(2), to cover minterm 15.

We have now found five prime implicants that cover all of the minterms of both f_1 and f_2. Prime implicants 4, 6, 12, 14(2, 8) and 4, 5, 6, 7(1, 2) are associated with function f_1 only. Prime implicant 13, 15(2) is associated with function f_2 only. Finally, prime implicants

	f_1									f_2				
	3	4	5	6	7	11	12	13	14	3	5	11	13	15
4, 5, 12, 13(1, 8) a		X	X				X	X						
4, 5, 6, 7(1, 2) b		X	X	X	X									
*4, 6, 12, 14(2, 8) c f_1		X		X			X		X					
3, 7(4) d	X				X									
3, 11(8) e	X					X								
3, 11(8) f										X		X		
13, 15(2) g f_2													X	X
11, 15(4) h												X		X
5, 13(8) i											X		X	
5, 13(8) j $f_1 f_2$			X					X			X		X	
3, 11(8) k	X					X				X		X		

Minterms covered by
essential prime implicant

Figure 3.5.22 Initial covering table for the functions $f_1(W, X, Y, Z)$ and $f_2(W, X, Y, Z)$.

68

		f_1 3	f_1 5	f_1 7	f_1 11	f_1 13	f_2 3	f_2 5	f_2 11	f_2 13	f_2 15
f_1	a		X			X					
	b		X	X							
	d	X		X							
	e	X			X						
f_2	f						X		X		
	g									X	X
	h								X		X
	i							X		X	
$f_1 f_2$	j		X			X		X		X	
	k	X			X		X		X		

Figure 3.5.23 Covering table after removal of the essential prime implicants.

		f_1 3	f_1 5	f_1 7	f_1 11	f_1 13	f_2 3	f_2 5	f_2 11	f_2 13	f_2 15
f_1	b		X	X							
	d	X		X							
f_2	g									X	X
	h								X		X
$f_1 f_2$	* j		X			X		X		X	
	* k	X			X		X		X		
	Minterms covered by prime implicants j and k	X	X		X	X	X	X	X	X	

Figure 3.5.24 The reduced covering table showing the secondary essential PIs.

5, 13(8) and 3, 11(8) are common to both f_1 and f_2. Using these PIs, the final minimum expression becomes

$$
\begin{aligned}
f_1(W, X, Y, Z) &= X\overline{Z} + \overline{W}X + X\overline{Y}Z + \overline{X}YZ \\
f_2(W, X, Y, Z) &= WXZ + X\overline{Y}Z + \overline{X}YZ
\end{aligned}
\qquad (3.5.27)
$$

These equations have thus been expressed in a form using a total of 5 distinct prime implicants which require a total of 13 literals to implement. If we had simply minimized each expression, one possible result would be

$$f_1(W, X, Y, Z) = X\overline{Z} + \overline{W}X + X\overline{Y} + \overline{X}YZ$$
$$f_2(W, X, Y, Z) = WXZ + X\overline{Y}Z + \overline{X}YZ \qquad (3.5.28)$$

which requires 6 distinct product terms using a total of 15 literals.

3.6 AN ANNOTATED BIBLIOGRAPHY

There are numerous books that discuss the general topics covered in this chapter. An exhaustive list of these would be out of the question; however, three excellent and very readable books are those by Hill and Peterson, Mano, and Roth.

HILL, J. F., and G. R. PETERSON, *Introduction to Switching Theory and Logical Design*, 3rd ed., Wiley, New York, 1981.

MANO, M. M., *Digital Logic and Computer Design*, Prentice-Hall, Englewood Cliffs, N.J., 1979.

ROTH, C. H., *Fundamentals of Logic Design*, 2nd ed., West Publishing, St. Paul, 1979.

A very extensive bibliography of works dealing with all aspects of digital systems can be found in Muroga. This book also gives excellent discussion of many theoretical topics in switching theory. Muroga also discusses a number of topics that are important in the design of VLSI circuits that are not usually found in switching theory texts.

MUROGA, S., *Logic Design and Switching Theory*, Wiley-Interscience, New York, 1979.

Two other references of note, dealing with the general topic of Boolean and switching algebra, are the books by Miller and Harrison. Both of these texts present ideas and concepts in a very rigorous, mathematical fashion and so should be considered "advanced" texts (especially Harrison). These books are recommended for the more precocious reader only.

HARRISON, M. A., *Introduction to Switching and Automata Theory*, McGraw-Hill, New York 1965.

MILLER, R. E., *Switching Theory*, Wiley, New York, 1965.

On more specific topics, the derivation of the postulates that describe a Boolean algebra can be found in the original paper by Huntington.

HUNTINGTON, E. V., "Sets of Independent Postulates for the Algebra of Logic," *Trans. Am. Math. Soc.*, vol 5 July 1904, pp. 288–309.

There are many examples of proof by algebraic manipulation presented in Chapter 2 of the book by Givone.

GIVONE, D. D., *Introduction to Switching Circuit Theory*, McGraw-Hill, New York, 1970.

The use of Karnaugh maps for the simplification of switching functions was first described in 1953 by M. Karnaugh. Virtually all of the books mentioned above discuss this method. When discussing maps of more than four variables, most texts use the form shown in Figure 3.5.9(b). Roth's book (mentioned above), however, uses the form shown in Figure 3.5.9(a); in fact it appears that this form of map is due to Roth.

KARNAUGH, M., "The Map Method for Synthesis of Combinational Logic Circuits," *Comm. Electronics*, November 1953, pp. 593–599.

Simplification procedures based on the Quine-McCluskey algorithm are also discussed in most of the texts cited above. However, methods other than this algorithm do exist. These methods are generally more efficient than Quine-McCluskey when carried out on a computer. This is usually true because with these methods it is not necessary to find all of the prime implicants. Chapter 10 of Dietmeyer and chapter 4 of Nagle et al. give two approaches which seem to be typical of the non-Quine-McCluskey algorithms. Nagle's procedure, however, only finds the prime implicants and not a minimal cover. A paper by Rhyne et al. describes a similar procedure which appears to be quite efficient. This paper also gives a nice list of the various approaches to this simplification problem. Muroga, cited on page 70, describes a very interesting, but quite different, approach, in Chapter 4. He refers to this method as the "branch-and-bound" method.

DIETMEYER, D. L., *Logic Design of Digital Systems*, Allyn & Bacon, Boston, 1978.

NAGLE, H. T., Jr., B. D. CARROLL, and J. D. IRWIN, *An Introduction to Computer Logic*, Prentice-Hall, Englewood Cliffs, N.J., 1975.

RHYNE, V. T., P. S. NOE, M. H. MCKINNEY, and U. W. POOCH, "A New Technique for the Fast Minimization of Switching Circuits," *IEEE Trans. Comp.*, Vol. C-26, No. 8, August 1977, pp. 757–764.

Finally, several of the books mentioned above also describe simplification procedures that can be applied to circuits having multiple outputs. Two excellent and readable references are those of Hill and Peterson and of Givone.

3.7 PROBLEMS

3.1. Prove the following algebraically. Identify the postulates or theorems used at each step.
(a) $a + a \cdot b = a$
(b) $a \cdot b + \bar{a} \cdot c + b \cdot c = a \cdot b + \bar{a} \cdot c$
(c) $a \cdot b + \bar{a} \cdot b = b$
(d) $\bar{a} + a \cdot b = \bar{a} + b$

3.2. Construct the truth tables for the AND, OR, and NOT operations in a four-element Boolean algebra having the elements 0, 1, a, and b.

3.3. Assuming an n element Boolean algebra, how many functions are there on m variables?

3.4. Determine which of the following equations are valid:
(a) $\bar{a}c + \bar{a}b + bc + ab + a\bar{c} = a + b + c$
(b) $AB + A\bar{C} + \bar{A}C = AC + BC + A\bar{C}$
(c) $\bar{B}D + CD + \overline{ABC} + ABC = \bar{B}D + \bar{A}CD + ABC$

3.5. Write the dual of each of the following expressions:
(a) $a + bc$
(b) $\bar{a}b + \bar{c}(d + e)$
(c) $a\bar{b} + c\bar{d}e$
(d) $(a + b)(\bar{a} + cd)$

3.6. Using De Morgan's theorem, take the complement of each of the expressions in Problem 3.5.

3.7. Write each of the expressions in Problem 3.5 in a

    (a) Sum of products form (SOP) and a

    (b) product of sums form (POS).

    Use the distributive postulates to accomplish this.

3.8. For each of the given functions, (i) write the function as a minterm list, (ii) write the function as a maxterm list, (iii) write the function in canonical minterm form, and (iv) write the function in canonical maxterm form.

    (a) $f(a, b) = a + b$

    (b) $f(a, b, c) = a + b\bar{c}$

    (c) $f(a, b, c) = ab + \bar{a}c + bc$

3.9. Convert the following expressions to a product of sums form:

    (a) $\bar{a}b\bar{c} + a\bar{b} + b\bar{c}d$

    (b) $\bar{X}Y + \bar{X}\bar{Z} + XYZ$

    (c) $wxy + \bar{w}\bar{y}\bar{z} + \bar{x}\bar{y} + wz$

    (d) $abc + \bar{a}bd + bd + \bar{c}d$

3.10. Convert each of the following expressions to a sum of products form:

    (a) $(\bar{a} + b)(a + \bar{b} + c)$

    (b) $(\bar{x} + \bar{y} + z)(w + y)$

    (c) $(\bar{x} + y)(w + z)(x + y + z)$

3.11. Algebraically reduce each of the following expressions to a minimal SOP expression:

    (a) $\bar{a}b + bc + abc + \bar{a}b\bar{c}$

    (b) $\bar{w}\bar{x}z + xyz + w\bar{y}x + wy\bar{z} + \bar{x}\bar{y}\bar{z} + \bar{x}z$

    (c) $xz + \bar{x}\bar{y}z + yz$

3.12. Plot each of the given functions in a Karnaugh map.

    (a) $f(a, b, c, d) = \bar{a}bc + a\bar{d} + b\bar{c}$.

    (b) $g(x, y, z) = \Sigma\, m(0, 1, 4, 6, 7)$.

    (c) $h(A, B, C, D) = (A + \bar{B} + C)(A + B + \bar{D})(\bar{B} + \bar{C})$.

    (d) $f(A, B, C, D, E)$ is 1 if the number represented by the 5-tuple (A, B, C, D, E) is even or is divisible by 3.

    (e) $G(a, b, c, d) = \Pi\, M(0, 1, 3, 4, 5, 7, 10, 11, 12, 13)$.

3.13. List all of the prime implicants for the functions given in Problem 3.12.

3.14. For each of the Karnaugh maps shown in Figure P3.14, write an expression for the function implemented in minimal sum of products form.

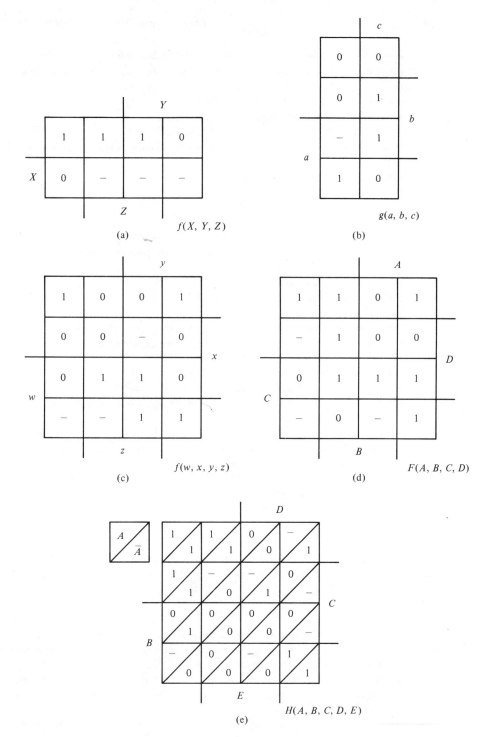

Figure P3.14

3.15. Repeat Problem 3.14, writing the expression in minimal product of sums form.

3.16. Using Karnaugh maps, find a minimal SOP expression for the following functions:
  (a) $f(a, b, c, d) = \Sigma\, m(2, 3, 6, 9, 12, 13, 14) + d(0, 11)$
  (b) $f(W, X, Y, Z) = \Pi\, M(3, 4, 5, 8, 10, 11, 12)$
  (c) $G(A, B, C) = \Sigma\, m(0, 1, 2, 4, 6)$
  (d) $h(a, b, c, d, e) = \Sigma\, m(7, 17, 23, 25, 28, 29, 30, 31)$

3.17. Write the prime implicants in product of literals form corresponding to the minterm list forms shown.
  (a) 8, 9, 10, 11(1, 2), on four variables
  (b) 0, 2, 8, 10(2, 8), on four variables
  (c) 5, 13, 21, 29(8, 16), on five variables
  (d) 0, 1(1), on four variables
  (e) 0, 2, 4, 6, 16, 18, 20, 22(2, 4, 16), on five variables

3.18. Write the prime implicants in minterm list form corresponding to the product of literals forms shown.
  (a) $\bar{a}cd$, on five variables
  (b) $b\bar{c}$, on four variables
  (c) $b\bar{c}d$, on five variables
  (d) $a\bar{b}$, on two variables

3.19. Using the Quine-McCluskey algorithm find a minimal cover for each of the following functions:
  (a) $F(A, B, C, D) = \Sigma\, m(0, 1, 2, 3, 5, 7, 10, 12)$
  (b) $h(A, B, C, D) = \Sigma\, m(0, 1, 2, 4, 5, 7, 8, 10) + d(3, 11, 15)$
  (c) $f(w, x, y, z) = \Sigma\, m(0, 1, 7, 8, 10, 12, 14, 15) + d(2, 5)$
  (d) $f(a, b, c, d, e) = \Sigma\, m(3, 4, 6, 7, 12, 18, 19, 20, 22, 23, 24, 25, 26, 27, 28, 29, 30, 31)$
  (e) $g(A, B, C, D, E) = \Sigma\, m(0, 1, 2, 3, 4, 6, 8, 11, 12, 27, 28) + d(9, 16, 17, 18, 19)$

3.20. The following functions yield cyclic covering tables. After finding the prime implicants, use the Petrick algorithm to find a minimal cover for each function.
  (a) $f(w, x, y, z) = \Sigma\, m(0, 1, 5, 7, 8, 12, 14, 15)$
  (b) $g(w, x, y, z) = \Sigma\, m(3, 4, 6, 7, 11, 12, 13, 15)$
  (c) $h(a, b, c, d) = \Sigma\, m(0, 4, 5, 7, 8, 10, 11, 14, 15)$

3.21. Find all of the possible solutions for the example given in Section 3.5.6.

3.22. Apply the multiple-function version of the Quine-McCluskey algorithm described in Section 3.5.6 to the implementation of the three functions f, g, and h given in Problem 3.20.

3.23. Let $f(\mathbf{x})$ be a switching function on the n variables $x_1, x_2, \ldots, x_n$ and let $f^d(\mathbf{x})$ be the dual of $f(\mathbf{x})$. Prove that

$$f^d(\mathbf{x}) = \bar{f}(\bar{\mathbf{x}}) = \bar{f}(\bar{x}_1, \bar{x}_2, \ldots, \bar{x}_n)$$

3.24. A switching function $f(\mathbf{x})$ is said to be *unate* if there exists an expression for $f(\mathbf{x})$ such that each of the variables appears in either complemented or uncomplemented form but not both. For example, $f(a, b, c) = a + \bar{b}c$ is unate, but $g(a, b) = \bar{a}b + a\bar{b}$ is not, since a and b appear in both complemented and uncomplemented form. Prove that all prime implicants of a unate function have a minterm in common and therefore that the minimal SOP expression for a unate function is unique.

Gates and the Design of Switching Circuits

4

4.1 INTRODUCTION

Probably the earliest hardware for carrying out digital computation was a simple adding machine built by Blaise Pascal for his father, a bookkeeper, in 1645. This machine and the famous calculating engine of Charles Babbage, circa 1822, were constructed using gears, cams, levers, and the like. These machines, and their successors, were basically decimal machines in which each digit was represented by one of ten possible positions of a ten-toothed gear or some similar mechanism.

Although the computations these machines performed could be carried out in the binary number system, there was no need to do so, since the machines were mechanical. However, the use of relays and, later, vacuum tubes and, still later, transistors, all of which are basically switches having two states, required that binary numeration be used for computing devices. These binary switches are generally of two types: bilateral and unilateral. A bilateral switching device, such as a relay contact or a simple switch, allows information to flow in two directions. A unilateral device, such as a transistor, restricts information flow to only one direction. Switching functions can be physically implemented using either type of device. By far the most common devices used today for the implementation of switching functions are unilateral.[1] Such devices are referred to as *gates*. In what follows, we will define a consistent gate symbology and show how to use these devices for the implementation of switching functions.

---

[1] Bilateral devices are, however, used extensively in VLSI circuits and will be discussed further in Chapter 8.

4.2 GATE SYMBOLOGY

Experience has shown that the symbology used in the design of large-scale digital systems is most important in conveying information about the operational characteristics of the system as well as the logical intent of the designer. In any digital circuit, there are two points of view from which the circuit may be analyzed: the *logical* (or mathematical) and the *physical*. The logical point of view considers only the 1 versus 0 behavior of switching variables and functions. The physical point of view considers the actual voltage levels used to implement the switching variables. These voltage levels are, of course, what one would observe on an oscilloscope and would thus indicate how the circuit is actually operating. The physical and the logical points of view coincide when one voltage level is used to represent a logical 1 and another, quite different, voltage level is used to represent a logical 0. Thus, it is important that the symbology used to represent digital circuits be capable of conveying both physical and logical information.

This need has led to the particular set of symbols and the standard for their usage that was adopted for general use by the Department of Defense in February 1962. This is MIL-STD-806B, which is now being used in some form or another by most of the digital integrated circuit manufacturers in this country. More recent versions are in existence but, at this time, have generally not met with as great a degree of acceptance as has standard 806B. More will be said about these standards at the conclusion of Section 4.2. In what follows, a symbology standard will be described which is completely compatible with standard 806B but has been extended to reflect current industrial usage.

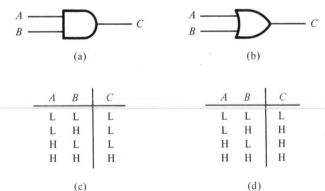

A	B	C
L	L	L
L	H	L
H	L	L
H	H	H

(c)

A	B	C
L	L	L
L	H	H
H	L	H
H	H	H

(d)

Figure 4.2.1 The gate symbols for (a) the AND and (b) the OR functions with their respective physical truth tables (c and d). (H = high voltage; L = low voltage.)

4.2.1 The Symbols and Their Meaning

Basically, *a gate is a physical device, electronic, mechanical, or otherwise, which implements a logical operation.*[2] As described above, the symbol that represents a function must show not only the physical behavior of the gate but, also, the logical function desired

---

[2] In all that follows, we will assume the use of electronic gates in which voltages are used to represent the logical values.

by the designer. The symbols used for the AND and OR functions are shown in Figure 4.2.1. These symbols are *distinctive symbols*, in that their shape corresponds to the intended logical operation performed by the respective gate. Consider, first, the symbol for the AND gate shown in Figure 4.2.1(a). This symbol represents a device that has the following physical behavior: the output, C, is a high voltage if input A is a high voltage *and* input B is also a high voltage. In a similar manner, the symbol for the OR gate, shown in Figure 4.2.1(b), is interpreted physically to have a high voltage on output C if either input A is a high voltage *or* input B is a high voltage. These two interpretations can be summarized in the *physical truth tables* given in Figure 4.2.1(c) and (d).

The logical interpretation of these gates depends on how we associate voltages with logical 1s and 0s. There are obviously two ways in which this can be done, namely:

$$1 = \text{high voltage}$$
$$0 = \text{low voltage}$$

or

$$1 = \text{low voltage}$$
$$0 = \text{high voltage}$$

Assume, for the moment, that a high voltage corresponds to a 1 and a low voltage corresponds to a 0. If we substitute these values into the physical truth tables given in Figure 4.2.1, we obtain the logical truth tables shown in Figure 4.2.2(a) and (b). These tables are, of course, the same as derived in Section 3.2, which defined the AND and OR switching functions. If we take the alternative point of view, that a 1 corresponds to a low voltage and a 0 to a high voltage, then the logical truth tables that result from this substitution are shown in Figure 4.2.2(c) and (d). Writing the logical function implemented by tables (c) and (d), we find that the AND gate now realizes the function $C = A + B$, or, rewriting

A	B	C
0	0	0
0	1	0
1	0	0
1	1	1

(a)

A	B	C
0	0	0
0	1	1
1	0	1
1	1	1

(b)

A	B	C
1	1	1
1	0	1
0	1	1
0	0	0

(c)

A	B	C
1	1	1
1	0	0
0	1	0
0	0	0

(d)

Figure 4.2.2   Logical truth tables resulting from the two possible assignments of voltages to logic levels for the physical gates of Figure 4.2.1. (a) AND gate and (b) OR gate for $1 = \text{high}$ and $0 = \text{low}$; (c) AND gate and (d) OR gate for $1 = \text{low}$ and $0 = \text{high}$.

$A(L)$ ——————| $C(L)$

Figure 4.2.3 An OR gate that is physically the same as the AND gate of Figure 4.2.1(a).

in a different form, $\overline{C} = \overline{AB}$; and we find that the function implemented by the OR gate now is $C = AB$, or $\overline{C} = \overline{A} + \overline{B}$. Thus, we see that *a physical gate can implement several different logical functions, depending only on how we associate the voltage levels with logical values.*

In order to avoid confusion, we need to indicate how the voltage levels are to be assigned to the logic levels. We will do this by using the notation $X(H)$ or, simply, X to indicate that a 1 corresponds to a high voltage and $X(L)$ or $X\_$ to indicate that a 1 corresponds to a low voltage. Note that in all that follows, if no indication is given, it will be assumed that 1 is associated with a high voltage, as is customary. We will refer to the assignment of a logical 1 to a low voltage as being *asserted low*; the reverse situation would be termed *asserted high*.

Consider next the gate shown in Figure 4.2.3. The physical interpretation of this gate symbol is that the output, C, is low if either input A is low *or* input B is low. The "bubbles" (small circles) on the inputs and outputs of the gate are used to indicate that a low voltage is expected for assertion, and the OR symbol is used to indicate the physical OR of voltage levels. It is easily verified that the resulting physical truth table is the same as the one shown in Figure 4.2.1(c). Since the physical truth tables are identical for the gates shown in Figures 4.2.1(a) and 4.2.3, the two symbols must represent the same physical gate. However, the logical functions performed are quite different. If, for example, we now assume that all inputs and outputs to the gate of Figure 4.2.3 are considered to be a logical 1 when the voltage is low, as indicated by the labels $A(L)$, $B(L)$, and $C(L)$, then the logical truth table becomes, by simply substituting 0 for H and 1 for L in the table of Figure 4.2.1(c), the same as Figure 4.2.2(c). Thus we have implemented the logical function $C = A + B$!

Let us now consider the example shown in Figure 4.2.4. The gate symbol itself indicates—if we ignore the correspondences associated with the labels A, B, and C, which are, after all, arbitrary (the designer knows what these are supposed to be)—that the output is high if either input is low; in other words, the OR operation. Now let us look at the logical function implemented by this gate. To do this, first construct the physical truth table for the gate; this is shown in Figure 4.2.4(b). The logical truth table can now be constructed from the physical truth table by replacing the lows and highs with the logical 1s and 0s

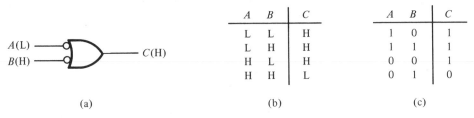

A	B	C
L	L	H
L	H	H
H	L	H
H	H	L

A	B	C
1	0	1
1	1	1
0	0	1
0	1	0

(a)   (b)   (c)

Figure 4.2.4 A gate with mixed logical conventions. (a) Mixed logic symbol. (b) Physical truth table. (c) Logical truth table.

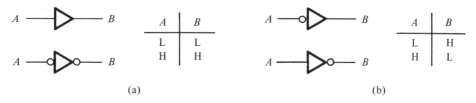

Figure 4.2.5   Equivalent symbols for the buffer (a) and the level shifter (b), with their respective physical truth tables.

assumed by the notation used in Figure 4.2.4(a), namely, that A is a logical 1 when low, or A is asserted low; that B is asserted high; and that C is asserted high. Figure 4.2.4(c) is the resulting logical truth table, from which the switching function realized is

$$C = A + \overline{B} \qquad (4.2.1)$$

Consider for a moment how the $\overline{B}$ entered this equation. By the definition of the gate symbol, the output is asserted high if either input is asserted low. The input signal called B is assumed to be a logical 1 when it is high, as indicated by the notation B(H). The corresponding gate input is a logical 1 when its voltage is low. Thus the logical interpretations of the two points in the circuit are complementary, and so B appears in the output function as $\overline{B}$. On the other hand, the driving signal A and the corresponding gate input are both asserted low, and so no complementation arises. The same holds true for the output, except that in this case both are asserted high. From this discussion we observe the following rule:

Rule 1.   A logical complementation will arise at any time when the assertion levels on opposite ends of a line are different.

Before proceeding to analyze more complex circuits, we need to introduce a new gate type called a *buffer*. Figure 4.2.5(a) shows the logic symbols associated with this gate and gives the corresponding physical truth table. We will generally refer to a buffer with a ''bubble'' on either the input or the output as a *level shifter*[3] or a NOT gate. Figure 4.2.5(b) shows the associated symbols for the level shifter and the corresponding physical truth table. The buffer is generally used to amplify a signal so that it can serve as an input to many more gates than would be physically possible otherwise. The level shifter is generally used to shift an assertion level from one value to another so that either a logical inversion is implemented, which produces the NOT or complementation operation, or avoided, as needed. Figure 4.2.6 shows the use of level shifters for these two functions. If we were to remove the level shifter of Figure 4.2.6(a), input A would appear at the output uncomplemented. Similarly, removal of the level shifter, as shown in Figure 4.2.6(b), would cause A to be complemented.

---

[3] The term ''level shifter'' is used here to indicate a gate that converts one logic level to another. This term may also be used, in other contexts, to indicate a device for shifting one voltage level to another because of conflicting electrical or electronic requirements.

(a)   (b)

Figure 4.2.6   Use of the level shifter (a) to create or (b) to remove a logical complementation or inversion.

4.2.2 Analysis of Mixed-Logic Circuits

When we use the ideas just presented, the analysis of rather complex circuits becomes a straightforward job. The basic analysis procedure consists of performing the AND function for AND gates and the OR function for OR gates, and complementing a function whenever an assertion-level mismatch occurs. Figure 4.2.7 shows a moderately complex gate network realizing some switching function. The problem is to write the switching expression for the function implemented. The analysis is done by writing the function implemented at the output of each gate without reference to any bubbles at these outputs. Complementations are generated wherever mismatches occur on the inputs. For example, the output of gate 1 realizes the function $A + B$ regardless of the fact a bubble occurs at this output. When this output is used as an input to gate 2, it appears uncomplemented, because the input to gate 2 also has a bubble (recall rule 1). However, it appears complemented at the input to gate 3, because there is a logical mismatch at that input. Continuing the process of writing equations on a line-by-line basis, the function realized becomes

$$F = [(A + B)C][\overline{(A + B)} + \overline{D}] + DE \tag{4.2.2}$$

A very important observation should be made at this point. If an oscilloscope is connected to some point in the circuit, the voltage behavior of a correctly operating circuit can be predicted by knowing the logical function and the assertion level of the gate output driving the point under test. For example, the output of gate 4 in Figure 4.2.7 realizes the logical function

$$[(A + B)C](\overline{A}\overline{B} + \overline{D}) = (A + B)C\overline{D} \tag{4.2.3}$$

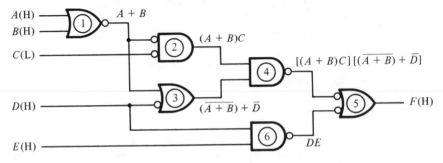

Figure 4.2.7   Analysis of a mixed-logic circuit.

(a)   (b)

Figure 4.2.8   NAND (a) and NOR (b) equivalent symbols.

Thus, this output will be a low voltage if C is asserted and D is not asserted (or is negated) and either A or B is asserted. In terms of the physical assertion levels, this output is low if C is low and D is low (negated) and either A or B, or both, are high.

If we peruse a 7400 series TTL[4] data book, we will notice very quickly that although there are a lot of different gates, none is shown in the form of gate 2, 3, or 5 of Figure 4.2.7. In fact, no single gate package exists which implements gate 3. However, it is easily verified by examination of the physical truth tables that gate 2 is equivalent to the so-called NOR gate and gate 5 is equivalent to the so-called NAND gate, both of whose symbols appear in the data book. Figure 4.2.8 shows the physically equivalent representations for these two gates. The name "NOR" comes from the function implemented by the gate if one assumes that all inputs and outputs are asserted high. This function is

$$C = \overline{(A + B)} = \overline{A}\,\overline{B} \tag{4.2.4}$$

which is the "NOT of the OR of A and B," or NOR. Similarly for the NAND, whose function is

$$C = \overline{(AB)} = \overline{A} + \overline{B} \tag{4.2.5}$$

It is important to note, however, that if the output is assumed to be a logical 1 when low while the inputs are taken as a 1 when high, then the logical functions implemented in the two cases are the OR and the AND, respectively.

It will be useful later to be able to convert between physically equivalent mixed-logic AND and OR symbols as done in Figure 4.2.8. The general conversion process is easily accomplished by use of the following rule:

Rule 2.   To convert a mixed-logic AND gate symbol to a physically equivalent mixed-logic OR gate symbol, change the AND symbol to an OR symbol, place bubbles on all signal lines in the OR symbol that did not have bubbles in the AND symbol, and remove all bubbles on signal lines in the OR symbol that had bubbles in the AND symbol. Conversion of a mixed-logic OR to a mixed-logic AND is done in exactly the same manner.

As an example, Figure 4.2.9(a) shows a three-input mixed-logic AND gate symbol that is to be converted to a physically equivalent OR gate symbol. Applying rule 2 to this AND gate results in the OR gate symbol of Figure 4.2.9(b). The reader should verify, by constructing physical truth tables for each symbol, that the two gates, although not performing the same *logical* function, are *physically* the same gate.

---

[4] TTL, or transistor-transistor logic, is the most common technology used today for the implementation of simple to moderately complex logical functions. Other technologies, such as MOS, or metal-oxide-semiconductor, and CMOS, or complementary MOS, are typically used for very complex circuits.

Figure 4.2.9 Two physically equivalent gate symbols. (a) A mixed logic AND gate symbol. (b) OR symbol physically equivalent to part (a).

(a)   (b)

4.2.3 Synthesis of Switching Functions Using Mixed Logic

Suppose we are given a switching function such as

$$Z = \overline{EF}(AB + \overline{C} + \overline{D}) + GH \tag{4.2.6}$$

and are asked to design a gate network that implements the function using only NAND gates and NOT gates (level shifters). Further assume that all signals are asserted high. Our job now is to carry out a design and draw the circuit diagram in a manner that indicates the exact logical form of the equation. The design procedure, starting with an equation, is very straightforward and may be summarized by the following steps:

Step 1. Ignoring logical complementations, lay out a circuit implementing the equation with AND and OR gates only. The result for this example is shown in Figure 4.2.10.

Step 2. Affix "bubbles," or assertion-level indicators, to each gate to produce the physical gate required by the problem constraints. In this case, NANDs are required. Figure 4.2.8(a) shows the two equivalent symbols that can be used for this gate. Figure 4.2.11 shows the result of this step.

Step 3. Add level shifters as necessary to either create or remove logical complementations. Figure 4.2.12 shows the result of this final step.

If one examines the TTL data books, one will observe that, in addition to NANDs, NORs, and NOTs, there are also AND and OR gates, although not in as great a variety. A logical question, then, is, If such gates are available, why should one be concerned with NANDs and NORs—why not implement everything using AND, OR, and NOT gates? The primary

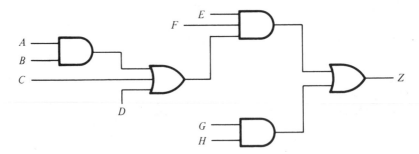

Figure 4.2.10 Result of synthesis step 1, which yields the function $Z = (AB + C + D)EF + GH$.

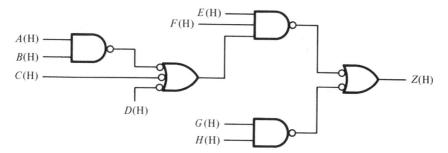

Figure 4.2.11   Result of synthesis step 2, which yields the function $Z = (AB + \overline{C} + D)EF + GH$.

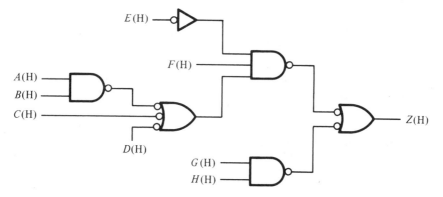

Figure 4.2.12   Final realization of Equation (4.2.6).

reason stems from the fact that computers and their related memories and peripherals generally require a large number of signals that are asserted low, i.e., that cause some significant action to occur when low. In addition, the NAND and NOR gates are generally faster than AND and OR gates. Finally, the NAND gate (and also the NOR gate) is a universal gate in that all functions can be implemented with this gate only (refer to Problems 4.7 and 4.8). For these reasons, implementations of switching functions using NANDs, NORs, and NOTs are usually preferable to implementations using ANDs, ORs, and NOTs.

4.2.4 Converting Ill-Formed Circuits to Standard Form

It would certainly be nice if everyone designing digital systems in the real world would adhere to the symbology usage just described, since it would certainly make schematic diagrams easier to read, understand, and maintain. In fact, most integrated circuit manufacturers and original equipment manufacturers (OEMs) do use the symbology presented above. Unfortunately, others do not. In many textbooks, technical periodicals, and hobby magazines, one may find circuits drawn as shown in Figure 4.2.13, for example. We will

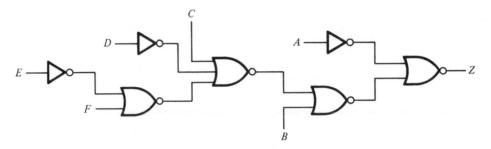

Figure 4.2.13 An ill-formed circuit.

refer to circuits like this as being ill-formed. The difficulty of analysis is obvious: if an oscilloscope is connected to Z, what conditions of voltage levels on A, B, C, D, E, and F will cause Z to be a high voltage (or did the designer want it to be a low voltage in the first place?)? Asked another way: What switching function of the primary input variables is Z, and what function did the designer want to implement? The analysis can clearly be carried out, but it is complicated by the fact that there are assertion-level mismatches on almost every line! De Morgan's theorem will have to be applied repeatedly to such circuits to answer the question, and this is a process highly susceptible to error.

What is needed for the analysis of ill-formed circuits like that shown in Figure 4.2.13 is a simple procedure to redraw the circuit so as to make it readable. Fortunately, this is quite straightforward. To illustrate the procedure, assume that all signals in Figure 4.2.13 are asserted high. The procedure is as follows:

Step 1. Convert the gate that generates the output to a physically equivalent gate so that no assertion-level mismatch occurs at the output. In the case at hand, we need to convert the OR symbol at the output to a physically equivalent AND symbol so as to remove the logical mismatch at the output. The result of this step is shown in Figure 4.2.14.

Step 2. Convert the gates driving the output gate so that no level mismatches occur between the gates. In this case, no mismatches occur, and so no change is necessary.

Step 3. Continue on succeeding levels converting the gates so that mismatches in assertion levels are eliminated or minimized. Note that no physical gate can be changed or

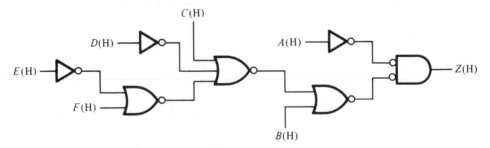

Figure 4.2.14 First-level conversion of Figure 4.2.13.

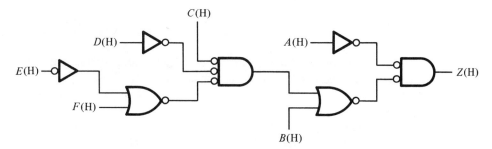

Figure 4.2.15 A readable circuit in standard form equivalent to the circuit of Figure 4.2.13.

added, since this would change the original physical circuit. Figure 4.2.15 shows the result of this step.

The principal objective of this process is to move the mismatches as close to the inputs as possible. Since expressions implemented at various points in the circuit are simpler as we get closer to the inputs, any application of De Morgan's theorem will be easier if this is done. We will refer to circuit diagrams drawn in this fashion as being given in *standard form*. It is now easy to determine that the logical function implemented by the original circuit is

$$Z = A[B + \overline{C}D(F + \overline{E})] \qquad (4.2.7)$$

4.2.5 Some Notes on Other Symbols Usage and Other Standards

The symbols used here are those of MIL-STD-806B. Although the H and L tags on the signal lines are not a part of standard 806B, they are used extensively in industry. Unfortunately, the literature, including the semiconductor manufacturers, does not use the tagged symbols to any great extent. Rather, the usual usage is to use a complementation indicator (usually an overbar) to indicate an asserted low signal and to use no tag to indicate an asserted high. Thus

$$A(L) = \overline{A}$$

and

$$A(H) = A$$

Although most IC houses use this convention, only a few, Texas Instruments included, remove any possible ambiguity by specifying their devices' behavior with a physical truth table and not a table of 1s and 0s.

There are two other symbols that are part of standard 806B and appear commonly in the literature which have not, as yet, been discussed. The first is the Exclusive OR symbol, shown in Figure 4.2.16. The logical function realized by this symbol is

$$C = A \oplus B = \overline{A}B + A\overline{B} \qquad (4.2.8)$$

Figure 4.2.16 The Exclusive OR gate.

"Bubbles" may appear in this symbol in exactly the same way as with any other gate symbol and carry the same physical meaning as before. The second symbol not mentioned thus far is actually two: the *wired* OR and *wired* AND. Figure 4.2.17 shows these two symbols. Actually, these symbols do not represent gates at all! They represent a logical function generated by physically soldering the outputs of two circuits—outputs *A* and *B* or outputs *D* and *E*—together to form a single signal, called *C* or *F*, respectively. These connections occur most frequently on devices having an "open-collector" output. An open-collector output is just the collector voltage of a transistor having an uncommitted collector. Figure 4.2.18 shows two such gates connected together in this fashion. It is easy to see that *C* will be a high voltage, +5 V, if both transistors *A* and *B* are off—the AND operation. Alternatively, *C* will be a low voltage, or at ground, if either transistor *A* or transistor *B* is on—the OR operation. The symbol selected to depict this wired logic should be based on the function actually wanted by the designer.

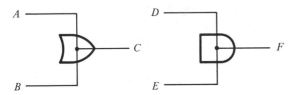

Figure 4.2.17 The "wired" OR and AND symbols.

The symbol used for the wired OR operation, unfortunately, does not follow the standards which we have just described. The appropriate symbol for such an interpretation would be as shown in Figure 4.2.19. Such a symbol is not used, however.

Other standards do exist, and so also do other forms for the symbols. Notable is the current IEEE Std. 91-1973 (ANSI-Y32.14-1973). This standard modifies the interpretation of the bubble in a somewhat confusing, although consistent, way and adds a set of gate symbols in addition to those given above. The Appendix describes some of these changes.

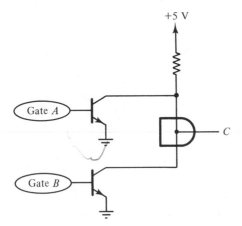

Figure 4.2.18 Two open-collector gates connected in a wired-logic arrangement.

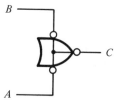

Figure 4.2.19 An appropriate, but unused, symbol for the wired OR.

This standard has not been generally accepted by industry. It is for this reason that we do not use it in this text. It is unfortunate that acceptance of this standard has been slow, since it very clearly shows both the physical and the logical behavior of a design.

Both the symbology standard used here and Std. 91-1973 are well suited for describing small to medium-size systems. As integrated circuits become more complex, however, this simple symbology becomes inadequate. What is needed is a symbology which clearly describes the function of such large-scale circuits. Such a notation has recently been introduced as IEEE Std. 91-1984. In this standard, the symbols are generally uniform, rather than distinctive, as used here, and the functions of the inputs and outputs are indicated by a special notation termed *dependency notation*. Since we will be dealing with rather simple circuits throughout this text, we will not introduce this standard here. A basic introduction to this new symbology and to dependency notation is given in the Appendix. Further references can be found in the bibliographies at the end of this chapter and at the end of the Appendix.

4.3 SWITCHING CIRCUIT DESIGN EXAMPLES

The design of computers and other large-scale digital systems is usually accomplished by breaking the system up into small portions, each of which is designed separately. By approaching the problem in this way, each component can be individually tested. Thus, when the final system is assembled, the likelihood of its functioning properly the first time is tremendously increased. The design of each of the smaller subsections follows a very well-defined procedure. First, the problem is specified by describing the specific function and operation of the subsystem. This specification, usually in written form, is then translated into a set of switching expressions which are implemented using gates and other digital components. The schematic diagrams that result are then used to build the component and finally test it for proper operation. In what follows, we will illustrate this design procedure by looking at a number of examples.

4.3.1 A Binary Adder

The heart of any computer is clearly its central processing unit (CPU), which is made up of circuitry that can perform arithmetic and logical operations on information. Among the arithmetic operations, addition is the most used. What we would like to do in this section is to design a piece of hardware that can be used to add two signed 2's complement numbers.

We can think of the adder as being a logic network having two sets of inputs, each

C_i	A_i	B_i	S_i	C_{i+1}
0	0	0	0	0
0	0	1	1	0
0	1	0	1	0
0	1	1	0	1
1	0	0	1	0
1	0	1	0	1
1	1	0	0	1
1	1	1	1	1

Figure 4.3.1 Truth table for a 1-bit binary adder.

consisting of n bits, and an $(n + 1)$-bit output, where the extra output bit is used to give the carry generated by the addition. For any reasonable value of n, we quickly see that a truth table representation for the n-bit adder is not feasible. For example, if $n = 8$, a reasonable size for most microprocessors, there will be 16 bits on the input—8 bits for each of the two numbers—meaning a truth table with $2^{16} = 65{,}536$ entries. Obviously, this approach leaves something to be desired. However, if we change our point of view and consider the process to be one of adding two 1-bit numbers, the problem becomes very simple, indeed.

Let us consider what happens in the addition of two n-bit numbers at the ith bit position. First, we add the two bits of the given numbers, A_i and B_i. Once this is done we must add to this result any carry that came from the next lower-order bit addition, i.e., the addition of A_{i-1} and B_{i-1}. The result of this addition will give a single bit for the sum and a single bit for the carry into the next higher position. Figure 4.3.1 gives a truth table for this single-bit addition. This table was first derived in Section 2.3. In this table, C_i is the carry in and C_{i+1} is the carry out of the ith bit position, and S_i is the sum bit for the ith position.

By plotting S_i and C_{i+1} in Karnaugh maps and finding a minimal SOP expression for each, we obtain the equations

$$S_i = \overline{A_i}\overline{B_i}C_i + \overline{A_i}B_i\overline{C_i} + A_i\overline{B_i}\overline{C_i} + A_iB_iC_i$$
$$C_{i+1} = A_iB_i + A_iC_i + B_iC_i \tag{4.3.1}$$

We could, of course, implement these equations directly in a "two-level" circuit.[5] Such an implementation would require five gates for S_i and four gates for C_{i+1}. However, by factoring S_i, we find, from the following equation, that it is equal to the Exclusive OR of A_i, B_i, and C_i:

$$S_i = \overline{(A_i\overline{B_i} + \overline{A_i}B_i)}C_i + (A_i\overline{B_i} + \overline{A_i}B_i)\overline{C_i}$$
$$= A_i \oplus B_i \oplus C_i \tag{4.3.2}$$

S_i can now be implemented using two Exclusive OR gates. C_{i+1} can be factored as well to produce a result which will also yield a simpler implementation. This is done as follows:

[5] The term "two-level" refers to using a group of ANDs at the input level to implement the product terms, then ORing these at the second level to generate the output.

$$C_{i+1} = A_iB_i + A_i\overline{B_i}C_i + \overline{A_i}B_iC_i$$
$$= A_iB_i + (A_i\overline{B_i} + \overline{A_i}B_i)C_i \qquad (4.3.3)$$
$$= A_iB_i + (A_i \oplus B_i)C_i$$

In this case, only three additional gates are required to implement C_{i+1}, since the Exclusive OR of A_i and B_i has already been implemented. The resulting gate realization is shown in Figure 4.3.2.

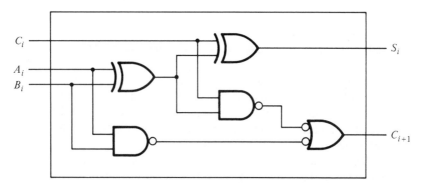

Figure 4.3.2   The implementation of the 1-bit adder.

In this realization we may note that the first exclusive OR generates the sum of A and B, while the second adds in the carry. Since the Exclusive OR gate, in each case, is performing "half" of the addition task, this gate is sometimes referred to as a *half adder*. The circuit shown in Figure 4.3.2 is then referred to as a *full adder*.

Now, our original problem was to design an *n*-bit adder, not just a 1-bit adder. By cascading these full adders so that the carry out from one becomes the carry in to the next higher bit position, adders of arbitrary length can be created. This *iteration* of circuit elements is shown in Figure 4.3.3. Note that for proper operation the carry in to the least significant bit is set to a 0 by tying it to a low voltage; specifically, ground.

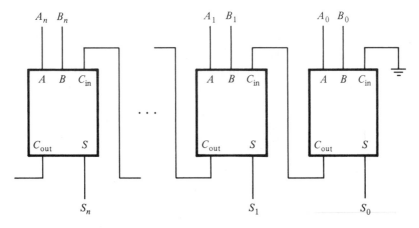

Figure 4.3.3   An iterative network of full adders which realizes an *n*-bit adder.

As we shall see in the next chapter, the "propagation" of a carry through this sequence of adders can take a great deal of time and thus slow down the addition process. One solution to this problem is to design 2-bit or 4-bit full adders, a process more complicated than the design of 1-bit adders but still tractable. These multibit adders can then be cascaded in the same manner as for the single-bit adder to produce an n-bit adder. If we assume that it takes as long to propagate a carry through one 4-bit adder as it does to propagate through a 1-bit adder, then the effective time to perform the addition process can be reduced by a factor of 4. We will discuss this design in Section 4.4.

4.3.2 The Comparison of Two Binary Numbers

Another operation that is performed quite often in computation is that of comparing two numbers. What we want to do is to determine whether number A is greater than, less than, or equal to number B. One way of doing this is to subtract the two numbers and look at the result. Another approach, and the one we will take here, is to design a specific circuit which makes this comparison directly.

The *algorithm*, or procedure, for performing this comparison can be described as follows. Given the numbers A and B, we begin the comparison by looking at the high-order bits. If the high-order bit of A is 1 and that of B is 0, then A is greater than B. If the high-order bit of B is a 1 and that of A is a 0, then A is less than B. However, if the high-order bits are the same, either both 0 or both 1, we must look at the bits of the next higher order. By continuing this bit-by-bit comparison from left to right, we will eventually determine the ordering of the two numbers.

Using this algorithm, we can design a 1-bit comparator and then, by cascading n of these, produce an n-bit comparator, just as we did in Section 4.3.1 for the n-bit adder. Each 1-bit comparator must have four inputs and two outputs. The four inputs are, first, the two bits to be compared, A_i and B_i; then a bit E_{in} that indicates whether the bits to the left of the ith bit are all equal; and a bit G_{in} that tells whether it has already been determined that A is greater than B. The two outputs are the "equals" and "greater than" indicators that exist after the ith bits are compared. We will refer to these as E_{out} and G_{out}, respectively. Figure 4.3.4 shows the truth table for this circuit.

Note that if E_{in} and G_{in} are both 0, then A has already been determined to be less than B. Thus the ith bits are irrelevant, or don't cares. The same situation occurs if $G_{in} = 1$ and $E_{in} = 0$, except that now A is greater than B. Only when all of the higher-order bits are the same ($E_{in} = 1$ and $G_{in} = 0$) do the ith bits matter. Observe also that

E_{in}	G_{in}	A_i	B_i	E_{out}	G_{out}
0	0	–	–	0	0
0	1	–	–	0	1
1	0	0	0	1	0
1	0	0	1	0	0
1	0	1	0	0	1
1	0	1	1	1	0

Figure 4.3.4 Truth table for a 1-bit comparator.

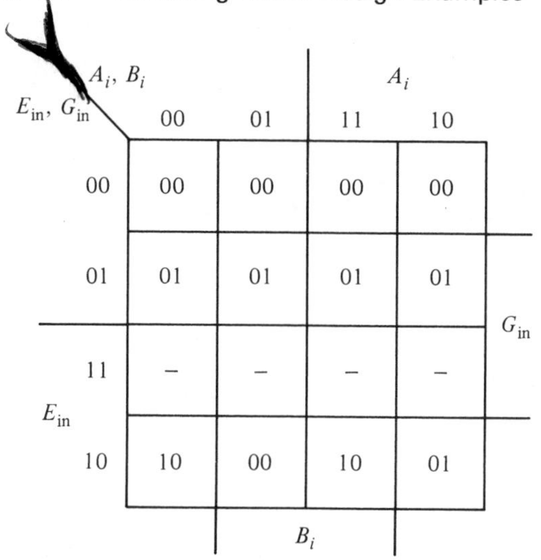

A_i, B_i
E_{in}, G_{in}
00 01 11 10
A_i
00 00 00 00 00
01 01 01 01 01
11 – – – –
10 10 00 10 01
G_{in}
E_{in}
B_i
E_{out}, G_{out}

Figure 4.3.5 The Karnaugh map from Figure 4.3.4.

$E_{in} = G_{in} = 1$ is not possible, since this would imply that A was equal to B and, simultaneously, greater than B. Plotting this truth table in the Karnaugh map of Figure 4.3.5, we arrive at the following design equations:

$$\begin{aligned} E_{out} &= E_{in}\overline{A}_i\overline{B}_i + E_{in}A_iB_i \\ &= E_{in}(\overline{A}_i\overline{B}_i + A_iB_i) \\ &= E_{in}\overline{(A_i \oplus B_i)} \end{aligned} \tag{4.3.4}$$

$$G_{out} = G_{in} + E_{in}A_i\overline{B}_i$$

A little thought shows that these equations make a great deal of sense, since the first is 1 only if $A_i = B_i$ and every pair of bits to the left are equal, and the second is 1 either if A has already been determined to be greater than B or if that determination occurs at this bit position. Figure 4.3.6(a) shows the 1-bit comparator implementation, and Figure 4.3.6(b) shows how these are cascaded to produce the n-bit comparator. The output from the least significant comparator gives the final comparison result: E and G.

4.3.3 Digital Multiplexers

It quite often happens, in the design of large-scale digital systems, that a single line is required to carry two or more different digital signals. Of course, only one signal at a time can be placed on the one line. What is required is a device that will allow us to select, at different instants, the signal we wish to place on this common line. Such a circuit is referred to as a *multiplexer* or *selector*.

Assume that we have four lines, I_0, I_1, I_2, and I_3, which are to be multiplexed on a single line, Y. Since there are four inputs, we will need two additional inputs to the multiplexer to select which of the I inputs is to appear at the output. Call these select lines S_1 and S_0.

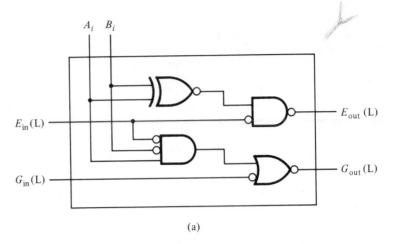

(a)

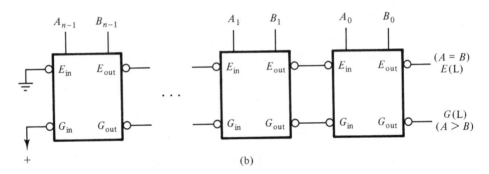

(b)

Figure 4.3.6   Design of an *n*-bit binary comparator. (a) An implementation of the 1-bit comparator. (b) An iterative array of 1-bit comparators forming an *n*-bit comparator.

Figure 4.3.7 shows a truth table for the resulting multiplexer. We can write the equation for Y directly from this table:

$$Y = I_0\bar{S}_1\bar{S}_0, + I_1\bar{S}_1S_0, + I_2S_1\bar{S}_0, + I_3S_1S_0 \tag{4.3.5}$$

Figure 4.3.8 shows the resulting gate implementation, assuming that all inputs and outputs are asserted high, and an appropriate symbol for the multiplexer.[6]

4.3.4 A Priority Encoder

In any computer system, there are a number of I/O devices that can communicate with the central processor. Each of these devices may request the attention of the central processor at any time. For example, when a user strikes a key on the computer's keyboard, the computer must respond by reading the value of the key depressed or else the information may be lost. Obviously, the processor cannot handle all of the requests simultaneously.

[6] The Appendix gives an IEEE standard symbol for this multiplexer, or MUX.

S_1	S_0	I_3	I_2	I_1	I_0	Y
0	0	–	–	–	0	0
0	0	–	–	–	1	1
0	1	–	–	0	–	0
0	1	–	–	1	–	1
1	0	–	0	–	–	0
1	0	–	1	–	–	1
1	1	0	–	–	–	0
1	1	1	–	–	–	1

Figure 4.3.7   A 4-line to 1-line multiplexer/selector.

There is a need, therefore, to somehow queue up the various requests and handle them one at a time. However, not all of the requests for service to the central processor have the same degree of urgency. For example, the human pressing a key on the keyboard will hold that key down for a hundred milliseconds or longer, whereas data found on a disk drive will be present for only a few microseconds or less. Obviously, if both requests are made at the same time, the disk needs to be taken care of before the key is read, since its information will vanish long before the human's finger is removed from the key. Thus, some mechanism is needed to identify the *priority* of the request for service. The basic idea is that each device which can request the service of the central processor is assigned a *priority level*. Then when a device wants service, it makes the request by asserting a line corresponding to its priority level. A piece of hardware, called a *priority encoder*, then

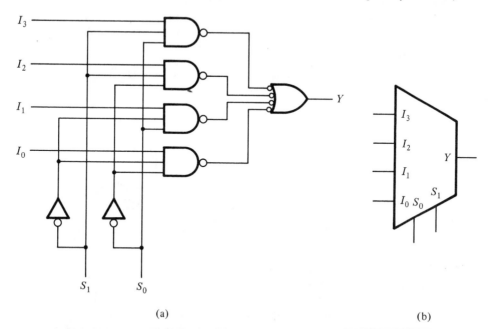

(a)

(b)

Figure 4.3.8   Gate implementation (a) of the 4-line to 1-line multiplexer/selector of Figure 4.3.7 and its schematic symbol (b).

P_3	P_2	P_1	P_0	Y_1	Y_0	R
0	0	0	0	–	–	0
1	–	–	–	1	1	1
0	1	–	–	1	0	1
0	0	1	–	0	1	1
0	0	0	1	0	0	1

Figure 4.3.9 The truth table for a four-level priority encoder.

determines which line is requesting service at the highest priority and generates a number corresponding to this priority.

Let us consider the design of a four-level priority encoder. Let the request lines be P_0, P_1, P_2, and P_3, where P_3 has the highest priority. Since there are four levels, we need two outputs to encode the various requesting levels. Let these two outputs be Y_1 and Y_0. If we encode (Y_1, Y_0) from (00) to (11) to represent the requests on lines P_0 to P_3, respectively, we will need one more output, R, to differentiate between no request and a level 0 request. Figure 4.3.9 shows the truth table for this priority encoder.

The necessary design equations can be derived directly from this truth table or by plotting the table in a Karnaugh map. In either case, it is easily verified that these equations become

$$Y_1 = P_3 + P_2$$
$$Y_0 = P_3 + \overline{P_2}P_1 \qquad (4.3.6)$$
$$R = P_3 + P_2 + P_1 + P_0$$

If we assume that the inputs and the output R are asserted low, which is the usual case for this function, and that the Y_1 and Y_0 are asserted high, then the resulting realization becomes as shown in Figure 4.3.10.

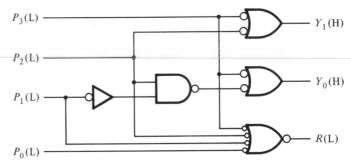

Figure 4.3.10 Implementation of the four-level priority encoder.

4.4 COMBINATIONAL LOGIC DESIGN USING ROMs AND PLAs

At the end of Section 4.3.1, we discussed the possibility of designing a 2- or 4-bit adder which could be cascaded to form an n-bit adder, as was done for the 1-bit adder designed there. The advantage of such a multibit adder is that it can speed up the addition process.

Let us consider how we would carry out the design of the 2-bit adder. This circuit would have five inputs, two for the number A, two for the number B, and one for the carry in. It would also have three outputs, two for the sum bits and one for the carry out. Let A_1 and A_0 be the bits of one of the numbers and let B_1 and B_0 be the bits of the second. The sum bits will be represented by S_1 and S_0. Finally, let C_{in} be the carry in and C_{out} be the carry generated by this addition. The truth table for this adder can be organized as two tables, one for $C_{in} = 0$ and one for $C_{in} = 1$. This is shown in Figure 4.4.1.

Inputs				$C_{in} = 0$			$C_{in} = 1$		
A_1	A_0	B_1	B_0	S_1	S_0	C_{out}	S_1	S_0	C_{out}
0	0	0	0	0	0	0	0	1	0
0	0	0	1	0	1	0	1	0	0
0	0	1	0	1	0	0	1	1	0
0	0	1	1	1	1	0	0	0	1
0	1	0	0	0	1	0	1	0	0
0	1	0	1	1	0	0	1	1	0
0	1	1	0	1	1	0	0	0	1
0	1	1	1	0	0	1	0	1	1
1	0	0	0	1	0	0	1	1	0
1	0	0	1	1	1	0	0	0	1
1	0	1	0	0	0	1	0	1	1
1	0	1	1	0	1	1	1	0	1
1	1	0	0	1	1	0	0	0	1
1	1	0	1	0	0	1	0	1	1
1	1	1	0	0	1	1	1	0	1
1	1	1	1	1	0	1	1	1	1

Figure 4.4.1   Truth table for a 2-bit adder.

Either by plotting the outputs in five-variable Karnaugh maps or by use of the Quine-McCluskey algorithm, the equations for the sum bits and the carries can be derived. These equations are

$$S_0 = \overline{C}_{in}\overline{A}_0 B_0 + \overline{C}_{in}A_0\overline{B}_0 + C_{in}\overline{A}_0\overline{B}_0 + C_{in}A_0 B_0$$

$$S_1 = \overline{C}_{in}(\overline{A}_1\overline{A}_0 B_1 + \overline{A}_1 B_1\overline{B}_0 + A_1\overline{A}_0\overline{B}_1 + A_1\overline{B}_1\overline{B}_0)$$
$$+ C_{in}(\overline{A}_1\overline{B}_1 B_0 + \overline{A}_1 A_0\overline{B}_1 + A_1 B_1 B_0 + A_1 A_0 B_1)$$
$$+ \overline{A}_1\overline{A}_0 B_1\overline{B}_0 + \overline{A}_1 A_0\overline{B}_1 B_0 + A_1\overline{A}_0\overline{B}_1\overline{B}_0 + A_1 A_0 B_1 B_0 \qquad (4.4.1)$$

$$C_{out} = A_0 B_1 B_0 + A_1 A_0 B_0 + C_{in}B_1 B_0 + C_{in}A_0 B_1 + C_{in}A_1 B_0$$
$$+ C_{in}A_1 A_0 + A_1 B_1$$

It will obviously take a large number of gates to implement these equations. The implementation of a 4-bit adder will be even more complex. What we would like is a single device that could be used to implement a wide range of complex functions. Fortunately, two such devices exist: read-only memories (ROMs) and programmable logic arrays (PLAs).

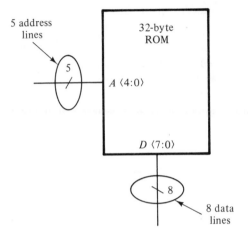

Figure 4.4.2 A 32-byte ROM.

4.4.1 Read-Only Memory (ROM)

We can think of a read-only memory as a table or dictionary that contains information. To look something up in this table we need a pointer or an index that identifies the location of a particular piece of information. This index is referred to as an *address*. This address is numeric and is generally selected to be uniquely associated with a particular piece of information. Most ROMs store information in 8-bit, or byte, quantities. Figure 4.4.2 shows a symbol for a ROM that stores 32 bytes of information. When the address lines, $A\langle 4:0\rangle$,[7] take on some value, say 00001, then the information stored in the ROM corresponding to this address will appear on the output lines, $D\langle 7:0\rangle$.

To see how such a device can be used to implement a switching function, and in particular the 2-bit adder, consider the truth table of Figure 4.4.1. The inputs to the adder represent minterms for the individual output functions. These minterms identify when a particular output function is 1 and when it is 0. Thus, if the values of the function are stored in the memory, then the inputs can be thought of as addresses which point to these values. Although ROMs usually store information in 8-bit bytes, we may associate a particular output function with a particular bit in the byte. Thus, letting $D\langle 0\rangle$ correspond to the carry out, C_{out}, and $D\langle 2\rangle$ and $D\langle 1\rangle$ correspond to the sum bits, S_1 and S_0, respectively, we can make the read-only memory implement the 2-bit adder if the information stored is as shown in the abbreviated table of Figure 4.4.3, which is just the information found in the truth table for the adder given in Figure 4.4.1.

Note, in this example, that we could implement five more functions of the input variables by using the ROM data lines that are not being used for the adder function. For example, we might let $D\langle 3\rangle$ be 1 whenever the numbers A and B were equal. In a similar fashion we might also implement the "greater than" and "less than" signals as well.

---

[7] The notation $A\langle 4:0\rangle$ is a shorthand notation meaning that there are five lines, labeled $A\langle 4\rangle$, $A\langle 3\rangle$, $A\langle 2\rangle$, $A\langle 1\rangle$, and $A\langle 0\rangle$, with $A\langle 4\rangle$ being the most significant and $A\langle 0\rangle$ the least significant.

	Address lines				ROM data output lines					
					Not used			S_1	S_0	C_{out}
C_{in} $A\langle 4 \rangle$	A_1 $A\langle 3 \rangle$	A_0 $A\langle 2 \rangle$	B_1 $A\langle 1 \rangle$	B_0 $A\langle 0 \rangle$	$D\langle 7 \rangle$	$\cdots$	$D\langle 3 \rangle$	$D\langle 2 \rangle$	$D\langle 1 \rangle$	$D\langle 0 \rangle$
0	0	0	0	0	–		–	0	0	0
0	0	0	0	1	–		–	0	1	0
0	0	0	1	0	–		–	1	0	0
0	0	0	1	1	–		–	1	1	0
$\cdots\cdots\cdots\cdots\cdots$						$\cdots\cdots\cdots\cdots\cdots\cdots\cdots$				
1	1	1	1	0	–		–	1	0	1
1	1	1	1	1	–		–	1	1	1

Figure 4.4.3   The contents of the ROM that implements the 2-bit adder of Figure 4.4.1.

This example illustrates that read-only memories can be used to implement very complex switching functions by storing in the ROM the value of the function corresponding to the assignments of the input variables. Thus, functions implemented by ROMs need not be minimized, since we are basically implementing the function from the minterm (or maxterm) list, i.e., we are implementing the function in canonical form.

Many different types of read-only memory exist. Some have the information stored in them at the time they are manufactured. These are said to be *mask-programmed*. Others can be programmed, or loaded with the required information, by the user. Such read-only memories are referred to as *programmable read-only memories* (PROMs). Programming of these ROMs generally requires special equipment to erase, if possible, any information that might be in the ROM and then store any new information required. These ROMs come in basically two types. One type cannot be erased and therefore can be programmed only once. The second type, *erasable programmable read-only memories* (EPROMs), can be used over and over again to store many different data sets. Generally, the process of erasing involves erasing every byte in the memory and usually takes several minutes. Once programmed, information can be read from EPROMs at computer speeds. We might say that EPROMs are "mostly" read-only memories. ROMs also come in a great variety of sizes. Modern EPROMs range in size from 4K (4096) bytes, denoted as 4K $\times$ 8, to 128K $\times$ 8 and larger. The nonerasable PROMs are generally very small, on the order of up to a few hundred bytes. The number of variables that a function can have and be implemented with ROMs is limited by the number of address lines on the ROM. Thus, functions of up to 16 variables can be implemented with a 64K $\times$ 8 EPROM.

4.4.2 Programmable Logic Arrays (PLAs)

Another device which can be used to implement complex functions on many variables is a *programmable logic array* (PLA). Figure 4.4.4(a) shows a simplified schematic for a PLA. A PLA consists of a set of AND gates, each input of which can be connected to any input of the PLA itself or the complement of any input, and a set of OR gates, whose inputs can be connected to any of the AND gate outputs. The outputs of the OR gates serve as outputs of the device. In this diagram, a single line is shown at the input of each of the AND gates. This line is used to represent n lines, each of which can be connected

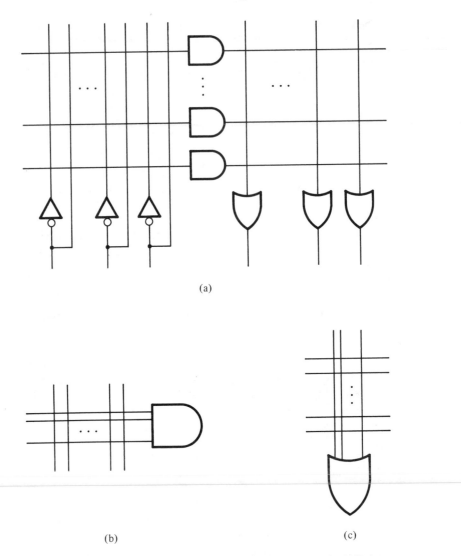

(a)

(b)   (c)

Figure 4.4.4 Programmable logic array. (a) Simplified diagram. (b) AND inputs. (c) OR inputs.

to a different device input or its complement. A similar situation exists for the OR gate inputs. This is shown in Figure 4.4.4(b) and (c). Thus arbitrary functions of the input variables can be implemented in *sum of products* (SOP) form with the PLA.

Consider, for example, the implementation of the adder outputs S_0 and C_{out} using a PLA having 5 inputs, 11 product terms, and 2 outputs. In order to implement these functions, we need to *program* the PLA. Programming a PLA consists of making connections between the device inputs—C_{in}, A_1, A_0, B_1, and B_0, in this case—and the AND gate inputs, as well as between the AND gate outputs, forming the product terms, and the OR gate inputs.

The required connections are shown in a *programming diagram*. A programming diagram is created by placing an X at the intersection of two lines that are to be connected. Thus, from the equations for S_0 and C_{in} given in equation group (4.4.1), we see that we need four product terms to form S_0 and seven product terms to form C_{out}. Note, in this case, that S_0 and C_{out} do not share a common product term, and thus a total of eleven AND gates are necessary for their implementation. Figure 4.4.5 shows the programming diagram used

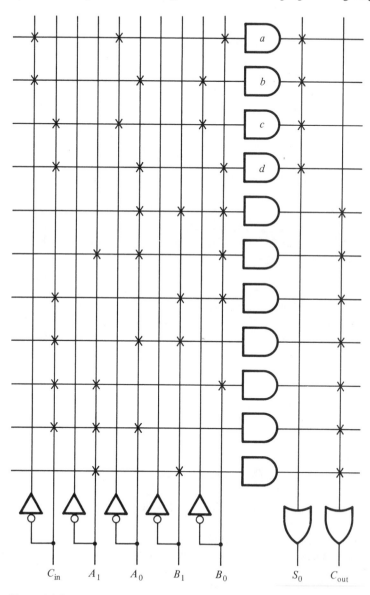

Figure 4.4.5 A PLA programming diagram for the 2-bit adder outputs S_0 and C_{out}.

to implement these two functions. In this diagram, for example, we see that S_0 is formed by ORing the outputs from the AND gates labeled a, b, c, and d. This is indicated by the X's at the intersection of these AND gate outputs and the inputs to the OR gate that forms S_0. The product term formed by AND gate a, $\overline{C}_{in}\overline{A}_1 B_0$, is indicated by the X's at the intersection of the input to this gate with inputs $\overline{C}_{in}$, $\overline{A}_1$ and B_0. The other product terms are indicated in a similar manner.

PLAs, like ROMs, exist in a multitude of different types. At present they are available in mask-programmed versions as well as in one-time-only programmable and erasable programmable versions. Sizes of these devices also vary greatly, but the devices usually have more than eight inputs and outputs. A typical example is the "field-programmable logic array" (FPLA) produced by Signetics, the 82S100, which has 16 inputs, 8 outputs, and 48 product terms. The number of product terms that can be implemented is the important factor in the use of a PLA device. Since the number of product terms in any such device is limited, it is important that the function to be implemented be in minimal sum of products form.

4.4.3 Some Comments on Implementation

The examples shown above clearly indicate that the implementing of switching functions using a PLA requires a somewhat different strategy from implementing them using a ROM. In the case of the ROM, we are actually implementing the truth table for the function and thus we need to express the function in terms of its minterms or maxterms, usually in the form of an index list. In the case of the PLA, however, we are implementing general product terms, not *just* minterms. Since PLAs have a limited number of AND gates, we need to find minimal SOP forms for the functions to be implemented. This requires the application of simplification procedures such as those described in Section 3.5 and, generally, the multiple-output simplification procedures given in Section 3.5.6.

Another important difference between ROMs and PLAs is that of speed. As we shall discuss in the next chapter, physical devices have associated *propagation delays*. This is the delay from the time an input changes until the output changes. It is generally true that PLAs have much shorter propagation delays than ROMs and so are best used in situations requiring a high speed of operation. ROMs, on the other hand, are usually less expensive.

4.5 AN ANNOTATED BIBLIOGRAPHY

An excellent discussion of the mixed-logic symbology presented in this chapter can be found in the books by Fletcher and by Prosser and Winkel. Some discussion can also be found in Kostopoulos. Both Kostopoulos and Fletcher also discuss at length the electrical characteristics of the various technologies used in integrated circuits today, including discussion of "wired logic" (e.g., open-collector logic).

FLETCHER, W. I., *An Engineering Approach to Digital Design*, Prentice-Hall, Englewood Cliffs, N.J., 1980.

KOSTOPOULOS, G. K., *Digital Engineering*, Wiley-Interscience, New York, 1975.

PROSSER, F. P., and D. E. WINKEL, *The Art of Digital Design: An Introduction to Top-Down Design*, 2nd ed., Prentice-Hall, Englewood Cliffs, N.J., 1987.

McCluskey explains in great detail the recent symbology IEEE Std. 91-1984 and uses it throughout his text. A short pamphlet by Mann gives a more formal definition of this new standard and a general discussion of its background.

MANN, F. A., "Overview of IEEE Std. 91-1984: Explanation of Logic Symbols," Texas Instruments, Inc., Carrollton, Tex., Publ. SDYZ001, 1984.

McCLUSKEY, E. J., *Logic Design Principles with Emphasis on Testable Semicustom Circuits*, Prentice-Hall, Englewood Cliffs, N.J., 1986.

Many authors discuss the use of PLAs and ROMs in the implementation of switching expressions. Such logic is quite often referred to as *programmable logic*. Chapter 8 of Fletcher is an excellent source for a thorough discussion of this topic in which numerous examples are given. McCluskey's text also describes the use of these devices, in Chapter 6. Other sources for coverage of this topic are the books by Mano, Dietmeyer, and Hill and Peterson.

DIETMEYER, D. L., *Logic Design of Digital Systems*, 2nd ed., Allyn & Bacon, Boston, 1978.

HILL, F. J., and G. R. PETERSON, *Introduction to Switching Theory and Logical Design*, 3rd ed., Wiley, New York, 1981.

MANO, M. M., *Digital Design*, Prentice-Hall, Englewood Cliffs, N.J., 1984.

One reference the reader should, without doubt, obtain is a digital integrated circuit catalog. These catalogs describe what devices are available, how they are packaged, and what their electrical characteristics are. TTL data manuals are readily available from electronics parts stores, computer stores, some bookstores, and the manufacturers themselves. TTL and CMOS devices are manufactured by most semiconductor firms today. Examples are Texas Instruments, Inc., Signetics Corp., Motorola, Inc., and National Semiconductor, Inc. The student interested in "playing" with some of these devices can find sources of supply at various electronic hobby stores and in the ads found in electronics and computer magazines. Experiments using these devices and the associated equipment can be found in manuals such as those of Williams, Wakerly, and Teng and Malmgren.

TENG, A. Y., and W. A. MALMGREN, *Experiments in Logic and Computer Design*, Prentice-Hall, Englewood Cliffs, N.J., 1984.

WAKERLY, J. F., *Logic Design Projects Using Standard Integrated Circuits*, Wiley, New York, 1976.

WILLIAMS, G. E., *Digital Technology—Lab Manual*, Science Research Associates, Inc., Chicago, 1977.

4.6 PROBLEMS

4.1 For each of the gates shown in Figure P4.1,
   (a) Construct the physical truth tables.
   (b) Construct the logical truth tables.
   (c) Write an expression for the logical function implemented.

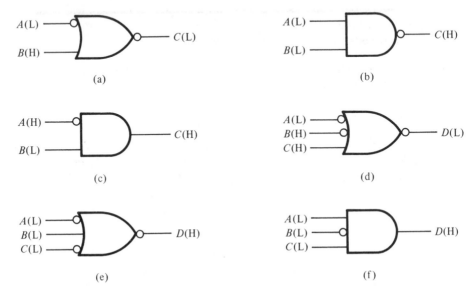

Figure P4.1

4.2. Which gates in Problem 4.1 are physically equivalent?

4.3. Construct the physical and the logical truth tables for each of the circuits shown in Figure P4.3.

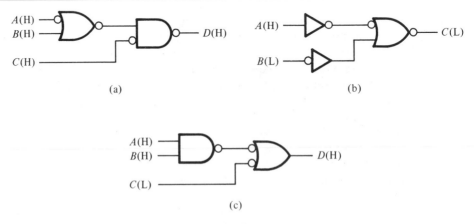

Figure P4.3

4.4. Write the function implemented by each of the circuits shown in Figure P4.4.

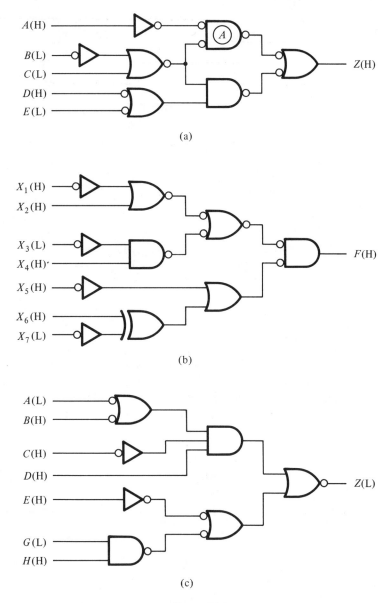

(a)

(b)

(c)

Figure P4.4

4.5. Suppose an oscilloscope probe were placed on the output of gate *A* of figure P4.4(a). What would the output of gate *A* look like if the inputs to the circuit are as shown in Figure P4.5?

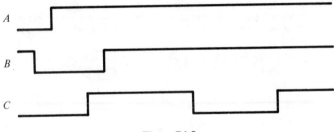

Figure P4.5

4.6. Construct alternative symbols physically equivalent to the gates shown in Figure P4.6.

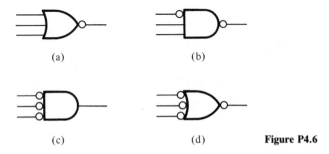

(a)   (b)

(c)   (d)   **Figure P4.6**

4.7. One of the reasons that NAND gates are used so extensively in the design of digital systems is that they can be used, alone, to implement any given switching function. Assuming all inputs are asserted high, show how to use the two-input NAND gate shown in Figure 4.2.8(a) to implement the NOT, the AND, and the OR operations.

4.8. Repeat Problem 4.7 using the NOR gate of Figure 4.2.8(b).

4.9. Using only the implication gate shown in Figure P4.1(c), implement the following functions and show a correctly drawn schematic. Assume that all inputs and outputs are asserted high.

(a) $g(a, b, c, d, e, f) = a + bc(d + ef)$

(b) $f(p, q, r, s, t, u, v, w) = (pq + rs)[t + v(u + w)]$

4.10. Repeat Problem 4.9, using NOR gates only.

4.11. Repeat Problem 4.9, using NAND gates only.

4.12. Redraw the circuits shown in Figure P4.12 in correct form and write the functions implemented by each.

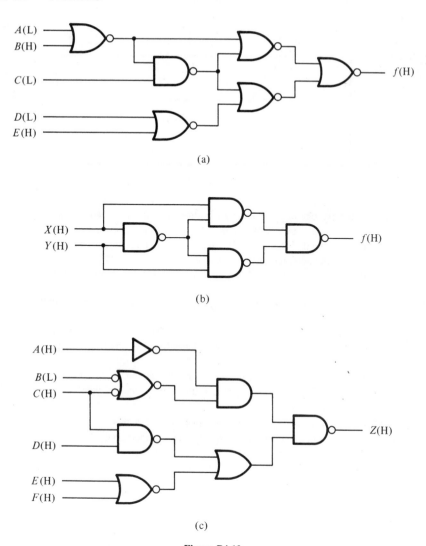

Figure P4.12

4.13. Suppose you are given the 4-line to 1-line multiplexer of Figure 4.3.8 and are told to implement the function $f(x, y, z) = x\bar{y} + y\bar{z}$ using *only* this circuit. Can this be done, and if so, how would you connect the asserted high inputs x, y, and z to the multiplexer to do the job?

4.14. Design a circuit using NANDs, NORs, and level shifters only that takes a 4-bit BCD number as an input and produces a 4-bit excess-3–coded number as an output. Assume that the inputs are asserted high and the outputs are asserted low.

4.15. Add an output to Problem 4.2 that is asserted low if the input is not a legal BCD number.

4.16. You are to design a one-digit BCD adder as follows. There are to be nine inputs, eight of which represent the two 4-bit BCD digits and the ninth of which is a carry into the adder. The output is five bits: four for the BCD sum digit and the fifth for the carry out. Figure P4.16

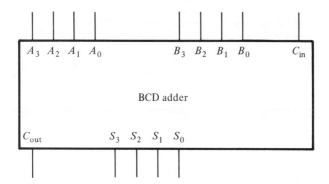

Figure P4.16

shows the circuit symbolically. Assume that all inputs and outputs are asserted high. (*Hint*: You may want to use the adder of Figure 4.3.3 as part of your circuit.)

(a) Using only NANDs, NORs, and level shifters, design the circuit and give a correctly drawn schematic diagram for your design.

(b) Given a 1024-byte ROM, indicate how you would program the ROM to implement the BCD adder.

(c) Assume you can purchase single integrated circuits (ICs) having either four 2-input NANDs, three 3-input NANDs, two 4-input NANDs, one 8-input NAND, four 2-input NORs, three 3-input NORs, or six level shifters or NOT gates. Assuming that the price of each of these ICs is 15 cents and the price of the ROM is $5, which of your designs, (a) or (b), is cheaper?

(d) Suppose, because of the cost of printed circuit boards, IC sockets, and the like, it costs an additional 50 cents per IC to implement either of your designs. Now which solution is the cheaper?

4.17. What would be the size specifications for a PLA that could be used to implement the BCD adder of Problem 4.16? Give the number of inputs, outputs, and product terms required.

Sequential Circuits

5

5.1 INTRODUCTION

Up to this point, we have dealt with switching networks whose outputs are functions only of the present input. It is possible for such networks to exhibit "memory," in the sense that the outputs can be made functions of not only the present inputs, but also some set of past inputs as well. Such systems are termed *sequential*, since the outputs may be functions of a sequence of past inputs. Basically, *sequential circuits* have memory because one or more of the outputs are "fed back" to serve as inputs to the network. Thus the next output will, somehow, be a function of the present inputs and the last output. To understand how this can happen, we must first introduce time as a variable in the system.

5.1.1 Delay in Gate Networks

We basically assumed, in our discussion of gates that a change in the output of the gate occurred at exactly the same instant of time that an input change occurred. This, of course, will not happen, since the gate is composed of electrical components that possess capacitance (among other things) to some degree. Since voltage cannot change instantaneously across a capacitor, the output of a gate cannot change simultaneously with the input. The time required for the output of a gate to change in response to a change in an input is referred to as *propagation delay*. Propagation delays for standard TTL (transistor-transistor logic) gates and other TTL devices vary but are usually in the range of 1 to 15 nanoseconds (ns).[1] Figure 5.1.1 shows how the input and output of a typical TTL NAND gate change in time.

---

[1] To put things in perspective, light travels approximately 30 cm (one foot) in one nanosecond.

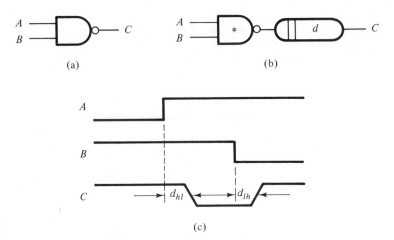

Figure 5.1.1 Propagation delay in a NAND gate. (a) The nonideal NAND gate. (b) An ideal NAND with delay. (c) Timing diagram.

Although it is usually the case that the delay for a high-to-low transition (d_{hl}) is different from the delay for a low-to-high transition (d_{lh}), for analysis purposes we may assume these to be the same. Figure 5.1.1(b) shows a model of the gate that can be used for purposes of analysis. This model consists of an ideal NAND gate, which has zero propagation delay, followed by a delay element. Using this model, we can examine the behavior of networks of gates as inputs change.

Consider the gate network shown in Figure 5.1.2, which realizes the function

$$f(A, B, C) = AB + \overline{A}C \qquad (5.1.1)$$

Suppose initially that inputs B and C are high and A is low, thus making the output high, or a logical 1. Now, suppose that at time t_1, input A goes from low to high. What happens at the output? To determine this, we need to follow the change in input A through the circuit to the output. We may assume, for this analysis, that the propagation delays through all of the gates are the same. Such an assumption, although strictly speaking not true, is good enough for our purposes. Now, when A goes high, lines x and y will both go low after a propagation delay d. The change in x will affect line z after another propagation delay d, at which time z will go high. Since y is low at this instant of time, the output, f, will stay high, as it should, from Equation (5.1.1).

Next, consider what happens at time t_2. When input A goes low, both of the lines x and y will go high after time interval d. Note that now both y and z are high, which means that the output, f, must go low after another interval d. At about this time, line z will go low, since x and C are now both high. Since z has gone low, the output must change once again and go high. All signals will now stay at these values. We see from this analysis that although the output, f, should, by Equation (5.1.1), have remained asserted, it has, in fact, changed for a brief period of time. This change is called a "glitch" and arises because

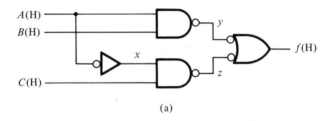

(a)

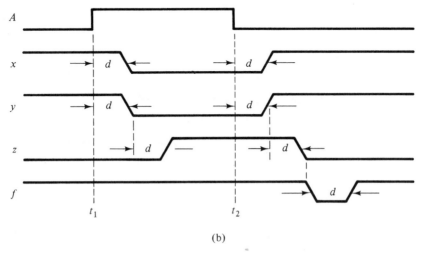

(b)

Figure 5.1.2   Generation of a "glitch" in a combinational circuit. (a) A realization of Equation (5.1.1). (b) Timing diagrams assuming B and C are both high throughout.

of the physical delays in a network.[2] Glitches can cause systems to fail and should, therefore, be avoided.

To see how glitches may be avoided, let us consider the cause of the glitch in the circuit of Figure 5.1.2. Basically, the glitch occurred because there were two separate paths from input A to the output, each having a different number of delays—one path, A-y-f, having two delays and the other, A-x-z-f, having three delays. This difference in path length causes the output to "see" A go low before it "sees" $\overline{A}$ go high. Thus, both product terms in Equation (5.1.1), as seen by the output, are 0. If, however, the *consensus* term BC (see Theorem 3.2.6 of Section 3.2) is added to Equation (5.1.1), the glitch will vanish, because $BC = 1$ throughout the various transitions on A and $\overline{A}$. With the consensus term, Equation (5.1.1) will become:

$$f(A, B, C) = AB + \overline{A}C = AB + \overline{A}C + BC \qquad (5.1.2)$$

[2] Glitches of this type are associated with what are generally referred to as "hazards," in this case, a static hazard, since the output was not supposed to change. The identification and elimination of hazards will be discussed in Chapter 6.

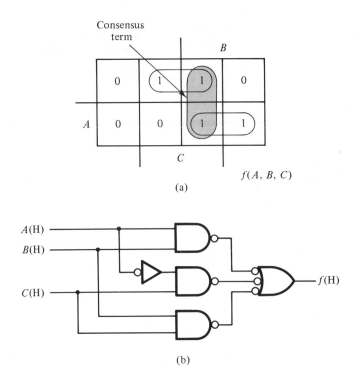

Figure 5.1.3   A glitchless implementation of Equation (5.1.1). (a) Added consensus term.
(b) Final realization.

Figure 5.1.3(a) shows how the consensus term is added in the Karnaugh map, and the resulting implementation is shown in Figure 5.1.3(b). This example illustrates the fact that the removal of consensus terms in logic circuits may cause undesired behavior.

5.1.2 Feedback

As mentioned above, a sequential circuit is one in which the output is a function of not just the present inputs but some set of past inputs as well. This form of "memory" is created in any system where outputs are "fed back" through a delay to the inputs. Effectively, the delay "remembers" some portion of the past history.

In the analysis of the combinational circuit of Figure 5.1.2, each gate had an associated delay which was considered individually when we analyzed the time behavior of the circuit. If we are to make the assumption that the combinational logic has been designed in such a way as to have no glitches, then a simplification can be made in the modeling process.[3] Since no glitches occur, we are interested only in the time it takes for a signal to propagate from the input to the output. Thus, we can combine all of the delays into a single, lumped

---

[3] It is generally possible to remove glitches in combinational networks by adding logic, as was done above when the consensus term was used to eliminate the glitch. Thus the assumption of a glitchless circuit is realistic.

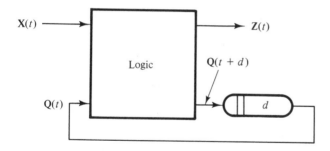

Figure 5.1.4 A model of a combinational circuit with feedback.

delay at the output. This is shown in the feedback model of Figure 5.1.4. In this figure, the symbol $X(t)$ represents a set of n inputs and the symbol $Z(t)$ a set of m outputs. The symbol $Q(t)$ is referred to as the *current state* of the system and is made up of a set of p outputs whose value will become $Q(t + d)$, the *next state*, at a time interval d from now. The delays shown in this model make it possible for us to separate the present state from the next state. Thus we see that the output of such a system is a function of the present input and the present state and that the next state is also a function of the present input and the present state.

A few simple, but important, observations can be made from the model shown in Figure 5.1.4. First, the delay represents memory in the system, since it holds, or remembers, the present state of the circuit while the circuit computes the next state. Second, whenever an input changes, both the outputs and the next state variables will change. With a change in the next state variables, one of two things can occur: either $Q(t) = Q(t + d)$ or $Q(t) \neq Q(t + d)$. In the first case, the system is *stable*: nothing changes, except possibly the Z's. In the second case, the system is *unstable*: in addition to possible changes in the outputs Z, the input state variables $Q(t)$ will change after the appropriate time delay. This change in the Q inputs may cause further changes in the next state variables. Two things can occur in such an unstable system: either the system will eventually become stable or it will continue to have changes forever. The latter case is usually an undesirable situation.[4]

The model shown in Figure 5.1.4 is an example of an *asynchronous* or *fundamental-mode sequential circuit*. As we shall see in the next section, such sequential circuits are extremely important in the design of memory elements for computers. However, because of the possibility of long-term instability in such systems, reliable design of large-scale asynchronous sequential circuits is very difficult, if not impossible. We will investigate the processes of analysis and design of such circuits in Chapter 6.

5.2 FLIP-FLOPS

In Section 5.1.2, we saw that a combinational circuit with feedback can be analyzed using the model given in Figure 5.1.4. In this section, we are going to use this model to examine a specific feedback circuit, called an *SR* flip-flop. This particular circuit forms the basis for all commonly used flip-flop types as well as computer memory. We will then examine

---

[4] This is, however, the exact behavior needed to produce oscillation for a computer's "clock."

the many different types of flip-flops that are available and define their operating characteristics. In succeeding sections, we will use these devices to design counters and sequential controllers.

Before proceeding, we should note that the term "flip-flop" is a generic term applied generally to electronic devices having two stable states. The flip-flop can be placed in one or the other of these states by applying various signals to its inputs. As we shall see, there are many types of flip-flops and many ways to control them.

5.2.1 The Simple SR Flip-Flop

Figure 5.2.1(a) shows a simple two-gate combinational circuit having one feedback signal called Q that is shown to be asserted high. The two inputs to the circuit, S and R, are both asserted low, as indicated in the figure. Figure 5.2.1(b) shows the circuit after the delays have been moved to the output. This figure is in the form of the sequential circuit model of Figure 5.1.3. The analysis of this circuit is easily carried out by writing the equation for the next-state variable $Q(t + d)$. This equation becomes

$$Q(t + d) = S + \overline{R}Q(t) \qquad (5.2.1)$$

On the basis of this equation, the timing analysis can be carried out as shown in Figure 5.2.2; remember that S and R are asserted low and Q is asserted high. Equation (5.2.1) states that if $S = 0$ and $R = 0$, then the next state will be equal to the present state: a

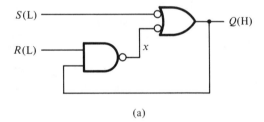

(a)

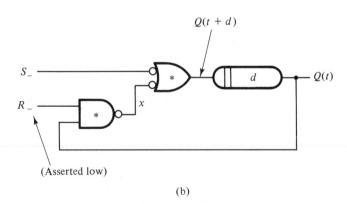

(b)

Figure 5.2.1 A simple *SR* flip-flop. (a) A two-gate circuit with feedback. (b) Model with delay.

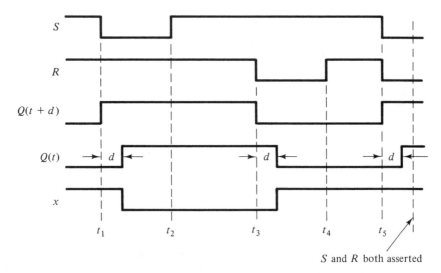

Figure 5.2.2   Timing diagram for the *SR* flip-flop model of Figure 5.2.1.

stable situation. In this situation, the output, Q, which can be either a 0 or a 1, will not change over time. Assume that $Q = 0$, and let S go from a 0 to a 1 (i.e., from a high to a low) at time t_1. The next state, from Equation (5.2.1), will thus be equal to 1, which means that after a delay of time interval d, the present state, $Q(t)$, will also be 1. When S returns to 0 at a time t_2, the output, Q, will have changed to 1 and will continue to be 1 thereafter. We say that the output has been *set* to 1 by assertion of the S input. Now suppose that R goes to 1 at time t_3. From Equation (5.2.1) we see that the next state will become 0, as will the present state after a delay of time interval d. When R returns to 0 at time t_4, the output, Q, will continue to have the value 0 from then on or until S changes again. We say that the output is *reset* to 0 by the assertion of the R input.

This circuit is called a *set-reset (SR) flip-flop*, since a momentary assertion of the S input sets the output to 1 and a momentary assertion of the R input resets the output to 0. Once the flip-flop is placed in some state, it will remember its state until the next input change. Thus the flip flop can "store" 1 bit of information.

If we look a bit closer at the *SR* flip-flop of Figure 5.2.1, we may observe, ignoring propagation delay differences between the two signals, that the line labeled x takes on the opposite value of the output Q as long as S and R are not both simultaneously asserted. If both S and R are asserted, as, for example, at time t_5 in Figure 5.2.2, then both Q and x will be high, as is easily verified by an examination of Figure 5.2.1 and the timing diagram of Figure 5.2.2. If we assume that this never occurs or is never allowed to occur, then we can think of the signal x either as $\overline{Q}$ or as $Q(L)$. Figure 5.2.3(a) shows the symbol we will use for this flip-flop under the condition that S and R are never asserted at the same time, and Figure 5.2.3(b) gives the defining physical truth table for the flip-flop. In this symbol, the preferred label for line x of Figure 5.2.1(a) is Q_- or $Q(L)$.[5]

---

[5] The simple *SR* flip-flop shown in Figure 5.2.1 is generally referred to in the IC data catalogs as an "*S-R* latch."

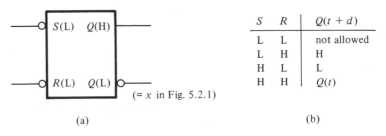

S	R	$Q(t + d)$
L	L	not allowed
L	H	H
H	L	L
H	H	$Q(t)$

(a)   (b)

Figure 5.2.3 A symbol for the *SR* flip-flop (a) and its truth table definition (b).

5.2.2 A Clocked SR Flip-Flop

In the basic flip-flop, changes in the output occur whenever either input changes. In a computer, separate operations occur at specific instances of time defined by an internal "clock." Thus, to use the basic flip-flop in the design of a computer, we must make certain that the outputs change only at very specific instances of time. This can be accomplished by adding a clock input to the flip-flop, as shown in Figure 5.2.4. In this flip-flop, the output Q will be unaffected by any change in the S and R lines as long as the clock, G, is negated. The output is allowed to change only when G is asserted. Note, however, that as long as G is asserted, the output will follow the changes in the S and R lines. This is an example of what we will refer to as a *latch-mode flip-flop*. Specifically, a latch-mode flip-flop is one whose outputs "functionally" follow the inputs for as long as the clock line is asserted. This means that the flip-flop basically becomes a simple combinational circuit in which the bistable nature of the device becomes transparent.

Another type of clocked flip-flop using the basic *SR* flip-flop circuit is the *D* flip-flop. This device has one input, D, and, of course, the clock. The output Q equals the input D whenever the clock line is asserted. Figure 5.2.5 shows the circuit for a *D*-type latch-mode flip-flop. This type of flip-flop is used extensively in the design of computers and other digital systems for the temporary storage of information and is often referred to simply as a *latch* or as a *transparent latch*.

In the latch-mode *SR* flip-flop, if the S and R lines change more than once while the G line is asserted, the output Q will also change more than once. This, quite often, is an undesirable characteristic. Suppose we wish the output to change only once during the period that the clock is asserted and to take on the value specified by the last input change. This can be done by using two *SR* flip-flops in tandem, as shown in Figure 5.2.6. Such

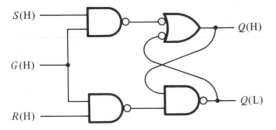

Figure 5.2.4 A clocked *SR* flip-flop.

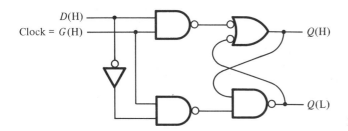

Figure 5.2.5   A clocked *D*-type flip-flop.

an arrangement is referred to as a *master-slave flip-flop*. In the master-slave *SR* flip-flop, the output of the slave flip-flop takes on the value of the output of the master while the clock is negated. When the clock is asserted, the slave flip-flop latches, or holds, this value while the master flip-flop changes to its new value. This new value is then passed to the output when the clock is once again negated.

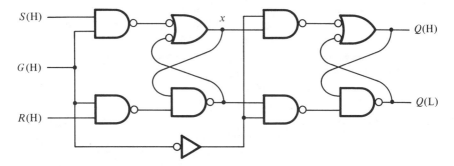

Figure 5.2.6   The master-slave *SR* flip-flop.

A rather serious problem, however, exists in the master-slave flip-flop. As we saw in Section 5.1, glitches can occur in combinational circuits because of differing propagation delays in the network. Suppose that the *S* and *R* inputs are both to be negated while the clock line is asserted. This, of course, should result in the output of the flip-flop remaining at its old value at the end of the clock pulse.[6] Figure 5.2.7 shows what can happen if a glitch should occur during this interval. We see from this figure that the glitch causes the output to change when it is supposed to stay the same. Such a situation could cause a large system to fail, with potentially catastrophic consequences. For this reason, master-slave flip-flops are not in general use today.

5.2.3 Edge-Triggered Flip-Flops

A type of flip-flop that avoids the glitch problem, and many similar noise-induced problems as well, is the *edge-triggered flip-flop*. In this flip-flop, the next output value is dependent only on the values of the inputs at the time when the clock line goes from low to high (or

---

[6] A *clock pulse*, as used here, is taken to mean a signal which is asserted for some period of time and then is negated for another period of time. Although this action usually occurs with regularity in a computer, regularity is not an essential feature for driving flip-flops.

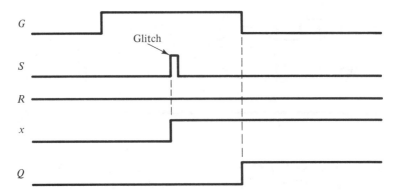

Figure 5.2.7 Timing diagram for a glitch-induced output error.

high to low) and is totally unaffected by any change on the inputs at any other time. Since the transition of a clock signal from low to high is usually very fast (a few nanoseconds), the likelihood that the inputs will change during this period is extremely small, indeed. In fact, the designer of a system using these devices should make sure that this *never* happens. The analysis and design of edge-triggered flip-flops will be discussed in Chapter 6. Figure 5.2.8 shows symbols used for the four basic types of edge-triggered flip-flops, along with their respective physical truth tables. Only two of these, however, are generally available as integrated circuits: the *D* and the *JK* flip-flops. Note that if there is a "bubble" at the clock input, then the output changes on a *negative*, or high-to-low, transition of the clock. If no bubble is present, then the output changes on the *positive*, or low-to-high, clock transition. In these truth tables, and all that follows, we will use an uppercase letter to refer to the next state and a lowercase letter to refer to the present state. Thus, $Q = Q(t + d)$ and $q = Q(t)$.

The characteristic behavior of each of these flip-flops is defined by the truth tables of Figure 5.2.8. It is sometimes convenient, however, to represent this behavior by the *characteristic equation* of the flip-flop. For example, Equation (5.2.1) is the characteristic equations for the *SR* flip-flop. The characteristic equations for the various flip-flops can be derived from the truth tables that define them. For example, for the *JK* flip-flop, assuming that inputs *J* and *K* and output *Q* are asserted high, as indicated in Figure 5.2.8(b), we see that

$$Q = q\overline{J}\,\overline{K} + J\overline{K} + \overline{q}JK = q\overline{K} + J\overline{K} + \overline{q}J$$
$$= q\overline{K} + \overline{q}J \qquad (5.2.2)$$

Each of these flip-flops have characteristics useful in different applications. The *D*, or *delay*, flip-flop, as mentioned earlier, is used extensively for temporarily storing information in a computer. A collection of *D* flip-flops might make up a "register" in the central processing unit of a computer. The *T*, or *toggle*, flip-flop is most often used for the design of counters, as we shall see in the next section. A quick examination of the truth tables for the various flip-flops shows that the *JK* is a combination of both the *T* and the *SR* flip-flops. Specifically, if the condition that the *J* and *K* lines are never asserted at the same

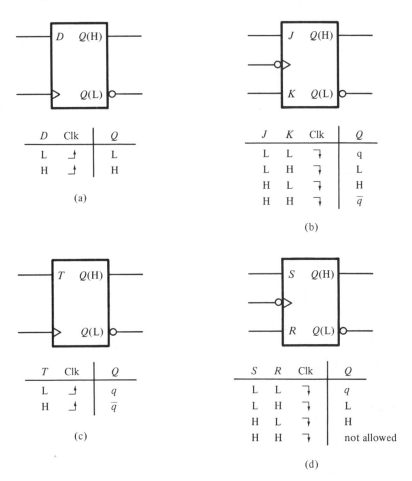

Figure 5.2.8 The four basic edge-triggered flip-flops: (a) *D* flip-flop; (b) *JK* flip-flop; (c) *T* flip-flop; (d) *SR* flip-flop.

time is maintained, then the *JK* is exactly equivalent to the *SR*, with *J* equal to the *S* input and *K* equal to the *R* input. If the *J* and *K* lines are tied together, the resulting single line is equivalent to the toggle flip-flop's *T* input. This, of course, helps to explain why only the *D* and the *JK* flip-flops are generally available as integrated circuits.

The defining truth tables of Figure 5.2.8 clearly and unambiguously show the behavior of the four basic flip-flop types. However, in the design process, a different version of these tables will be useful. In the design, or synthesis, process we must design logic that generates signals at the inputs to the flip-flops that will cause them to produce specific outputs on the next clock pulse. For example, suppose that the current output of a *JK* flip-flop is 1 and the next value is required to also be 1. What should the *J* and *K* inputs to the flip-flop be? From Figure 5.2.8(b), we see that two conditions on the inputs will cause the output to remain 1. The first is when $J = 1$ and $K = 0$, a condition causing the output to be set

q	Q	D
0	0	0
0	1	1
1	0	0
1	1	1

(a)

q	Q	J	K
0	0	0	-
0	1	1	-
1	0	-	1
1	1	-	0

(b)

q	Q	T
0	0	0
0	1	1
1	0	1
1	1	0

(c)

q	Q	S	R
0	0	0	-
0	1	1	0
1	0	0	1
1	1	-	0

(d)

Figure 5.2.9   Current state–next state truth tables for the various flip-flops: (a) *D* flip-flop; (b) *JK* flip-flop; (c) *T* flip-flop; (d) *SR* flip-flop.

to 1, and the second is when $J = 0$ and $K = 0$, a condition resulting in no change in the output. Thus, the value of J becomes a don't care and the value of K must be 0 for the output to remain a 1. The remaining combinations are determined in a similar manner. This "present-state–next-state" behavior is shown in the tables of Figure 5.2.9 for each of the basic flip-flops. These tables will be used extensively in the design of sequential circuits and should therefore be committed to memory.

5.3 COUNTERS

Consider the three–flip-flop circuit shown in Figure 5.3.1(a). The analysis of this circuit is most easily carried out by writing the flip-flop input equations, i.e., the equations for T_1, T_2, and T_3, and then, on the basis of the truth table defining the T flip-flop shown in Figure 5.2.8(c), constructing a table showing how the outputs, q_1, q_2, and q_3, change on each occurrence of the clock pulse. For example, if $T_2 = 1$, then the output of this flip-lop, q_2, will toggle, or change value, when the clock signal goes from a low to a high. If, on the other hand, $T_2 = 0$, then the output of the flip-flop will not change when the clock changes. The flip-flop input equations are easily seen to be

$$T_1 = 1$$
$$T_2 = q_1 \qquad\qquad (5.3.1)$$
$$T_3 = q_1 q_2$$

Using these input equations, we can determine the successive output values of each flip-flop. The three outputs at any given time, taken collectively, are referred to as the *state* of the machine. The table that shows how the outputs change and the circuit moves from state to state with each clock pulse is, therefore, referred to as a *state transition table*. This table is shown in Figure 5.3.1(b). As we indicated in Section 5.2.3, the current state of the machine is indicated with lowercase letters and the next state with uppercase letters.

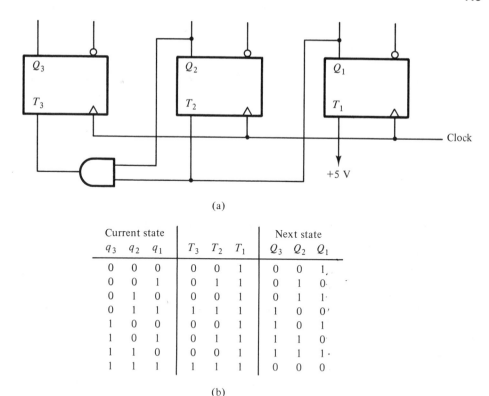

(a)

Current state						Next state		
q_3	q_2	q_1	T_3	T_2	T_1	Q_3	Q_2	Q_1
0	0	0	0	0	1	0	0	1
0	0	1	0	1	1	0	1	0
0	1	0	0	0	1	0	1	1
0	1	1	1	1	1	1	0	0
1	0	0	0	0	1	1	0	1
1	0	1	0	1	1	1	1	0
1	1	0	0	0	1	1	1	1
1	1	1	1	1	1	0	0	0

(b)

Figure 5.3.1   (a) A 3-bit binary counter made up of T flip-flops and (b) its state transition table.

Thus q_1 is taken as the present output of flip-flop 1, and Q_1 is taken as the next value of this output. (This practice will be followed in the remainder of this text.) From this figure, it is easily seen that the three outputs, q_1, q_2, and q_3, change in such a way as to count the clock pulses, at least to 7, and thus the circuit is a counter.

The sequence of states that a counter goes through can also be shown in a *state diagram*. The state diagram for the counter shown in Figure 5.3.1 is given in Figure 5.3.2. As can be seen from this figure, the state diagram consists of *nodes*, whose labels represent the state of the network at particular times. These nodes are connected to each other by *directed edges*, which show what state the system will go to on the next clock pulse. Such a diagram is very useful for visualizing the behavior of more general sequential circuits, as we shall see shortly.

Counters, of course, need not count in the sequence just given. We may want a counter to count in a Gray code sequence or some other sequence not representing a binary counting sequence. Suppose that we are required to design a counter that counts in the Gray code sequence, 000, 001 , 011, 010, 110, 111, 101, 100, 000, ... The question is, How do we proceed with the design? The analysis procedure that was used above began by deriving the flip-flop input equations. These were then used, in conjunction with the

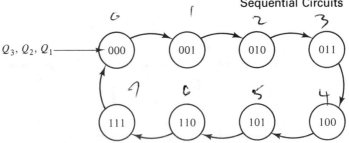

Q_3, Q_2, Q_1

Figure 5.3.2   The state diagram for the counter of Figure 5.3.1.

truth table that defined the flip-flops, to derive a state transition table. What we need to do for the design of a counter is exactly the reverse: derive the state transition table, and then, using the definition of the flip-flop to be used in the counter being designed, derive the flip-flop input equations. To carry out the process, let us assume that we are to do the design using *JK* flip-flops. Figure 5.3.3 shows the state transition table for the specified count sequence and the values that the *J* and *K* lines for each flip-flop must take on to generate the required next-state values. This table is usually referred to as an *excitation table*, since it gives the flip-flop input values necessary to cause the flip-flops to change state in a particular way. The values for *J* and *K* are easily obtained from the current state–next state tables given in Figure 5.2.9. For example, if the current state of flip-flop 1 is 0 and the next state is to be a 1, then, from Figure 5.2.9(b), we see that $K = 1$ and that *J* becomes a don't care. The resulting excitation table is shown in Figure 5.3.3.

The equations for the flip-flop inputs can easily be derived from the excitation table by plotting the *J* and *K* values in Karnaugh maps. These are shown in Figure 5.3.4, from which the flip-flop input equations are seen to be

$$J_3 = \overline{q}_1 q_2$$
$$K_3 = \overline{q}_1 \overline{q}_2$$

$$J_2 = q_1 \overline{q}_3$$
$$K_2 = q_1 q_3$$

$$J_1 = \overline{q}_2 \overline{q}_3 + q_2 q_3$$
$$K_1 = \overline{q}_2 q_3 + q_2 \overline{q}_3 = \overline{J}_1$$

(5.3.2)

Present state			Next state			Flip-flop inputs					
q_3	q_2	q_1	Q_3	Q_2	Q_1	J_3	K_3	J_2	K_2	J_1	K_1
0	0	0	0	0	1	0	–	0	–	1	–
0	0	1	0	1	1	0	–	1	–	–	0
0	1	1	0	1	0	0	–	–	0	–	1
0	1	0	1	1	0	1	–	–	0	0	–
1	1	0	1	1	1	–	0	–	0	1	–
1	1	1	1	0	1	–	0	–	1	–	0
1	0	1	1	0	0	–	0	0	–	–	1
1	0	0	0	0	0	–	1	0	–	0	–

Figure 5.3.3   The excitation table for a Gray code counter.

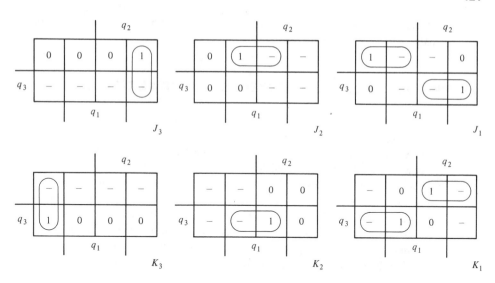

Figure 5.3.4 Derivation of the Gray code counter input equations.

Using equations (5.3.2), the physical implementation of the Gray code counter becomes as shown in Figure 5.3.5. Note the use of both the asserted low and asserted high flip-flop outputs to generate the various functions without the use of level shifters. The reader should verify that this counter does implement the specified Gray code counter by deriving the state transition table as was done in the previous analysis example. This table

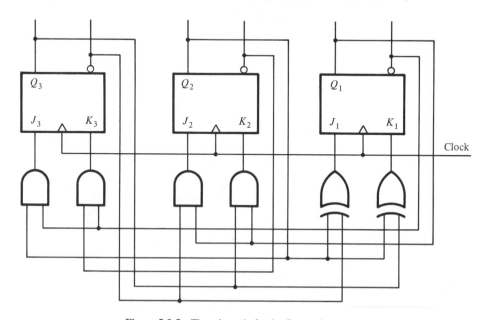

Figure 5.3.5 The schematic for the Gray code counter.

should look exactly like the excitation table of Figure 5.3.3 except that this time the don't cares will have values assigned to them.

5.4 SYNCHRONOUS, OR CLOCKED, SEQUENTIAL CIRCUITS

In Figure 5.1.4 a model for a sequential circuit was shown. This circuit was termed an asynchronous sequential circuit, since there was nothing to control the time at which the outputs change except for the propagation delays and the times at which the inputs change. Because it is generally not possible to control, with any degree of accuracy, the propagation delays inherent in the circuit, circuits of this type are of little use in the implementation of large-scale systems. However, if the feedback delays are replaced by clocked flip-flops,[7] a type of sequential circuit whose behavior is easily controlled is created. Such a circuit will be called a *clocked*, or *synchronous*, *sequential circuit*. In this section we will investigate clocked sequential circuits by first showing models that can be used for their analysis. We will then look at the problems associated with the analysis and design of such machines.

5.4.1 Models for Clocked Sequential Circuits

As indicated above, if we replace the delay elements in Figure 5.1.4 with some type of clocked flip-flop, a clocked, or synchronous, sequential circuit results. Such a circuit is also, sometimes, referred to as a *state machine*. The model that results when we replace the delays with D-type flip-flops is shown in Figure 5.4.1(a). This model is referred to as the *Mealy model*. It is easily seen from this figure that the next state, Q, is a function of the current state, q, and the current inputs, X.[8] This is also true of the outputs, Z. In other words, we may express Q and Z as follows:

$$Q = f(q, X)$$
$$Z = h(q, X)$$
$$(5.4.1)$$

An alternative model arises if we assume that the sequential circuit's outputs are functions only of the state of the machine. The counters of Section 5.3 are examples of this form of sequential circuit. Figure 5.4.1(b) shows the general model resulting from this assumption, when we further assume that D flip-flops are used in the feedback paths. This is referred to as the *Moore model*, in which

$$Q = f(q, X)$$
$$Z = h(q)$$
$$(5.4.2)$$

The models given in figure 5.4.1 use edge-triggered D flip-flops in the feedback paths. Since D flip-flops simply transfer the input of the flip-flop to the output when the

---

[7] The flip-flops cannot actually replace the physical delays in the circuit. What they do is to prevent changes on the inputs from causing changes in the feedback lines except at the point in time at which the flip-flop outputs change, and this is controlled by the clock.

[8] Remember that the symbols Q, q, X, and Z refer not to a single signal but to a collection of state variables (flip-flop outputs), inputs, and outputs.

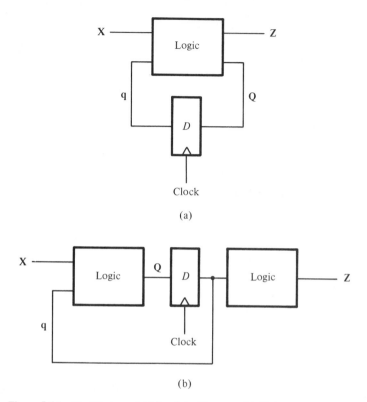

Figure 5.4.1   The Mealy model (a) and the Moore model (b) for sequential circuits.

asserted clock transition occurs, the next state, **Q,** of the system is, in fact, the value of the D flip-flop inputs, **D;** that is, **D = Q.** If, however, we were to use other flip-flop types, such as *JK*, the next state, or set of next flip-flop outputs, becomes a function of these inputs as defined by the tables in Figure 5.2.8. Thus, the outputs labeled **Q** in Figure 5.4.1 become the flip-flop inputs required to generate the next state. We will see examples of this in what follows.

5.4.2 The Analysis of Clocked Sequential Circuits

The analysis of the behavior of clocked sequential circuits requires that we determine the output equations (the **Z**'s) and the next-state equations (the **Q**'s), from which we can derive the state transition table, or *state table*, and a *state diagram* similar to the one we derived in Section 5.3. Consider, for example, the clocked sequential circuit shown in Figure 5.4.2. The next state equations and output equation are easily derived from this circuit. They are, respectively,

$$
\begin{aligned}
Q_1 &= \overline{q}_1 q_2 + q_1 \overline{q}_2 \qquad (= D_1) \\
Q_2 &= q_2 \overline{x} + \overline{q}_2 x \qquad (= D_2) \\
Z &= q_1 + \overline{q}_2 \overline{x}
\end{aligned}
\qquad (5.4.3)
$$

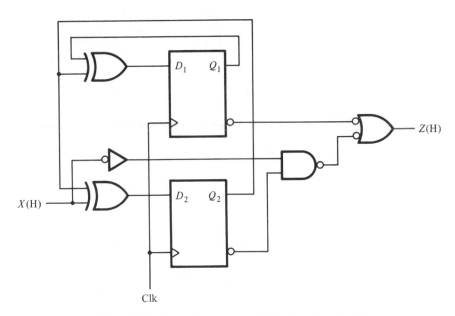

Figure 5.4.2   A synchronous sequential circuit to be analyzed.

We see that since Z is a function of both the state variables q_1 and q_2 and the input X, this circuit fits the Mealy model.

Although equations (5.4.3) completely defines the behavior of the sequential circuit shown in Figure 5.4.2, they are a little awkward to use in trying to determine the sequence of outputs that will be produced by a given sequence of inputs. By plotting these equations

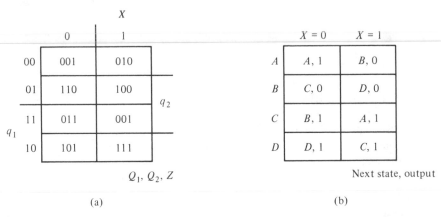

Figure 5.4.3   State tables for the circuit of Figure 5.4.2: (a) state transition table; (b) symbolic state table.

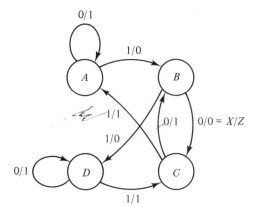

Figure 5.4.4 The state diagram corresponding to the state table of Figure 5.4.3.

in the form of a state table, the behavior of the system becomes a bit clearer. Figure 5.4.3(a) shows this plot.

Assuming that the goal of analysis is to be able to predict how the outputs of the sequential circuit will change as the inputs change, we need not know the specific values that the state variables take on as the inputs change. We only need to distinguish between states to predict the behavior of the machine. Thus, we may replace the specific values that the state variables can take on by simple labels. For example, we might make the following replacements for the states (q_1, q_2): $(00) = A$, $(01) = B$, $(11) = C$, and $(10) = D$. The resulting symbolic state table is shown in Figure 5.4.3(b).

An alternative representation to the state table is a *state diagram*. We use one of two forms for the state diagram, depending on the sequential circuit model being used. In both cases the state diagram is a *directed linear graph* in which the nodes represent the states of the machine and the edges represent the inputs required to move from one state to the next. In the Mealy model, the outputs are a function of both the input and the state, and therefore the outputs must be associated with the edges in the state diagram. In the Moore model, the outputs are functions only of the state, and so the outputs, for this model, are associated with the nodes. Figure 5.4.4 shows the state diagram for the sequential circuit of Figure 5.4.2. In this diagram the edges are labeled in the form X/Z. This state diagram defines the behavior of the system. For example, suppose that we start in state A and apply the input sequence 101101; the output sequence that results will be, from either the state table or the state diagram, 001001.

As a second example, let us analyze the sequential circuit shown in Figure 5.4.5. Our analysis objective is, once again, to produce a state table and a state diagram that identifies the behavior of this circuit. The circuit shown in this figure differs from the one shown in Figure 5.4.2 in that this circuit uses JK flip-flops instead of D flip-flops. Thus, the flip-flop input equations are not equivalent to the next state equations, as they would be for D flip-flops. However, we can easily obtain the next-state values by first writing the flip-flop input equations, or *excitation equations*, and then use the table defining the JK

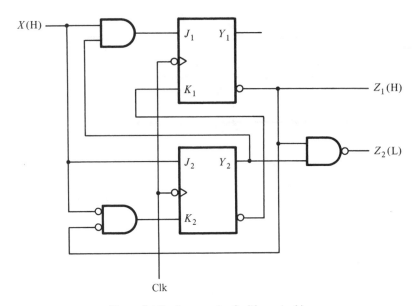

Figure 5.4.5 An example of a Moore machine.

flip-flop, given in Figure 5.2.8(b), to derive the next state values. The input and output equations for this circuit become

$$J_1 = Xy_2$$
$$K_1 = \bar{y}_2$$
$$J_2 = X$$
$$K_2 = \bar{X}y_1$$
$$Z_1 = \bar{y}_1$$
$$Z_2 = \bar{y}_1 y_2$$

(5.4.4)

A second difference between this example and that of Figure 5.4.2 now becomes apparent. That is, the outputs are functions of the state variables only, and so the Moore model is to be used for this analysis.

The inputs of equations (5.4.4) are plotted in the excitation tables of Figure 5.4.6(a); the resulting transition and state tables are also given in Figure 5.4.6. To see how the state table is derived from the flip-flop excitation tables, refer to Figure 5.2.8(b). Consider, for example, the first row of the excitation tables in Figure 5.4.6(a). For $X = 0$, J_1 and K_1 are 0 and 1, which causes the flip-flop to reset; and therefore $Y_1 = 0$, as shown in Figure 5.4.6(c). The same situation occurs when $X = 1$. On the other hand, when $X = 0$, J_2 and K_2 are 0 and 0, respectively. Thus, the flip-flop output should not change. Since the current state variable, y_2, is 0, the next-state value, Y_2, must also be 0. When $X = 1$, however, J_2 and K_2 are 1 and 0 and therefore $Y_2 = 1$. The remainder of the assigned table, Figure 5.4.6(c), is completed in a similar manner. Figure 5.4.6(d) shows the symbolic state table, where the states are assigned as follows: $A = (00)$, $B = (01)$, $C = (11)$, and $D = (10)$.

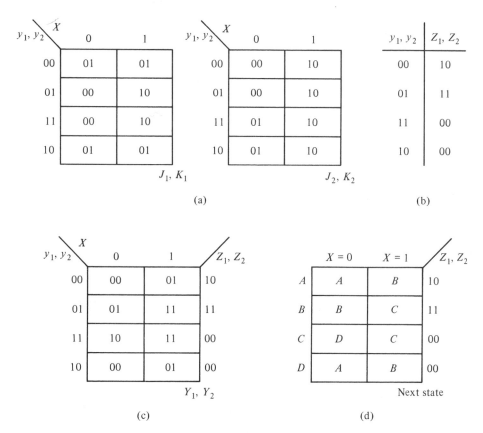

Figure 5.4.6   The flip-flop excitation and state tables for the sequential circuit of Figure 5.4.5: (a) the excitation tables; (b) the output table; (c) the state transition table; (d) the state table.

The output of a Moore machine is associated only with the value of the state variables and thus the outputs for this circuit are plotted along the right side of the state tables to correspond to the appropriate state variable values. Figure 5.4.7 shows the state diagram that corresponds to this state table. Note that the outputs are shown not on the edges of the graph, but in the nodes associated with the states that cause these outputs.

5.4.3 The Design of Clocked Sequential Circuits

As was the case with the design of counters in Section 5.3, the design of more general sequential circuits is just the reverse process of the analysis. There are basically five steps in this process:

Step 1. From the problem statement, obtain a state diagram and a state table.

Step 2. Assign a coding to the states to form a transition table.

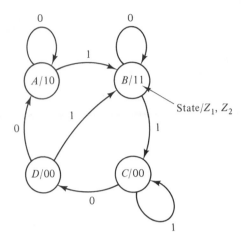

Figure 5.4.7   State diagram of the Moore machine of Figure 5.4.5.

Step 3. Specify a flip-flop to use and derive the flip-flop excitation tables from the transition table.

Step 4. Derive the flip-flop input equations and the circuit output equations from these tables.

Step 5. Draw a circuit diagram.

We can best illustrate this design process by an example.

Problem Statement 1

Design a sequential circuit having one input, X, and one output, Z. Z is to be 1 whenever the four most recent inputs are 1011, where the most recent input is the rightmost in the string. Overlapping of sequences is allowed so that the input sequence 1011011 will produce an output of 0001001. The input X is to be asserted low, and the output Z is to be asserted high.

We begin the design process by constructing a state diagram for a sequential circuit to meet these requirements. In doing this, we should try to associate a specific meaning with each state in the diagram. For example, in the state diagram of Figure 5.4.8, let state A correspond to the situation where we have seen no part of the input sequence. Then let state B be the state corresponding to seeing the first 1 in the sequence. C can correspond to 10, and D to 101. By the statement of the problem, a 1 on the input while the circuit is in state D should produce a 1 out, but what is the next state? Well, since this 1 can also be taken as the first 1 in the sequence, we should go to state B. This is the sequence of states that will result if the input sequence is the one desired. What happens, however, if this sequence is broken? For example, suppose the current input is 101, which would put us in state D, and then a 0 comes in. In this case, the desired sequence is broken; however, the last 1 of 101 becomes the first 1 of 10, which corresponds to state C. Therefore, an input of 0 in state D will cause us to go to state C. Continuing in this way, we arrive at the state diagram of Figure 5.4.8.

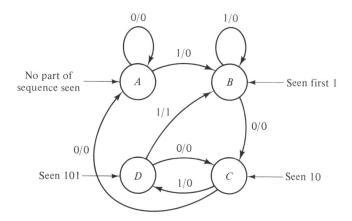

Figure 5.4.8   The state diagram for a sequence detector that detects the sequence 1011.

We have drawn this state diagram for a Mealy machine. There is actually nothing in the problem statement that would indicate a preference. It is generally true, however, that Moore machines usually require more states than the equivalent Mealy machine, as we shall shortly see. Thus, we will use the Mealy model whenever possible. More will be said about the model equivalences or lack thereof in Section 5.6.

The state table for this sequence detector can easily be constructed from the state diagram. This is done in Figure 5.4.9(a). In order to derive the flip-flop input equations, we first must have an assignment, or coding, for the states required in this problem. Since there are four states, we must have two state variables to distinguish each of the states. Call these Y_1 and Y_2. We may, at this stage, assign states to state variable values arbitrarily. Assume the assignment is made on (Y_1, Y_2) as follows: $(00) = A$, $(01) = B$, $(11) = C$, and $(10) = D$. On the basis of this assignment, the state transition table becomes as shown in Figure 5.4.9(b).

	$X = 0$	$X = 1$
A	$A, 0$	$B, 0$
B	$C, 0$	$B, 0$
C	$A, 0$	$D, 0$
D	$C, 0$	$B, 1$

Next state, output

(a)

y_1, y_2 \ X	0	1
00	000	010
01	110	010
11	000	100
10	110	011

Y_1, Y_2, Z

(b)

Figure 5.4.9   Flip-flop state tables (a) and state transition table (b) for the sequence detector of Figure 5.4.8.

Before we can derive the flip-flop input equations, we must specify the type of flip-flop to be used in the design. For this example, let us use *JK* flip-flops. The flip-flop excitation tables can now be derived from the state table using Figure 5.2.9(b) in exactly the same way as was done for the counter design example in Section 5.3. The resulting excitation tables are shown in Figure 5.4.10.

Now, from the state table of Figure 5.4.9 and the excitation tables of Figure 5.4.10, we may derive the flip-flop input equations and the network output equations. These equations become:

$$J_1 = \overline{X}y_2$$
$$K_1 = \overline{X}y_2 + X\overline{y}_2 = X \oplus y_2$$
$$J_2 = y_1 + X \qquad\qquad (5.4.5)$$
$$K_2 = y_1$$
$$Z = y_1\overline{y}_2X$$

Equations (5.4.5) yield the final circuit implementation shown in Figure 5.4.11. The reader should verify these results by deriving the state diagram of Figure 5.4.8 from the circuit diagram of Figure 5.4.11, using the analysis procedure of Section 5.4.2.

It should be appreciated that the complexity of the implementing equations depends

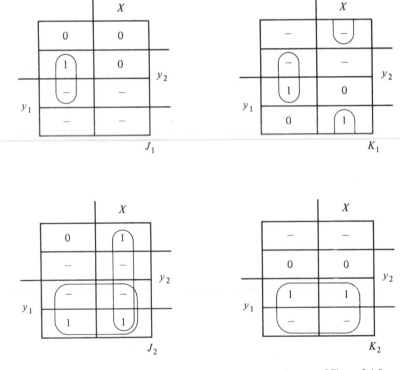

Figure 5.4.10 Flip-flop excitation tables for the sequence detector of Figure 5.4.8.

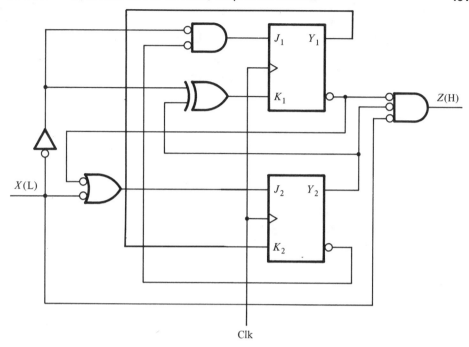

Figure 5.4.11   Implementation of the 1011 sequence detector.

on how this assignment is made. Ideally, we would like to assign states in such a way as to make the equations as simple as possible. This general *state assignment* problem is a very difficult problem and will not be discussed here. A number of references to this problem are given at the end of the chapter. Although the general problem is very difficult, there are a couple of rules of thumb that can be used which usually lead to fairly simple implementing equations. These rules, in order of importance, may be stated as follows:

Rule 1.   States that have the same next state for a given input should be adjacent. Priority should be given to states having common next states for the largest number of inputs.

Rule 2.   States that are the next state of a given state should be adjacent.

The rationale for rule 1 is that when two rows are adjacent and the state variable in the same column of these two rows is the same, this state variable will have a term which is independent of the state variable that differs between the two rows. Thus, the final equations will have simpler terms. The reason for the second rule is that when these states are adjacent they will differ in only one state variable. Thus, the state variables that are common may share common terms.

We can illustrate the application of these rules with the last example. Referring now to the state table of Figure 5.4.9(a), we see that by rule 1, states B and D should be adjacent, since they map into the same next states for each input: C for $X = 0$ and B for $X = 1$.

Further, by rule 2, we would like to make B and C adjacent. This leaves state A to be adjacent to either C or D. Applying rule 2, we see that A and D should be adjacent, because these are the next state of state C. The resulting state table and transition table are shown in Figure 5.4.12(a) and (b). The excitation tables for the JK flip-flops are shown in Figure 5.4.12(c). The resulting flip-flop input equations become

$$J_1 = \bar{y}_2$$
$$K_1 = y_2 + X$$
$$J_2 = X \qquad\qquad (5.4.6)$$
$$K_2 = \bar{X}$$
$$Z = y_1 y_2 X$$

These equations are clearly simpler than those of equation (5.4.5), and thus the rules of thumb have, indeed, done their job.

Let us consider another design example.

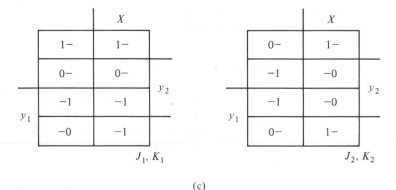

y_1, y_2		$X = 0$	$X = 1$
00	C	$A, 0$	$D, 0$
01	B	$C, 0$	$B, 0$
11	D	$C, 0$	$B, 1$
10	A	$A, 0$	$B, 0$

(a)

y_1, y_2	x 0	1
00	100	110
01	000	010
11	000	011
10	100	010

Y_1, Y_2, X

(b)

	X	
	$1-$	$1-$
	$0-$	$0-$
	-1	-1
	-0	-1

J_1, K_1

	X	
	$0-$	$1-$
	-1	-0
	-1	-0
	$0-$	$1-$

J_2, K_2

(c)

Figure 5.4.12   The tables resulting after applying the rules of thumb to the state table of Figure 5.4.9(a): (a) permuted state table; (b) state transition table; (c) excitation tables.

Problem Statement 2

Design a two-input (A and B), one-output (Z) sequential circuit (using a Mealy model) having the following characteristics: $Z = A(t)A(t - 1)$ until the B input becomes 1, at which time $Z = A(t) + A(t - 1)$. The next occurrence of a 1 on input B causes Z to switch back to the AND operation. Z continues to switch between the OR operation and the AND operation on each occurrence of 1 at input B. Assume that both inputs and the output are asserted high.

As before, we began the design process by constructing the state diagram, which is always the most difficult part. Let us begin by assuming that Z implements the AND and derive a portion of the state diagram corresponding to this part of the problem. To get ourselves started, let us hold $B = 0$ for a while and develop the state diagram based on changes in A. Again, let a meaning be associated with each state in the system. Referring now to Figure 5.4.13, let state S_0 correspond to the case in which the last value of A was 0. If a 0 occurs next on input A, then we will stay in this state and produce an output of 0. Next, if A becomes a 1, we need to go to a state corresponding to the case in which the last value of A was 1. Call this state S_1. An input of 1 on A while the circuit is in the S_1 state will, of course, take us back to S_1 and will produce an output of 1, since the last two inputs on A were 1. An input of 0 on A, on the other hand, will take us back to S_0. A similar argument can be made if we assume that Z is to implement the OR operation. In this case, S_2 will correspond to the case in which the last A was 0, and S_3 will correspond to that in which the last A was 1.

The next step in deriving the state diagram is to connect the two pieces just derived. Suppose we are in state S_0 and B becomes a 1. Two possibilities arise. First, A can be a 0, in which case the two inputs, A and B, are 01. The 1 on B must cause us to switch to the OR operation, and the 0 on A must cause us to go to the state in the OR diagram corresponding to the case in which the last A was 0. The resulting state is S_2. But what is the output? Should we make the output the AND operation, which is what it was, or

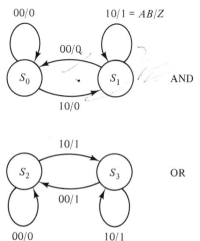

Figure 5.4.13 Partial state diagram for the function generator.

should we make the output the OR operation, which is what it is supposed to switch to? Nothing in the statement of the problem tells us what to do, so we may arbitrarily decide. Let us assume, then, that Z takes on the last operation required before the switch. Thus, the output in this case will be 0. Figure 5.4.14 shows this transition. Next, assume that A is 1 while B is 1 and, again, we are in state S_0. In this case, we must go to state S_3, which corresponds to the state in which the last A was a 1, and the output is to switch to the OR operation. On the assumption that the output is the function of the last two A inputs before the switch, the output here becomes a 0. Repeating these arguments starting in each of the other states results in the state diagram of Figure 5.4.14, which is the desired final diagram. The state table shown in Figure 5.4.15(a) is derived from this state diagram.

The next step in the design process is to assign states and reconstruct the state diagram on the basis of this assignment. Since there are four states, we need two state variables to code the states. Let the state variables be Y_1 and Y_2. Applying the rules of thumb given above, we see that by rule 1, states S_0 and S_1 should be adjacent, as should states S_2 and S_3. Application of rule 2 does not give us any further information, so let us assume that S_1 and S_2 are adjacent. Thus, we may assume a coding for the states (Y_1, Y_2) of $S_0 = (0, 0)$, $S_1 = (0, 1)$, $S_2 = (1, 1)$, and $S_3 = (1, 0)$. The resulting encoded state table is shown in Figure 5.4.15(b).

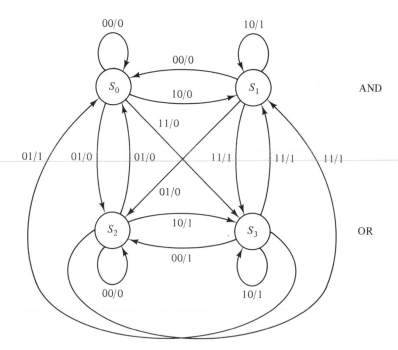

Figure 5.4.14 Final state diagram for the function generator.

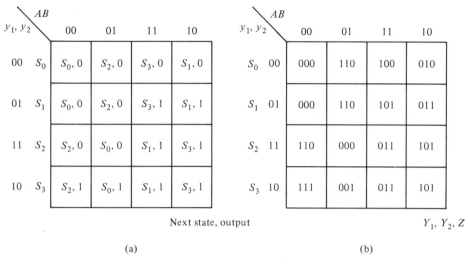

Next state, output Y_1, Y_2, Z

(a)   (b)

Figure 5.4.15   (a) State table for the function generator. (b) State transition table.

Assuming the use of T flip-flops, the flip-flop excitation tables become as shown in Figure 5.4.16, from which the flip-flop input equations become

$$T_1 = B$$
$$T_2 = \overline{A}\,\overline{B}\,\overline{y}_1 y_2 + \overline{A}\,\overline{B} y_1 \overline{y}_2 + \overline{A} B \overline{y}_1 \overline{y}_2$$
$$+ \overline{A} B y_1 y_2 + A \overline{B} \overline{y}_1 y_2 + A \overline{B} y_1 \overline{y}_2 \qquad (5.4.7)$$
$$+ A B \overline{y}_1 \overline{y}_2 + A B y_1 y_2$$
$$= A \oplus B \oplus C \oplus D$$

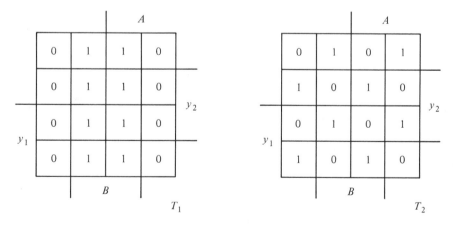

Figure 5.4.16   Flip-flop excitation tables.

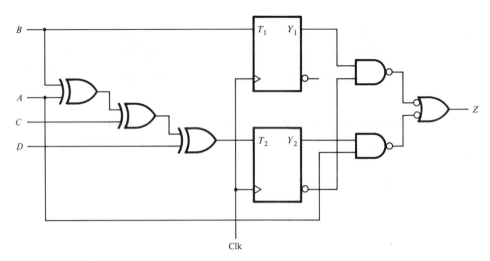

Figure 5.4.17 Final realization for the function generator circuit.

The output can be derived from the state tables of Figure 5.4.15 and is seen to be

$$Z = y_1\bar{y}_2 + Ay_2 \tag{5.4.8}$$

Figure 5.4.17 shows the final realization based on these equations.

It was mentioned earlier that the complexity of the realizing equations for a sequential circuit can depend heavily on the way in which the states are assigned. As an example, consider the state assignment for the problem just completed. Suppose that states S_0 and S_1 were as assigned above but that $S_2 = (10)$ and $S_3 = (11)$. The resulting assigned state table would appear as shown in Figure 5.4.18, from which it may be verified that

$$T_1 = B$$
$$T_2 = \bar{A}y_2 + A\bar{y}_2 = A \oplus y_2 \tag{5.4.9}$$
$$Z = y_1y_2 + Ay_1 + Ay_2$$

	A, B			
y_1, y_2	00	01	11	10
S_0 00	000	100	110	010
S_1 01	000	100	111	011
S_3 11	101	001	011	111
S_2 10	100	000	011	111

Y_1, Y_2, Z

Figure 5.4.18 An alternative state assignment and the resulting encoded state table.

Clearly, the equation for T_2 is much simpler than the one derived in the equations of group (5.4.7), although the equation for Z is slightly more complex. The resulting implementation will be somewhat simpler because of this.

5.5 THE SIMPLIFICATION OF SEQUENTIAL CIRCUITS

It can very easily happen that the designer of a sequential circuit will create a state diagram or table having more states than are actually required to implement the design. Since the number of flip-flops in the feedback path, as well as the complexity of the implementing equations, depends on the number of states, it is important that they be reduced to a minimum. In order to do this, we must introduce the concept of equivalent states. This is done in the following definition.

Definition 5.5.1.   Let p and q be two states of a sequential machine M. We will say that *p is equal to q*, or *p is indistinguishable from q*, if the output sequence produced by applying an arbitrary input sequence to the machine is the same regardless of whether we start the machine in state p or state q.

The relationship between states p and q specified in this definition is an example of an *equivalence relation*, which is defined as follows:

Definition 5.5.2.   Let $*$ be a relationship between elements of some set S. Then $*$ is an equivalence relation if

  (i) For all x in S, $x * x$, that is, x is related to itself (the reflexive property);
 (ii) If $x * y$, then $y * x$, that is, the order of the relation makes no difference (the symmetric property); and
(iii) If $x * y$ and $y * z$, then $x * z$ (the transitive property).

The equivalence relation we are most familiar with is the algebraic equality, represented by the symbol $=$. It is easy to verify that state equivalence satisfies all three of the properties in definition 5.5.2 and is, therefore, an equivalence relation (see Problem 5.27).

The definition of state equivalence just given does not help much in determining whether two states are equal, since we would have to test them against every possible input string of which we might conceive. Fortunately, this really is not necessary. Note that two states will be *distinguishable*, or *not equal*, if we can find at least one input string that produces, on the last input, two different outputs depending on the state we started in. For example, consider machine M, whose state table is given in Figure 5.5.1. We can see immediately, from the state table, that states A and B are not equal, since if we start in A and apply a 1 on the input X, we get an output Z of 0; but if we do the same thing starting in B, we get a 1 out. Thus A and B cannot be equal. However, what can we say about A and C? Well, the outputs are the same: for $X = 0$ the outputs are both 0, as they are, also, for $X = 1$. This does not, however, mean that the states are necessarily equal. Note

	X = 0	X = 1
A	B, 0	A, 0
B	F, 0	E, 1
C	D, 0	B, 0
D	B, 0	A, 0
E	C, 0	B, 1
F	A, 0	E, 0
G	E, 0	G, 0

Next state, output   **Figure 5.5.1** State table for machine *M*.

that when *X* = 0, the pair of states *AC* goes to the pair of states *BD* (*A* goes to *B* and *C* goes to *D*) and when *X* = 1, *AC* maps into *AB*. We say that *AC implies BD* and *AB*, or that *BD* and *AB* are *implied* by *AC*. Now if *B* and *D* are equal and if *A* and *B* are equal, then *A* and *C* must also be equal (why?). However, we have already observed that state *A* is not equal to state *B* and, therefore, state *A* cannot be equal to state *C* (why?)

By continuing this process for every possible pair of states, we can determine which pairs are equivalent and which are not. This search process can be simplified tremendously by making a table giving all pairs and listing, for each, the set of implied pairs. This set of implied pairs is called an *implication set*. Figure 5.5.2(a) shows the resulting *implication table*. If the members of a pair have different outputs for some input, they are not an equal pair and so we indicate this in the table by simply crossing out the entry. For example, the entry for *AB* is crossed out, since *A* and *B* have different outputs for an input of 1. On the other hand, if a pair map into a single state for each possible input and the outputs are the

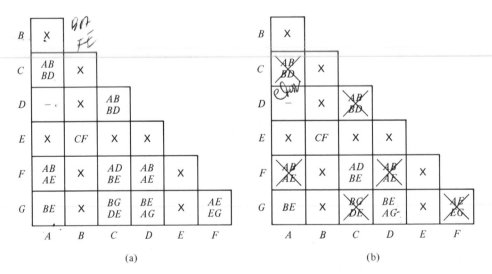

Figure 5.5.2 Implication tables for machine *M*: (a) initial; (b) final.

		$X = 0$	$X = 1$
(ADG)	A	$B, 0$	$A, 0$
(BE)	B	$C, 0$	$B, 1$
(CF)	C	$A, 0$	$B, 0$

Next state, output

Figure 5.5.3   A three-state machine equivalent to machine M.

same, as is true of AD, then the pair must be equal. This is shown in the table by a dash (—).

The identification of equivalent states proceeds as follows. Go to each table entry that is not crossed out and examine the implied pairs in the entry. If any of these pairs corresponds to a table entry that has already been crossed out, then the states corresponding to this position in the table cannot be equal and so the entry is crossed out. By making repeated passes through the implication table, eventually we reach a point where no further entries may be removed. The resulting implication table for machine M is shown in Figure 5.5.2(b).

The pairs corresponding to entries in the table that have not been crossed out are AD, AG, BE, CF, and DG. From this collection, we note that states A, D, and G are equal to each other: A equals G, and D equals G. So, too, state B is equal to state E and state C is equal to state F. We may now reduce the original table by replacing each occurrence of D and G by state A, each occurrence of state E by state B, and, finally, each occurrence of state F by state C. The resulting, reduced state table is shown in Figure 5.5.3.

Other techniques for carrying out this reduction process exist. Some of these can be found in the references cited at the end of the chapter. There are also methods, similar to that just described, for reducing machines which are not completely specified, i.e., machines which, for various reasons, may have unspecified, or don't care, next states or outputs. Such machines are commonly encountered in the design of very large digital systems such as the control unit for a computer. Usually, such machines arise out of very highly structured problems to begin with and, as a consequence, end up having, if not an absolute minimal number of states, at least a near minimal number. Because of this we will not discuss this process here. Procedures for reducing incompletely specified machines can be found in the references given at the end of the chapter.

5.6 MEALY-MOORE EQUIVALENCE
AND OTHER SEQUENTIAL CIRCUITS

We saw in Section 5.5 that there are two fundamental models for sequential circuits, the only difference between them being that the output in one is a function of the state variables only and in the other the output is a function of both the state variables and the current inputs. There must, of course, be an equivalence between these two models, since nothing fundamental can stop us from designing a system starting from either point of view. For example, a state diagram corresponding to the sequence detector of Section 5.4.3, assuming a Moore model instead of a Mealy model, is shown in Figure 5.6.1. Assuming that state A is the initial state in the Mealy version and state S_0 is the initial state in the Moore version, it can be verified that these two state diagrams will produce output sequences that are identical for identical input sequences.

To investigate the converison between models, consider, first, the conversion from a Moore machine to a Mealy machine. By assuming that the output associated with a state in the Moore machine can be associated with the *incoming* edges for an equivalent state in the Mealy machine, we will obtain the conversion shown in Figure 5.6.2(a). The reverse of this must give the Mealy to Moore equivalence. There is, however, a complication that arises in this case. If the outputs on all incoming edges of a state are the same, then this output becomes the output for the equivalent Moore state. If, however, the outputs are different, then the state must be "split" so that one Moore state will exist for each of the different incoming-edge outputs. Figure 5.6.2(b) shows this conversion. Application of this Mealy to Moore conversion process to the state diagram of Figure 5.4.8 yields the state diagram of Figure 5.6.1, where state B has been split into states S_1 and S_4. The remaining states correspond as follows: $A = S_0$, $C = S_2$, and $D = S_3$.

Starting with Figure 5.6.1, we see that a conversion to the Mealy equivalent yields a state diagram having five states rather than the four of the original machine. However, it can be verified by the simplification procedures described in Section 5.5 that states S_1 and S_4 in this transformed state diagram are equal, thus yielding the original four-state Mealy machine.

Although Mealy and Moore machines are equivalent in the sense just described, there are some important timing differences which should be noted. Since the output of a Moore machine is a function of the state only, the output must be stable, i.e., unchanging, as long as the state is fixed. Thus changes in the inputs between state changes cannot affect the output. In the Mealy machine, on the other hand, the outputs are functions of both the inputs and the state variables and so the output will change whenever either the input

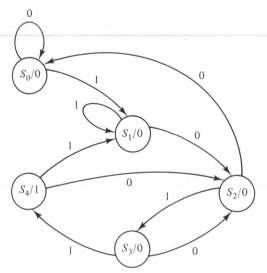

Figure 5.6.1 A Moore version of the 1011 sequence detector of Figure 5.4.8.

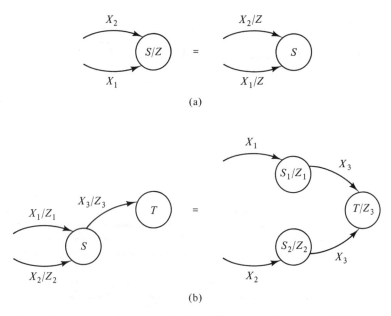

Figure 5.6.2 State equivalence in the Mealy-Moore conversion process: (a) Moore to Mealy transformation; (b) Mealy to Moore transformation.

changes or the state changes. These two cases are shown in Figure 5.6.3. From this figure, we see that *the output of a Moore machine is always valid*, except for the time required for all signals to settle down immediately after the state change. Alternatively, *the output of a Mealy machine is valid only at the instant that the state changes*. Even though the timing between these two models is quite different, they will produce the same results if the output is interpreted at the correct point in time.

We have mentioned asynchronous sequential circuits and have spent a good deal of time examining synchronous sequential circuits in this chapter. The principal difference between these two circuits is that the state changes and output changes are dependent on level changes in all of the inputs in the asynchronous case and only on a single clock input in the synchronous case. The term "level change" is used to mean that an input which has been at one voltage level for some period of time changes to another level and stays at this level for another period of time. A type of sequential circuit which is intermediate to these two is the *pulse-mode* circuit. This is basically an asynchronous sequential circuit in which one or more of the inputs are assumed to be "pulses." A pulse is defined, in a rather imprecise way here, as an assertion of an input which is long enough to allow the gates to see the change but short enough to be negated by the time any state changes *caused* by the input change are seen on the feedback paths. Obviously, pulses of this type are hard to control, and so pulse-mode circuits using "real" pulses are seldom encountered. A more practical variation on this theme is to use flip-flops in the feedback paths in a manner similar to clocked sequential circuits to control the times of state change. The incoming pulses thus appear as not one but many clocks. Circuits of this type have important application

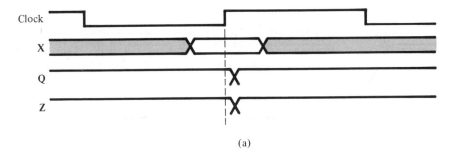

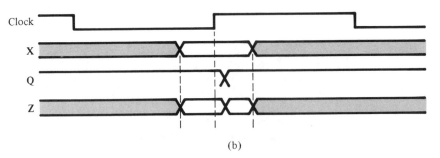

Figure 5.6.3 Timing differences between the Mealy and Moore models. (a) Moore machine timing. (b) Mealy machine timing.

in many problems. We will examine these two additional varieties of sequential circuits in Chapters 6 and 7.

5.7 AN ANNOTATED BIBLIOGRAPHY

Most texts dealing with digital design and switching theory discuss the analysis and synthesis of clocked sequential circuits. Very readable discussions can be found in the books by Mano, Hill and Peterson, Friedman, and Roth.

FRIEDMAN, A. D., *Fundamentals of Logic Design and Switching Theory*, Computer Science Press, Inc., Rockville, Md., 1986.

HILL, J. F., and G. R. PETERSON, *Introduction to Switching Theory and Logical Design*, 3rd ed., Wiley, New York, 1981.

MANO, M. M., *Digital Logic and Comptuer Design*, Prentice-Hall, Englewood Cliffs, N.J., 1979.

ROTH, C. H., *Fundamentals of Logic Design*, 2nd ed., West Publishing, St. Paul, 1979.

Hill and Peterson also give an excellent description of two commonly encountered methods for reducing the number of states: partitioning of the state set, and implication tables (the method described here). Friedman's book extends the concept of partitioning to reduce machines which are incompletely specified, i.e., those having state tables containing don't cares. Another good source for state reduction is the book by Kohavi. Finally, a paper by Meisel describes an algorithm that will generate a minimal machine equivalent to a given incompletely specified machine.

KOHAVI, Z., *Switching and Finite Automata Theory*, 2nd ed., McGraw-Hill, New York, 1978.

MEISEL, W. S., "A Note on Internal State Minimization in Incompletely Specified Sequential Networks," *IEEE Trans. Electronic Computers*, August 1967, pp. 508–509.

There are generally three approaches to the state assignment problem: rules of thumb, standard state tables and adjacencies, and partitions. Most of the texts on switching theory that discuss this problem give a set of rules of thumb. One of these is the book by Rhyne. McCluskey, Hill and Peterson, and Muroga are examples of texts that describe methods using standard state tables and state adjacencies. The use of partitions to predict the state variable dependencies is a method described at great length in Kohavi and Dietmeyer. The book by Harrison (the one described earlier as for precocious students only) also discusses this method.

DIETMEYER, D. L., *Logic Design of Digital Systems*, Allyn & Bacon, Boston, 1978.

HARRISON, M. A., *Introduction to Switching and Automata Theory*, McGraw-Hill, New York, 1965.

McCLUSKEY, E. J., *Introduction to the Theory of Switching Circuits*, McGraw-Hill, New York, 1965.

MUROGA, S., *Logic Design and Switching Theory*, Wiley-Interscience, New York, 1979.

RHYNE, V. T., *Fundamentals of Digital System Design*, Prentice-Hall, Englewood Cliffs, N.J., 1973.

Friedman discusses a procedure for converting Mealy machines to Moore machines and vice versa, in Chapter 5. A discussion of this machine conversion similar to that given here is also given in Chapter 11 of Hill and Peterson.

5.8 PROBLEMS

5.1. For each of the circuits shown in Figure P5.1, complete the timing diagram indicated. Assume that each gate has a propagation delay d which is much less than the time between changes of any of the signals.

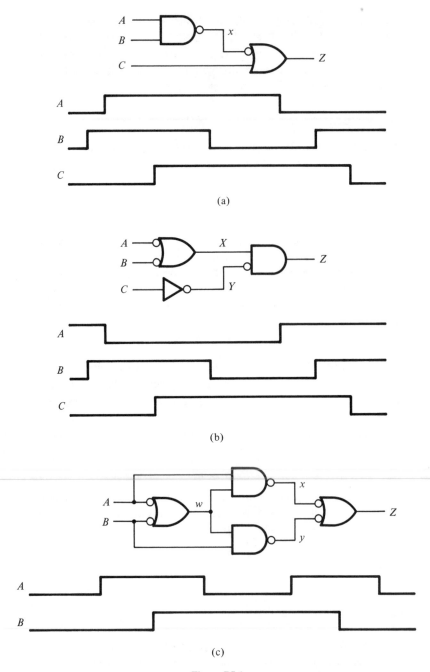

(a)

(b)

(c)

Figure P5.1

5.2. Complete the timing diagrams for the circuit shown in Figure P5.2.

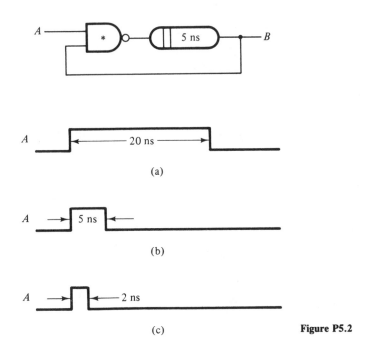

(a)

(b)

(c)   **Figure P5.2**

5.3. Redo the timing analysis of the NAND gate *SR* flip-flop shown in Figures 5.2.1 and 5.2.2, assuming that each gate has a delay of *d* seconds associated with it.

5.4. Repeat Problem 5.3 for the flip-flop shown in Figure P5.4.
   (a) First use the timing diagram shown in Figure 5.2.2.
   (b) Next, use the same timing but invert inputs *S* and *R*.

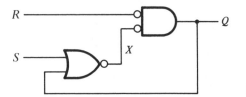

Figure P5.4

5.5. Complete the timing analysis for the circuit shown in Figure P5.5. Assume all gates have the same delay and that it is much less than the switching time of input *A*.

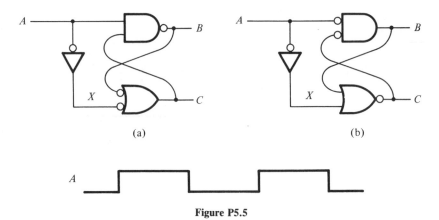

(a)   (b)

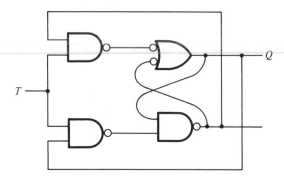

Figure P5.5

5.6. Assume that each gate in the circuit shown in Figure P5.6 has a propagation delay of d seconds. Assuming, further, that input T has been low for a very long time and then goes high for an interval x before going low again, construct a timing diagram for output Q assuming the following values for x:

(a) $x \gg d$

(b) $3d < x < 4d$

(c) $2d < x < 3d$

(d) $d < x < 2d$

(e) $x < d$

Figure P5.6

5.7. Derive the characteristic equation for the flip-flop of Figure P5.6.

5.8. Complete the indicated timing diagrams for the circuit shown in Figure P5.8. Assume that the propagation delay of the flip-flops is d.

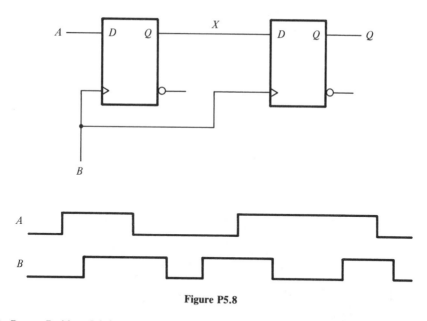

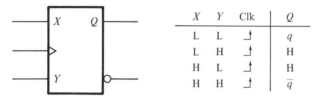

Figure P5.8

5.9. Repeat Problem 5.8 for the circuit shown in Figure P5.9.

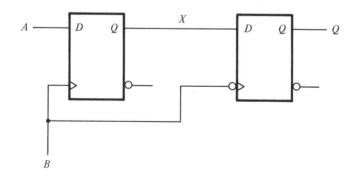

Figure P5.9

5.10. Derive the characteristic equations for each of the flip-flops shown in Figure 5.2.8.

5.11. Derive the characteristic equation for the XY flip-flop defined in Figure P5.11.

X	Y	Clk	Q
L	L	⌐↑	q
L	H	⌐↑	H
H	L	⌐↑	H
H	H	⌐↑	$\bar{q}$

Figure P5.11

5.12. Construct state diagrams for the counting sequences generated by the circuits shown in Figure P5.12.

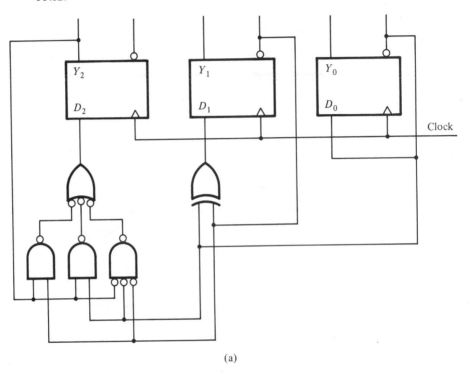

(a)

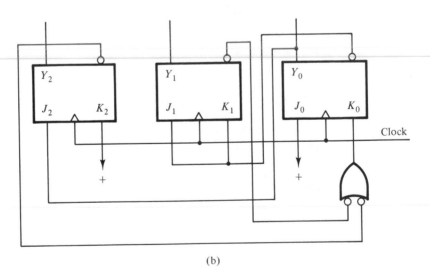

(b)

Figure P5.12

5.13. Using the flip-flops indicated, design a counter that counts in the sequence 0, 1, 5, 2, 6, 0, ... :

(a) T flip-flop

(b) D flip-flop

(c) JK flip-flop SR

5.14. Assuming that all gates and flip-flops have the same propagation delay d, derive a state diagram for the circuit shown in Figure P5.14.

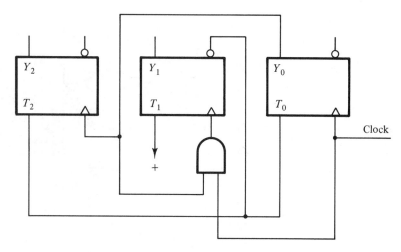

Figure P5.14

5.15. Derive the state diagrams for the clocked sequential circuits shown in Figure P5.15. Show the state diagram in the proper Mealy or Moore form.

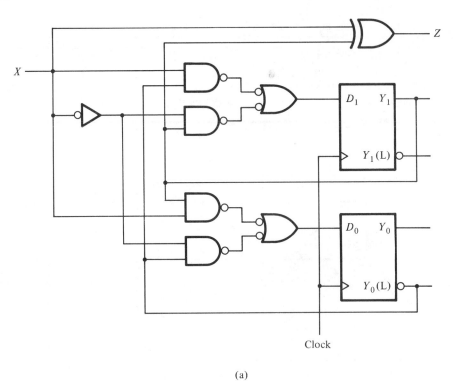

(a)

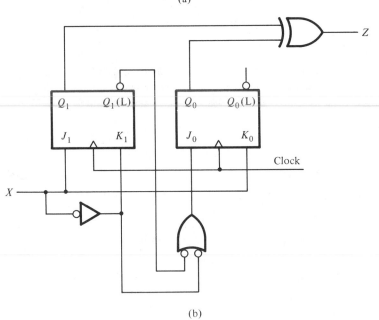

(b)

Figure P5.15

5.16. Show how you would convert from each of the following flip-flops into the alternative flip-flop indicated:

(a) D to T

(b) T to D

(c) D to JK

(d) JK to D

(e) T to JK

5.17. For the state diagram shown in Figure P5.17, what is the output sequence generated by the input sequence $x = 10111001$, assuming that you start in state S_0?

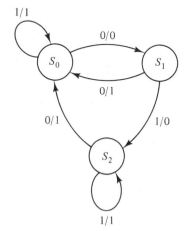

1/1

Figure P5.17

5.18. Design a circuit using D flip-flops that implements the machine whose state diagram is given in Figure P5.18.

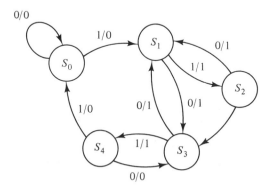

Figure P5.18

5.19. Repeat Problem 5.18 using JK flip-flops.

5.20. Based on your design for Problem 5.18, reconstruct the state diagram implemented, showing all possible states. What happens if on turning on the power to your design the flip-flops happen to start in one of the states not found in Figure P5.18?

5.21. Construct a Moore machine whose output is 1 if the last five inputs were 11010. Use JK flip-flops in your design.

5.22. Repeat Problem 5.21, but construct a Mealy machine this time.

5.23. A 3-bit counter is to be designed on the basis of the following specifications. If an input X is 1, the counter is to count in the sequence 000, 001, 010, 011, 100, 101, 110, 111, . . . If $X = 0$, the counter is to count in the sequence 000, 001, 100, 010, 000 . . . Design this counter using JK flip-flops.

5.24. Design a one-input (X), two-output (Z_1, Z_0) circuit whose outputs represent the number of 1s that have appeared in the last three inputs.

5.25. Using the XY flip-flop of Problem 5.11, design a sequential circuit to implement the machine whose state diagram is given in Figure P5.17.

5.26. A certain sequential circuit has the following next state equations:

$$Y_1 = \bar{x}y_1 + x\bar{y}_1$$
$$Y_0 = xy_1 + \bar{x}\bar{y}_1\bar{y}_0 \tag{P5.26.1}$$

The characteristic equation for a JK flip-flop is $Y = y\bar{K} + \bar{y}J$. If we plug this into equation pair (P5.26.1), we obtain

$$Y_1 = \bar{x}y_1 + x\bar{y}_1 = y_1\bar{K}_1 + \bar{y}_1J_1$$
$$Y_0 = xy_1 + \bar{x}\bar{y}_1\bar{y}_0 = y_0\bar{K}_0 + \bar{y}_0J_0 \tag{P5.26.2}$$

Solve equation pair P5.26.2 for K_1, J_1, K_0, and J_0 in terms of the variables x, y_1, and y_0. (*Hint:* Note that if $Y_i = f = g$, then $f \oplus g = 0$, and that the flip-flop inputs take on the general form, for example, $K_0 = c_0\bar{X}\bar{y}_1\bar{y}_2 + \cdots + c_7Xy_1y_2$.)

5.27. Show that the concept of indistinguishable states as defined in Definition 5.5.1 is an equivalence relation.

5.28. Find a minimal state machine equivalent to the machine whose state table is given in Figure P5.28.

	$X = 0$	$X = 1$
A	B, 1	C, 1
B	E, 0	G, 0
C	C, 1	A, 0
D	D, 0	B, 1
E	F, 1	C, 1
F	H, 0	D, 0
G	G, 0	F, 1
H	F, 1	C, 1

Figure P5.28

5.29. For each of the state diagrams shown in Figure P5.29, convert to the alternative form; i.e., convert the Mealy machine in part (a) to a Moore machine and convert the Moore machine of part (b) to a Mealy machine.

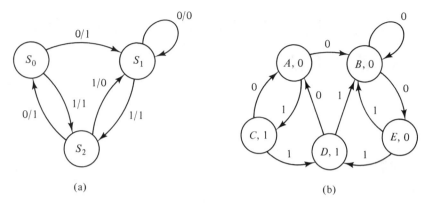

(a)

(b)

Figure P5.29

Asynchronous Sequential Circuits

6

6.1 INTRODUCTION

In Chapter 5, we introduced the idea of a combinational circuit with feedback. Such a circuit was termed a *sequential circuit*, since its outputs were dependent on some past sequence of inputs. In circuits of this type the output behavior is controlled by the physical delays of the various gates used to implement the circuit. As was pointed out in that chapter, a simple model for such circuits can be derived if we assume that the circuit is "glitch" free. This model was shown in Figure 5.1.4 and is repeated here as Figure 6.1.1. Since the delays in these circuits are not generally controllable, designing reliable circuits of this type is usually difficult. By replacing the delays with flip-flops, as was done in the model of Figure 5.4.1, we were able to completely control the feedback paths and therefore generate sequential circuits that performed predictably and reliably. There is, however, a "Catch-22," and that is that the circuit operates reliably only if the flip-flops operate reliably as well. The flip-flops themselves can only be modeled as shown in Figure 6.1.1. Thus, if the clocked sequential circuit is to work properly, we must develop techniques for designing reliable flip-flops.

In this chapter, then, we will investigate the analysis and design of sequential circuits that basically must be modeled as in Figure 6.1.1. These circuits will be termed *asynchronous sequential circuits*, since their output behavior is controlled by the changes in the input variables, which are, generally, not synchronized in any way. As was pointed out in Section 5.1.2, two possible things can occur in circuits of the type shown in Figure 6.1.1. First, if

$$Y_i(t + d) = y_i(t)$$

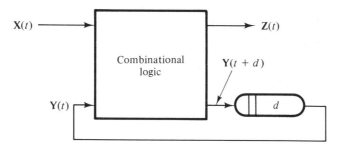

Figure 6.1.1   A model of a combinational circuit with feedback.

for all i, then the circuit will be *stable*; i.e., as long as the inputs do not change, neither the outputs nor the secondary variables will change. Second, if, for some combination of inputs and secondaries,

$$Y_i(t + d) \neq y_i(t)$$

for some i, the circuit is *unstable*; i.e., the present value of the secondary variable will change after propagation delay time d. If an unstable situation occurs, one of two things will happen. Either this change can result in another change, and so on indefinitely, or, after some finite number of changes, a stable state will be reached. In general, the inputs in an asynchronous circuit can change at any time, including times during which the outputs are changing. This, of course, can make analysis extremely difficult, because we are never sure of the input values at any given instant of time. If, however, we assume that no input makes a change until all of the outputs and state variables are stable, and then that only one input changes at a time, circuit analysis becomes considerably easier. Asynchronous sequential circuits in which this assumption is made are generally referred to as *fundamental-mode sequential circuits*.[1] In what follows we will be interested only in asynchronous sequential circuits that operate in this fundamental mode. We will, of course, be interested in flip-flops, but the techniques to be developed are applicable to many other circuits as well.

Before we begin our investigation of asynchronous circuits, a little historical and physical perspective is in order. Circuits of the type to be examined in this chapter have a very desirable characteristic, namely, that they can process information at a speed limited only by the longest propagation delay path in the implementation. Since current technology allows the design of gates having propagation delays of the order of several picoseconds (1 picosecond $= 10^{-12}$ seconds) and since it is possible to construct all combinational functions using only two levels of gates—an AND level to implement the minterms, followed by an OR level—it is conceivable that computers could be designed that can perform basic operations very fast, indeed. In fact, a computer having an arithmetic unit based on this idea was designed and built at the University of Illinois in the mid-1950s. At the time, Illiac II, as it was named, was the fastest computer in the world. Many much faster computers exist today, even though they are not asynchronous. Their greater speed is principally due to improvements in the technology used to implement the hardware. As we shall soon see,

---

[1] Another term that is encountered in the literature is *level-mode*, although this is not as common.

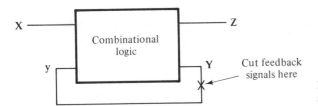

Figure 6.2.1 Asynchronous circuit analysis model.

designing large-scale asynchronous systems is, if not impossible, fraught with tremendous problems of reliability. Most of these problems can be handled by the manner in which the requisite functions are implemented, i.e., the form of the equations. Unfortunately, some of these problems can be managed only by controlling the feedback delays—which was the reason for inserting the flip-flops in the feedback paths in the first place!

6.2 THE ANALYSIS PROBLEM

As was the case in the analysis of clocked sequential circuits, the analysis of asynchronous sequential circuits involves determining how the outputs of the circuit change with changes in the inputs. These outputs are functions of the *primary variables*—the inputs—and of the signals associated with the feedback path delays. These feedback variables are usually referred to as the *secondary variables* and represent the state of the system at any instant of time. The process of identifying the circuit output behavior is not significantly different from what we saw in the last chapter. However, since there are no flip-flops in the feedback paths, identification of the state, or secondary, variables may not always be easy.

In general, the analysis of an asynchronous sequential circuit begins by identifying and "cutting" all of the feedback paths at the output of the gates that drive these feedbacks so that the resulting circuit is purely combinational.[2] The gate outputs which are cut will be labeled with a capital letter to indicate that this is the *next* value of the secondary variable, and all of the gate inputs that are driven by this line will be labeled with a lowercase letter to indicate the *current* value of the secondary variable. This conforms to the labeling of current state versus next state that was used in Chapter 5. Figure 6.2.1 shows this general model. The reader should compare it with Figure 6.1.1. We may now write equations for the outputs and the next value of the secondary variables in terms of the current secondary variables and the inputs. The equations for the secondary variables are referred to as the *excitation equations*. Plotting the excitation equations for the secondary variables in a Karnaugh map results in an *excitation table* (or matrix) that shows how the secondaries change with changing inputs. Using the excitation matrix, we can easily predict the output behavior. Since each entry in the excitation table represents the *next* value of the secondary variables, each entry in the table can be either stable or unstable. By circling the stable states, we produce what is sometimes referred to as a *transition table* (or matrix). This table makes it easy to identify the transitions from one stable state to another caused

---

[2] Although it is not always easy to identify a minimal set of feedback paths, cutting more paths than necessary will not change the analysis outcome, but will only make it more difficult by adding secondary variables.

by a single variable change on the inputs, as well as output changes produced by these input changes.

6.2.1 Derivation of the Excitation Table

Perhaps the best way to explain the process of fundamental-mode sequential circuit analysis is with an example. Consider, therefore, the circuit shown in Figure 6.2.2(a). As indicated above, the first task is to identify the feedback paths. The circuit clearly has only one feedback path, as indicated. On cutting this path at the output of the gate that drives it, we create the secondary variable Y, as shown in the figure. We also produce the purely combinational circuit shown in Figure 6.2.2(b). The *excitation equation* for this secondary variable is now easily derived:

$$Y = DG + y(D + \overline{G}) \qquad (6.2.1)$$

We may observe also from Figure 6.2.2(a) that the output is simply equal to the secondary variable and so

$$Z = Y \qquad (6.2.2)$$

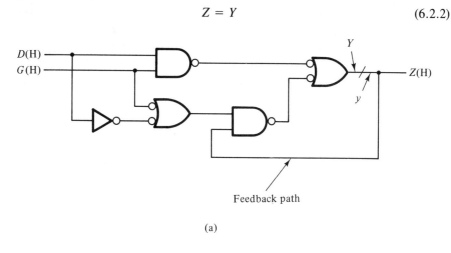

(a)

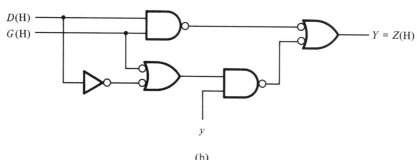

(b)

Figure 6.2.2   The analysis of a simple latch-mode D flip-flop. (a) Circuit with a single feedback path. (b) Feedback path cut to produce a combinational circuit for analysis.

If Equation (6.2.1) is now plotted in a Karnaugh map, we obtain the *excitation table* shown in Figure 6.2.3(a). Since the output, in this case, is a function of the secondary variable only, we have a Moore-type sequential circuit. To identify the behavior of this circuit on the basis of the excitation table, we first must identify the stable states. Recalling that a state is "stable" if the next value of the state variables is the same as the current value, we can identify those circuit states which are stable, in this example, by finding the entries in the table for which $Y = y$. Since the rows of the table are labeled according to the value of y, an entry in the table corresponds to a stable state if the plotted value for Y is the same as the row label. The stable states for this example are shown by the circled entries in the *transition table* of Figure 6.2.3(b).

To see how the transition table can now be used to understand the behavior of the circuit, let us assume that the inputs are $(D, G) = (0, 0)$. The main question is, What is the output Z? To answer this question, we must know the *total state* of the machine, where we define the total state as the value of both the primary variables, D and G, and the secondary variable, y. The total state thus corresponds to one entry in the transition table. Since we know the value of the inputs, we know that this entry must be one of the two in the column labeled $(D, G) = (0, 0)$ in Figure 6.2.3(b). In this case, both of these entries are stable. For the moment, let us assume that we are in the stable total state $(D, G, y) = (0, 0, 1)$ as indicated by the asterisk (*) in Figure 6.2.4(a). Now, unless an input changes, we will stay in this state indefinitely, and the output will be $Z = 1$.

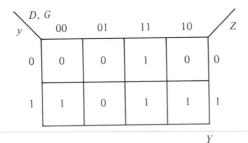

(a)

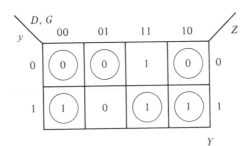

(b)

Figure 6.2.3   The excitation table (a) and transition tables (b) for the circuit of Figure 6.2.2.

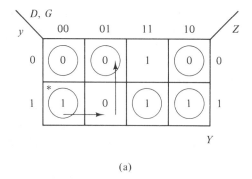

(a)

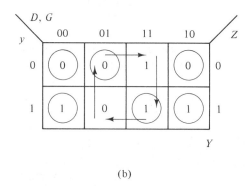

(b)

Figure 6.2.4 Some of the many possible transitions in the circuit of Figure 6.2.2. (a) Transition from total state (0, 0, 1) to (0, 1, 0). (b) The output follows the input, D, as long as G is asserted.

Suppose that input G changes from a 0 to a 1. This corresponds to moving one column to the right, as indicated in Figure 6.2.4(a). This state, however, is unstable, since it requires that the next value of the state variable Y be 0 when the present value y is 1. Thus, after some delay, y will change to a 1; this change corresponds to moving up one row in the transition table, as shown by the arrows in the figure. The resulting total state $(D, G, y) = (0, 1, 0)$ is stable, with an output of $Z = 0$. Thus, we have moved from a stable state in which the output is 1 to a stable state in which the output is 0 by holding $D = 0$ and changing G from a 0 to a 1. As shown in Figure 6.2.4(b), if we now hold G at 1 and change D to a 1, we will move to total state $(D, G, y) = (1, 1, 1)$ with an output of 1. If D changes back to a 0, we will move back to total state $(D, G, y) = (0, 1, 0)$. This loop is shown in the figure. In this loop we see that as long as the G input stays 1, the output will follow the input; i.e., if $G = 1$, then $Z = D$.

Continuing in the above manner, we can trace out all of the paths that lead from one stable state to another in the transition table. Simply stated, the process of tracing the paths involves changing an input variable, which causes the circuit to move over a column and then, depending on whether the resulting state is stable or not, move up or down within this column until a stable state is encountered. By carrying out this analysis for all of the possible paths, we can see that the circuit shown in Figure 6.2.2 is just the *latch-mode D flip-flop* discussed in Section 5.2. The behavior of this flip-flop can be described in words as having an output which is a stable 1 or 0, regardless of what input D does, as long as $G = 0$ and which takes on the value of the input when $G = 1$. The reader should trace out each of the possible transitions from one stable state to another to verify this behavior.

6.2.2 Race Conditions and Cycles

In the above analysis, we tacitly assumed that only one input changed at any instant of time. From a physical point of view, this seems to be a very good assumption, since even if "chance" causes the signals D and G, in the last example, to change at exactly the same instant of time, their effects on the outputs will occur at different times, because the propagation delays from the two inputs to the outputs will be different. A similar argument holds for two or more secondary variables changing at the same instant of time. If, however, it is required that two secondary variables change simultaneously, then the behavior of the circuit may not be completely predictable.

Consider the "fragment" of the transition table shown in Figure 6.2.5(a). Suppose that the circuit corresponding to this table is sitting in the stable state $(A, B, y_1, y_0) = (0, 1, 0, 0)$ and input B changes from a 1 to a 0. The circuit will move to total state $(A, B, y_1, y_0) = (0, 0, 0, 0)$, which shows a next state value of $(Y_1, Y_0) = (1, 1)$. This situation requires that *both* state variables change from $(0, 0)$ to $(1, 1)$! From a physical point of view, the simultaneous changing of two signals in a circuit is highly unlikely—one is bound to change slightly ahead of the other. Such a condition, where two or more variables are required to change at the same time, is termed a *race condition*. So the question is, How do we determine what happens next? In this case, we cannot say what happens for certain, but we can predict one of two outcomes. Suppose that variable y_1 changes first, so that (y_1, y_0) goes from a value of $(0, 0)$ to $(1, 0)$. This corresponds to moving to the bottom

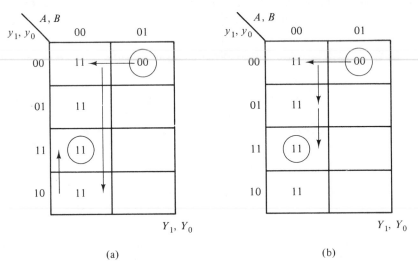

Figure 6.2.5 An example of a noncritical race. (a) y_1 changes before y_0. (b) y_0 changes before y_1.

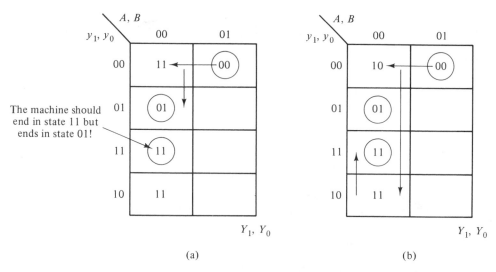

Figure 6.2.6   An example of a critical race. (a) Incorrect response. (b) Desired response.

row of the transition table, as shown in Figure 6.2.5(a). Since the resulting state is still unstable, requiring (y_1, y_0) to be equal to $(1, 1)$, the circuit has entered a *cycle*, a condition in which we move from one unstable state to another. The circuit must now move up a row in the table, as shown in the figure, and thus end in the stable state $(A, B, y_1, y_0) = (0, 0, 1, 1)$. Note that this transition is not a race, since the next change requires only a single variable change (y_0 changes from a 0 to a 1; y_1 is already a 1 at this point). This final stable state is where the transition table for the circuit indicates we were supposed to end (we needed to go from $(y_1, y_0) = (0, 0)$ to $(1, 1)$).

Suppose, on the other hand, that y_0 changes first, so that (y_1, y_0) goes from $(0, 0)$ to $(0, 1)$, corresponding to moving down one row in the transition table, as shown in Figure 6.2.5(b). The circuit now ends up in the total state $(A, B, y_1, y_0) = (0, 0, 0, 1)$, which is also unstable, requiring (y_1, y_0) to be equal to $(1, 1)$. The circuit has thus entered another cycle and must move down one more row to the stable state $(A, B, y_1, y_0) = (0, 0, 1, 1)$, where it will remain until another input change occurs.

We note in this particular example that regardless of the outcome of the race, the circuit will always end up in the same stable state. Such a race condition is referred to as a *noncritical race*, since the race outcome is not critical in determining the final stable state. Consider, on the other hand, what might happen in the case illustrated in Figure 6.2.6(a). In this case, if y_0 changes first, the circuit will not end up in total state $(A, B, y_1, y_0) = (0, 0, 1, 1)$, but rather in $(0, 0, 0, 1)$, since this is also stable. In this case, the circuit can end up in one of two different states, depending upon the outcome of the race. Such a race condition is termed a *critical race*.

Obviously, critical races must be avoided in designing such circuits if predictable and reliable behavior is to be guaranteed. We shall see in Section 6.5 that it is always possible to eliminate race conditions in asynchronous sequential circuits, although there may be a cost in terms of additional hardware. In fact, it is easy to see, in the example

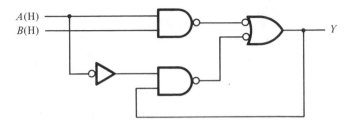

Figure 6.2.7   A circuit with a static hazard.

shown in Figure 6.2.6(a), that if the entry in the flow table at total state $(A, B, y_1, y_0) = (0, 0, 0, 0)$ had been changed from $(Y_1, Y_0) = (1, 1)$ to $(1, 0)$, as shown in Figure 6.2.6(b), y_1 would have been forced to change first and the circuit would have worked correctly. In this case, the secondary variables would have gone through a *cycle* of $(0, 0) \rightarrow (1, 0) \rightarrow (1, 1)$, which is the final stable state.

6.2.3 Static and Dynamic Hazards

Before looking at some more complex analysis examples, let us consider another situation that can cause asynchronous circuits to operate improperly. Consider the circuit shown in Figure 6.2.7. Proceeding as was done above, the excitation equation for this circuit becomes

$$Y = AB + \overline{A}y \tag{6.2.3}$$

Figure 6.2.8(a) shows the excitation table, and Figure 6.2.8(b) gives the corresponding transition table. Equation (6.2.3) has a *consensus* term that is missing, namely, By. As we demonstrated in Section 5.1, when such a term is missing in a combinational circuit, a "glitch" can occur, causing the output—the secondary variable, in this case—to momentarily change value. Such a situation is referred to as a *static hazard*, since the output

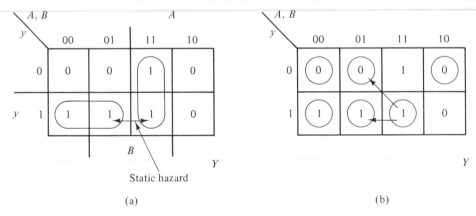

(a)   (b)

Figure 6.2.8   Excitation matrix (a) and transition matrix (b) for circuit of Figure 6.2.7. Transition matrix shows possible glitch-induced transitions.

is supposed to remain constant, or static, while an input changes. This particular example, in fact, is the circuit analyzed in Figure 5.1.2 with the output fed back to one of the inputs. As was shown in that example, the output, in this case y, will momentarily go to a 0 when input A goes from a 1 to a 0 if both B and y are initially 1. Looking at the transition table of Figure 6.2.8, this means that if the circuit is initially in the stable total state (A, B, y) $= (1, 1, 1)$, as shown in Figure 6.2.8(b), and A goes from a 1 to a 0, it is possible that the circuit could end up in stable state $(A, B, y) = (0, 1, 0)$ rather than the one intended, $(0, 1, 1)$. Whether this actually occurs or not depends on the relative delays through each of the gates and can only be determined by a complete timing analysis of the circuit. However, such a hazardous situation can *always be avoided by including the consensus terms*. Another way of putting this is that all static hazards can be avoided if all blocks of 1s in the excitation tables for each of the secondary variables are connected by the redundant consensus terms. For this example, this means adding the consensus term as shown in Figure 6.2.9(a), which results in the circuit realization shown in Figure 6.2.9(b).

Another hazard associated with combinational circuits which can cause asynchronous

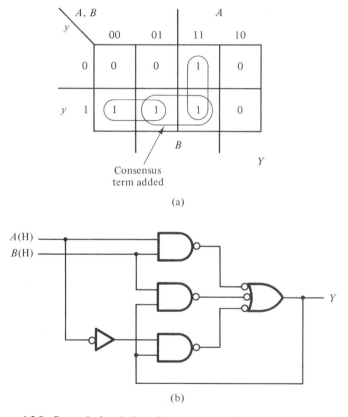

(a)

(b)

Figure 6.2.9 Removal of static hazard by connecting adjacent 1s with the consensus term. (a) Excitation table with added consensus term. (b) Added gate to remove "glitch."

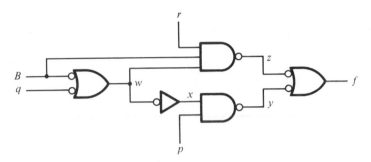

Figure 6.2.10   An example of a circuit having a dynamic hazard.

circuits to fail is the *dynamic hazard*. In a dynamic hazard, an output may change several times for a single change in an input. Dynamic hazards, like static hazards, can be eliminated logically by simply rearranging the form of the equation or by adding consensus terms. Figure 6.2.10 shows a typical circuit that has a dynamic hazard. The unsimplified equation for the output of this circuit is

$$f(B, p, q, r) = rB(\overline{B} + \overline{q}) + pqB \tag{6.2.4}$$

Notice, in this equation, that B appears in the term $B\overline{B}$ as well as in a term by itself. This is a characteristic of the dynamic hazard. Figure 6.2.11 shows the effect of this hazard on the output. It is assumed, in this figure, that all gate delays are the same and that inputs p, q, and r are all "high." Dynamic hazards are caused by the occurrence of three or more paths from an input to the output, each path having a different delay. By refactoring the implementing equation to eliminate these multiple paths, the dynamic hazard can be removed. In this case, all we need do is to write the equation as

$$f(B, p, q, r) = r\overline{q}B + pqB \tag{6.2.5}$$

and implement the circuit accordingly.

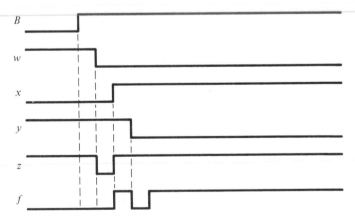

Figure 6.2.11   Timing diagram showing the effects of the dynamic hazard. Assume signals p, q, and r are all high.

6.3 ANALYSIS OF THE 7474 EDGE-TRIGGERED D FLIP-FLOP

In Chapter 5, we introduced the concept of an edge-triggered clocked flip-flop whose outputs were unaffected by the inputs except at the time that a transition occurred on the clock line. These flip-flops were used throughout Chapter 5 for controlling the feedback paths in clocked sequential circuits. As we shall see in Chapter 9, these flip-flops are also very important for use in the temporary storage of information in large-scale digital systems, such as computers. The two most commonly encountered flip-flops of this type are the 7474 and 74LS76 edge-triggered *D* and edge-triggered *JK*, respectively. The symbols for these devices are shown in Figure 6.3.1, along with their defining truth tables. Each of these flip-flops has two "asynchronous" inputs, *S* (set) and *C* (clear). These inputs are referred to as asynchronous because they cause the flip-flop output to be set to a 1 or cleared to a 0 regardless of the state of the other inputs. This is shown in the defining truth tables.

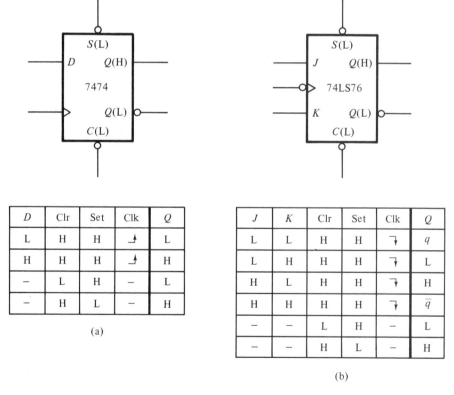

D	Clr	Set	Clk	Q
L	H	H	⤴	L
H	H	H	⤴	H
–	L	H	–	L
–	H	L	–	H

(a)

J	K	Clr	Set	Clk	Q
L	L	H	H	⤵	q
L	H	H	H	⤵	L
H	L	H	H	⤵	H
H	H	H	H	⤵	$\bar{q}$
–	–	L	H	–	L
–	–	H	L	–	H

(b)

Figure 6.3.1   Flip-flops with asynchronous presets and clears. (a) *D* flip-flop; type 7474. (b) *JK* flip-flop; type 74LS76.

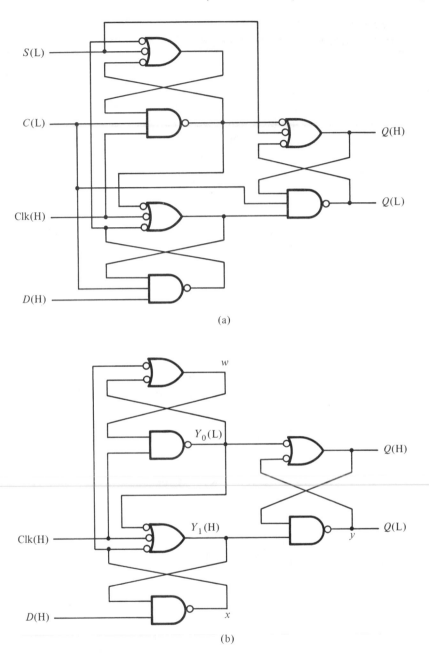

(a)

(b)

Figure 6.3.2 The implementation of the 7474 edge-triggered D flip-flop. (a) Implementation including the asynchronous set and clear. (b) Implementation with asynchronous set and clear tied high.

If we were to look up the 7474 in a 7400 series TTL (transistor-transistor logic) data book (or catalog), we would see the logic circuit given in Figure 6.3.2(a). What we would like to do, now, is to apply the analysis procedures given in Section 6.2 to this circuit and verify that the flip-flop behaves as specified. Let us begin our analysis by observing how the set and clear lines affect the output Q. It is easily seen that if S is asserted low and C is high, the outputs of all of the OR gates are high, thus making the circuit output, Q, high, or 1, regardless of the values of D and C1k. Conversely, if S is high and C is low, the output of all of the AND gates will be low, causing Q to go low, or to 0, again without regard to the value of the other inputs. If, however, both S and C are high, the output Q will be affected by the inputs D and C1k only. Thus, to investigate the dynamic behavior of this circuit, we will assume that both of the asynchronous inputs S and C are tied to a high voltage. (What happens if both S and C are low?) Figure 6.3.2(b) shows the resulting simplified circuit.

To begin our analysis of the circuit of Figure 6.3.2(b), we need to first identify the secondary variables corresponding to the feedback loops. It is not difficult to see that there are three such loops, which, when cut, produce a combinational circuit. The resulting secondary variables are labeled Y_0, Y_1, and Q, the circuit output, in this figure. With this choice for the secondaries, the excitation equations become

$$Q = y_0 + y_1 q$$
$$Y_1 = y_0 + \overline{C} + D y_1 \qquad\qquad (6.3.1)$$
$$Y_0 = C(y_0 + y_1 D)$$

Plotting these equations and circling the stable states produces the transition matrix shown in Figure 6.3.3.

Using the transition table, it is easy now to verify the behavior of the 7474. Before doing this, however, let us first examine the circuit to see whether there are any races or hazards. Checking for race conditions is a very straightforward process. A race condition will occur if two or more of the state variables q, y_1, and y_0 are required to change at the same instant of time. To identify a race, then, we start in a stable state and examine the entries in the same row to which the circuit can move, assuming that only one input changes at any given instant of time. If any of these entries requires more than one state variable to change, we have found a race. Consider, for example, the stable total state (C, D, q, y_1, y_0) = (1, 1, 0, 0, 0) shown as a in Figure 6.3.3. Only two possible moves are allowed from here. The first is to the stable total state (1, 0, 0, 0, 0), called b in the figure, and the other is to total state (0, 1, 0, 0, 0), marked c. The second of these two is unstable, requiring the state variables (q, y_1, y_0), to change from (0, 0, 0) to (0, 1, 0), which produces a final stable state, which is state d in the figure. Since only one of the state variables must change in this example, no race occurs.

Since races can also occur in cycles, we must check all of the possible cycles to make sure that each step in the cycle requires only a single variable change. Consider, for example, the cycle that occurs when we start in the stable total state (0, 1, 0, 1, 0), marked d in Figure 6.3.3 (which is the state we ended in, in the above example). If C now changes from a 0 to a 1, the circuit will move right to e in the column labeled (C, D) = (1, 1). This requires the state variables to change from (q, y_1, y_0) = (0, 1, 0) to (0, 1, 1), which

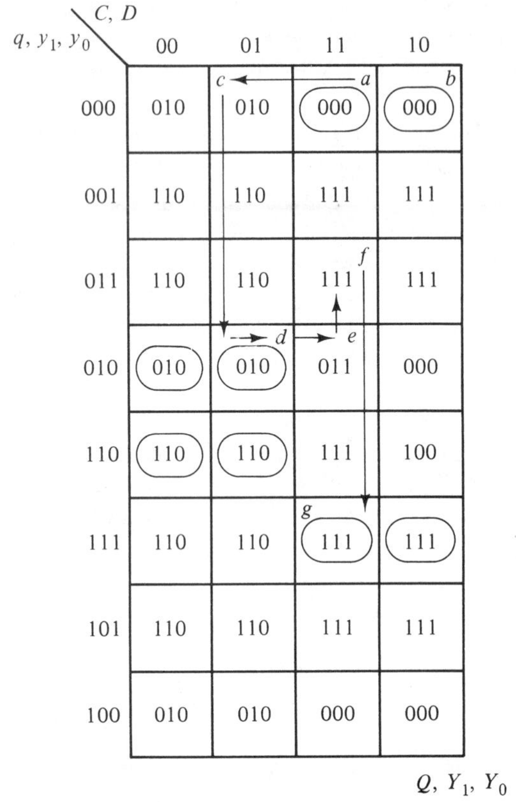

Figure 6.3.3 Checking for race conditions in the 7474 D flip-flop.

forces the system to move up one row to f. But the entry in this row requires the variables to change again from (0, 1, 1) to (1, 1, 1), which takes us down three rows to the stable total state (1, 1, 1, 1, 1). Since each step in this cycle requires only a single secondary variable change, no race condition exists. Proceeding in this manner with each of the other stable states in the transition matrix, we can verify that *this circuit is race-free*.

To determine whether the circuit is free of static hazards, we need only check the excitation tables for each of the state variables to determine whether groups of adjacent 1s are connected by overlapping groups of 1s. Figure 6.3.4 shows the excitation tables for each of the state variables. Note that all groups of 1s overlap, so that no glitch can occur on any of the state variables and so *no static hazard exists in this circuit*.

Finally, we note that since none of the equations has the characteristic form shown in Equation (6.2.4), which indicates the presence of a dynamic hazard, *this circuit is free of dynamic hazards* as well.

Analysis of the functional behavior of this circuit can now be carried out by tracing the various paths through the transition table which the circuit can follow. Take, for example, the situation where C and D are both 0 and the output $Q = 1$. If C goes high, the truth table for this flip-flop, given in Figure 6.3.1, indicates that the output is to change to a 0. This corresponds to the path shown in Figure 6.3.5 starting in a total stable state (C, D,

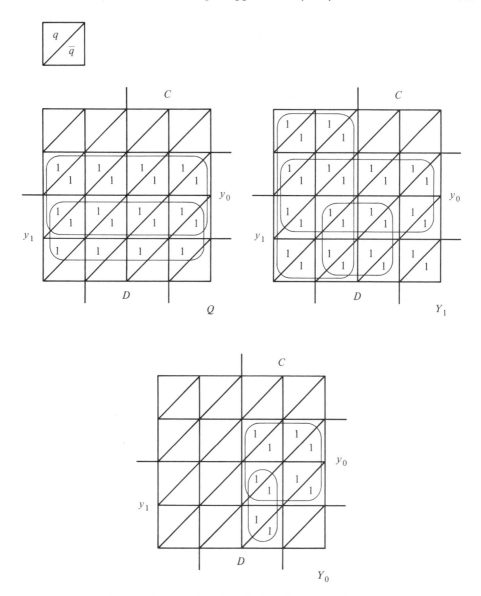

Figure 6.3.4 Checking the excitation tables for static hazards.

$q, y_1, y_0) = (0, 0, 1, 1, 0)$, marked *a* in the figure, and ending at *d*, which is total state $(1, 0, 0, 0, 0)$. Observe the three-step cycle required to get to the final stable state: *a* to *b*, *b* to *c*, and *c* to *d*. If, now, *C* returns to 0, the circuit will move over to the unstable total state $(0, 0, 0, 0, 0)$, or entry *e*, and then down to the final stable state $(0, 0, 0, 1, 0)$, entry *f*. The reader can trace through the many other possible paths in the circuit in a similar fashion.

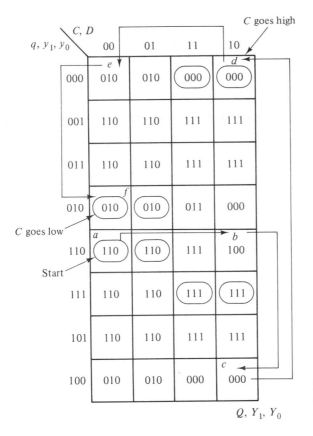

Figure 6.3.5 An example transition causing the output to change on a clock edge.

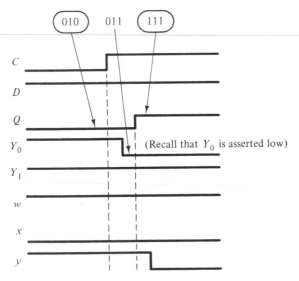

Figure 6.3.6 A timing diagram for the 7474, showing a cycle in the state variables.

Before leaving this example, it might be helpful to examine the time behavior of this circuit. Consider, for example, the path taken by the circuit in going from stable total state $(C, D, q, y_1, y_0) = (0, 1, 0, 1, 0)$ to $(1, 1, 1, 1, 1)$, which is shown as the path $d, e, f,$ g in the transition matrix of Figure 6.3.3. Figure 6.3.6 shows the sequence of changes occurring during this path transition. The timing diagrams shown in this figure are similar to what one might see on an oscilloscope and verify the circuit behavior as predicted by the transition matrix. Note in this example that output $Q(H)$ changes before the output $Q(L) = y$. This is typical of flip-flops having both asserted high and asserted low outputs present.

6.4 *THE SYNTHESIS OF ASYNCHRONOUS SEQUENTIAL CIRCUITS*

The synthesis process for asynchronous sequential circuits begins by creating the *primitive flow table*, a table analogous to the state table of synchronous sequential circuit design. As was the case in clocked circuits, the next steps involve, possibly, the simplification of the table followed by assigning values to the required state variables. In Chapter 5 we paid little attention to how the variables were assigned, since the assignment could not affect the operation of the circuit, although it could affect the complexity of implementation. In the case of asynchronous circuits, however, this somewhat cavalier attitude must be avoided. Here the state assignment is critical in creating circuits that are race-free. We also did not worry too much about glitches in Chapter 5, since the state could change only on a clock edge and by the time this occurred any glitches in the circuit would be gone. In the present case, however, glitches can cause the circuit to malfunction and thus must be avoided if we are to design reliable asynchronous circuits.

It is clear from this that the design of asynchronous sequential circuits must be attended with a great deal more care than was required for synchronous sequential circuits. Although the design processes are similar in most respects, the problems of races and hazards require some modification in the procedure. Perhaps the best way to describe the process of designing asynchronous sequential circuits is with an example. Consider, then, the following problem.

Design Problem

There are many applications where we would like to be able to turn a clock on and off using a manual switch. Usually a clock consists of pulses occurring at some fixed rate. We might be tempted to solve this problem simply by ANDing the switch with the clock. The difficulty is that since the switch is not synchronized to the clock, we might turn the switch on in the middle of a pulse and in so doing produce an output pulse shorter than that required by whatever system is being driven by the clock. Similarly, our switch might turn off in the middle of a pulse. Thus, what we want is a circuit that will produce nothing but complete pulses as long as the switch is on, regardless of when it was turned on or off. We will refer to this circuit as a *gated oscillator*.

6.4.1 Derivation of the Primitive Flow Table

One of the most direct ways in which we can begin the design process is to construct a timing diagram showing the various ways in which the inputs can change and showing also the output desired for each combination of inputs and each sequence of combinations. Figure 6.4.1 shows a typical timing diagram for the gated oscillator. A state in this timing diagram will be taken as a unique combination of the inputs and the associated outputs. Thus we begin the design process by identifying the sequence of states encountered in the timing diagram. We will arbitrarily start at the state labeled 0 in Figure 6.4.1, in which $(C, G, Z) = (1, 0, 0)$. (We are assuming here that all assertion levels are high.) The next state in the timing diagram occurs when C goes low. We will call this state 1. When C goes back high, we have returned to state 0. Now if G goes high while C is high, we move to a state not yet encountered. We will call this state 2. Proceeding through the timing diagram and introducing new states as necessary, we produce the sequence of states shown in the figure.

    The next step in the design process is to plot the sequence of state changes in a table similar to a state table called a *primitive flow table*. A primitive flow table has a single stable state associated with each row. This stable state is circled. The other entries in the row show to what stable state the circuit is to move for each of the possible changes in the inputs. Figure 6.4.2(a) shows the completed primitive flow table generated from the timing diagram of Figure 6.4.1. To illustrate how this table is derived from the timing diagram, consider the first row of the table, which corresponds to the stable state 0, circled in the figure. In the timing diagram, we move from state 0 to state 1 when the inputs change from $(C, G) = (1, 0)$ to $(0, 0)$. Thus, an uncircled 1, indicating an unstable state, is entered in the first row under the column labeled $(C, G) = (0, 0)$. We now create a new row having a circled 1, corresponding to a stable state, in this column, the second row in Figure 6.4.2. This will be the state we end up in after C changes from a 1 to a 0. The transition from stable state to stable state in this flow table is exactly equivalent to the way we move around in a transition table.

    Consider next the move from state 1 to state 0 in the timing diagram. When this occurs we must place an entry in the primitive flow table indicating an unstable 0 in the column headed $(1, 0)$. This, of course, means that we will return to the stable state 0 in the first row. In the timing diagram, the next change is from state 0 to a new state labeled

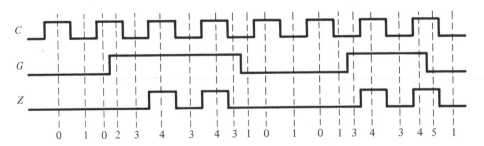

Figure 6.4.1 Typical timing diagram for the gated oscillator.

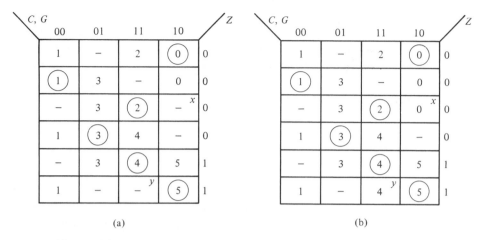

Figure 6.4.2 Primitive flow table derived from the timing diagram of Figure 6.4.1: (a) table derived directly from timing diagram; (b) added missing transitions.

2 corresponding to an input of $(C, G) = (1, 1)$. Thus, we make an entry of an uncircled 2, indicating an unstable state, in the first row in the column labeled $(C, G) = (1, 1)$. We must now create a new row having a stable state 2 in this column. This becomes the third row, in which a circled 2 appears, again indicating a stable state.

If we continue to create new rows in this way on the basis of the sequence of states described by the timing diagram of Figure 6.4.1, we will end up with the primitive flow table for the gated oscillator shown in Figure 6.4.2(a). Note, in this table, that a number of transitions have not occurred in the timing diagram. For example, no transition has occurred in the first row from the stable state 0 to the column headed $(C, G) = (0, 1)$. Such a transition would require that (C, G) change from $(1, 0)$ to $(0, 1)$, which means that both inputs would have to change *simultaneously*, a situation we have assumed all along cannot happen, or, at worst, is highly unlikely. Thus, we will not worry about this entry and will simply take it as a don't care. Two other entries in the table cannot be dismissed so lightly. These are labeled x and y in Figure 6.4.2. Neither of these entries requires both inputs to change simultaneously, and so both are possible, although no transitions into these states have occurred in the sample timing diagram. These entries can be handled in one of two ways: either fill in the entry with a reasonable value, or leave the entry as a don't care.

Filling in entries in the primitive flow table that do not occur in the sample timing diagram with ''reasonable'' values is usually not difficult. Consider, for example, the situation that exists if we start in stable state 2 and have the inputs change from $(C, G) = (1, 1)$ to $(1, 0)$, corresponding to moving from stable state 2 to the position labeled x in the figure. This situation happens when gate G goes high while C is high and then goes low before C goes back low. We may consider this a kind of glitch and simply ignore such spurious changes by forcing the circuit to go back to state 0 so that the output will not be affected, as shown in Figure 6.4.2(b). A similar situation exists if we start in stable state 5 and the inputs change from $(1, 0)$ to $(1, 1)$, the entry labeled y in the figure. This corresponds to G dropping in the middle of a clock pulse C and then going high again before the clock goes away. Again we have a glitch on G, and again, we ignore it by causing the system to return, in this case, to state 4, so that the output stays 1. The final primitive flow table with all of the possible transitions shown is given in Figure 6.4.2(b).

6.4.2 The Reduced Flow Table

The next step in designing an asynchronous sequential circuit is to reduce the primitive flow table to a table having as few rows as possible. We would like to do this because the flow table will eventually become the excitation table, in which the number of rows determines the number of state variables and, therefore, the complexity of the implementation.

The process of reducing the primitive flow table involves "merging" sets of two or more rows into a single row. We may *merge* two rows of the primitive flow table if, when the state labels in corresponding columns are matched up, each pair contains either two like entries or at least one don't care. When two rows are merged, a stable entry and an unstable entry become stable and two unstable entries stay unstable. For example, the first and second rows of the primitive flow table of Figure 6.4.2(b), corresponding to stable state 0 and 1, respectively, can be merged. The resulting row would have a stable 1 in the first column, an unstable 3 in the second column, an unstable 2 in the third column, and a stable 0 in the last column. Continuing to compare the rows in pairs, we can see that the rows corresponding to stable states 0 and 2 as well as those corresponding to stable states 1 and 2 can merge. We can describe all of this by a *merger diagram* showing which rows can merge and what the output value is that is associated with each row in the primitive flow table. Such a merger diagram is shown in Figure 6.4.3. In this diagram, the row state is given inside the circle and the output corresponding to this row is shown adjacent to the circle. We refer to the circled entries as *nodes*. Nodes which correspond to rows that can be merged are connected by a line, or edge. Thus, nodes 0 and 1 are connected, as are 0 and 2 and 1 and 2. The remaining connections in the merger diagram can be verified by observing, in Figure 6.4.2, that the corresponding rows can merge.

In general, a set of rows in a primitive flow table can be merged into a single row if the set is *strongly connected* in the merger diagram. "Strongly connected" means that

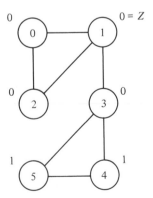

Figure 6.4.3 Merger diagram for the primitive flow table of Figure 6.4.2.

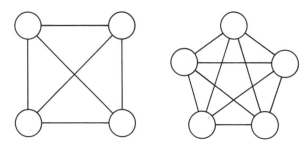

Figure 6.4.4 Strongly connected groups of four and five states.

each state in the set can be merged with every other state in the set. For example, Figure 6.4.4 shows strongly connected merger diagrams for four and five states. Since our objective is to reduce the primitive flow table to a minimum number of rows, we would like to find the strongly connected subsets of rows in the merger diagram and combine these into a single row in the merged flow table. Referring now to Figure 6.4.3, we see that the group of states 0, 1, 2 and the group 3, 4, 5 are both strongly connected and so each may be merged into a single row to produce the merged two-row flow table shown in Figure 6.4.5(a).

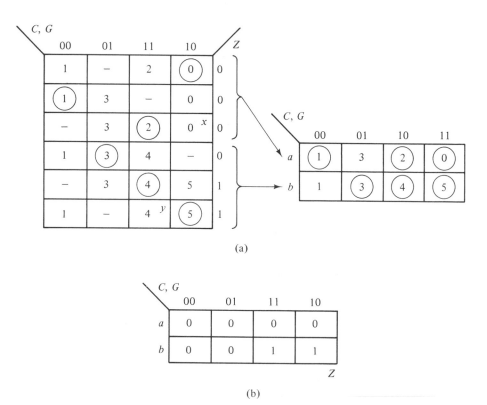

(a)

(b)

Figure 6.4.5 (a) Developing the merged flow table. (b) The output matrix.

Since the outputs in the merged flow table are associated with the individual stable states and not the separate rows (we merged the rows without regard to the associated outputs, after all), we must also derive an output matrix to accompany the merged flow table. On the surface, this is easy enough to do: simply place the output associated with each stable state in the corresponding position of the output matrix. For now, let us also set the output at unstable state entries to the value of the output associated with the corresponding stable state. Figure 6.4.5(b) shows the resulting output matrix.

6.4.3 Generation of the Excitation Table and the Final Circuit

Once we have obtained the merged flow table, we need to convert it to a transition matrix from which we can derive the equations for the secondary variables and thus implement the design. To get the transition matrix, we must first determine the number of state variables required and then assign them values for each row in the flow table. In the present example, this is easy. Since there are only two rows, we need only one state variable. Call this variable y. Figure 6.4.6 shows the resulting transition matrix, and Figure 6.4.7 shows the excitation and output matrices for the secondary variable y. The equations for the secondary variable and the output can now be derived from the excitation matrix and the output matrix:

$$Y = \overline{C}G + yC + yG \qquad (6.4.1)$$

$$Z = yC \qquad (6.4.2)$$

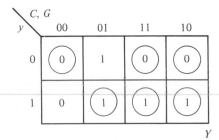

Figure 6.4.6 The resulting transition matrix.

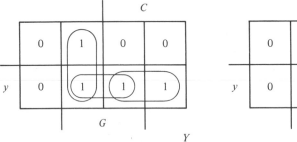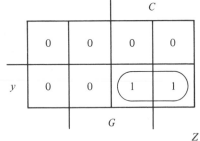

Figure 6.4.7 The final excitation and output matrices.

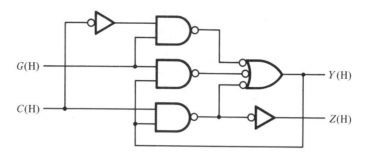

Figure 6.4.8   Final realization of the gated oscillator.

We have included the term yG in Equation (6.4.1) to prevent a static hazard. Figure 6.4.8 shows the resulting circuit, which realizes the gated oscillator required by the problem statement.

6.4.4 Merging When Multiple Choices Exist

Before we proceed with our discussion of the design process, let us pause for a moment and take a closer look at the merging process. As indicated in the last example, the merger diagram is set up by comparing pairs of rows. As pointed out above, two rows can be merged if in each column either the state labels are the same or one or both entries are don't cares. By comparing all possible pairs of rows in this way, the merger diagram is created. The next problem is to identify the largest strongly connected subsets of these rows which can merge into a single row. In the example just given, this choice turned out to be unique. This, usually, is not the case. The question then is, Given multiple ways in which rows can be merged, how do we select from the various possibilities?

In general, the objective of merging is to reduce the number of rows in the flow table to a minimum. Thus, if a merger diagram shows that multiple choices are possible, each producing the same minimal number of rows in the merged flow table, we must base our choice on some other criteria. Observe that if a set of rows are merged all of which have the same output, then the output will be a function only of the state variables corresponding to the merged row. If, on the other hand, rows are merged which have different outputs, then the output will be not only a function of the state variables, but also a function of the inputs, as was the case in the above example. Thus we may conclude that when a choice for merging is present we should *select a merger that reduces the complexity of the output function*. Consider, for example, the merger diagram shown in Figure 6.4.9. In this diagram there are two possible mergers: (0, 1, 2) and (3, 4), on the one hand, and (0, 1) and (2, 3, 4), on the other. The first of these might produce the merged flow table shown in Figure 6.4.9(a), from which we easily observe that the output Z is just equal to the state variable y used to encode the rows of the flow table. The second possible merger, shown in Figure 6.4.9(b), might produce the merged flow table and the corresponding output table shown, from which we see that the output Z is equal to yB. In the first case, no extra hardware will be required to implement the output Z. In the second case, we will need to add an extra gate to implement this output.

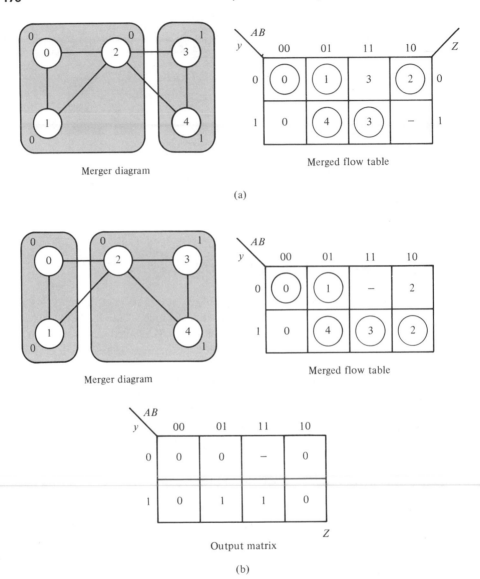

Merger diagram

Merged flow table

(a)

Merger diagram

Merged flow table

Output matrix

(b)

Figure 6.4.9 An example of multiple choices for row merging: (a) merging rows with common outputs; (b) merging rows with mixed outputs.

6.5 METHODS FOR AVOIDING RACES

The example designed above turned out to be a fairly simple circuit. Unfortunately, other asynchronous designs involve problems that did not appear in the last example. One such is the problem of assigning states so that races are avoided. As was mentioned earlier, races can always be eliminated. However, this may require adding states to the merged flow table.

Let us begin this discussion with an example. Consider the merged flow table shown in Figure 6.5.1(a). Since there are four rows in this table, two state variables are required to encode the corresponding four states. The problem now is to assign values to the state variables so that no races will occur. Recall that a race condition occurs if two or more state variables are required to change at the same instant of time. Thus, if the merged flow table indicates that the circuit is to move from a row with an unstable state to a row that

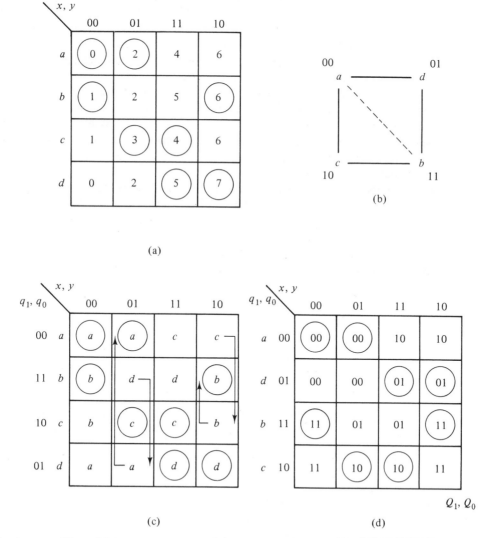

Figure 6.5.1   An example of avoiding races by creating cycles. (a) Merged flow table. (b) Adjacency diagram. (c) State table. (d) Resulting transition matrix.

is stable, the assignment of the state variables for these two rows must differ in only one bit. We will refer to these rows as being adjacent. For example, consider the column labeled $(x, y) = (0, 0)$ in Figure 6.5.1(a). In this column, row c must be adjacent to b so that when the circuit moves from stable state 3 to stable state 1, no race will occur whenever y changes from a 1 to a 0 and x is 0. Similarly, rows a and d must be adjacent. Examination of the column $(x, y) = (1, 1)$ shows that rows a and c as well as rows b and d must also be adjacent. These required adjacencies will be shown in an *adjacency diagram*, in which rows that must have assignments differing in only one variable are connected by a *solid line*. Figure 6.5.1(b) shows the resulting required adjacencies.

Now consider the remainder of the merged flow table. First examine the column labeled $(x, y) = (1, 0)$. Note here that there are two unstable states 6. To move directly to the stable state 6 from either of these unstable states would require that both rows a and b and rows c and b be adjacent. Since rows b and c are already required to differ in only one bit, we need worry only about the adjacency of rows a and b. In this case, we do not *have* to make these two rows adjacent, since we could create a cycle by making the unstable state 6 in row a go, first, to the unstable state 6 in row c and then to the stable state 6 in row b. Thus the adjacency of rows a and c is not essential. This is shown by the *dotted* line in the adjacency diagram. An identical argument holds for the adjacencies required in column $(x, y) = (0, 1)$. Figure 6.5.1(b) shows the completed diagram and an assignment for the state variables that will produce the required adjacencies. Figure 6.5.1(c) shows the flow table in terms of the rows and explicitly shows the necessary cycles. This table is basically equivalent to the state table of Chapter 5. Using the state assignments indicated, Figure 6.5.1(d) shows the resulting transition matrix from which the equations for the secondary variables may be derived.

Consider next the merged flow table of Figure 6.5.2(a). The adjacency diagram for this flow table is shown in Figure 6.5.2(b). It should be fairly clear there is no way that,

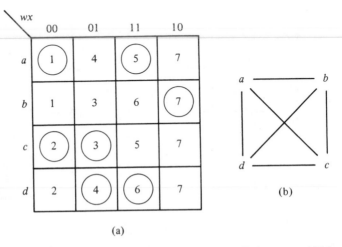

(a)

Figure 6.5.2 A flow table in which cycles cannot be used to eliminate races. (a) Merged flow table. (b) Adjacency diagram.

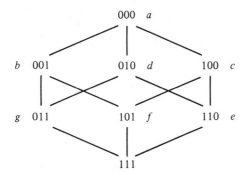

Figure 6.5.3 A map of adjacent assignments.

using only two state variables, we can make row a simultaneously adjacent to rows b, c, and d. However, if we use three state variables instead of two to encode the rows of the flow table, we may be able to accommodate all of the required adjacencies by creating cycles. Figure 6.5.3 shows all possible adjacencies for each of the eight assignments on the three variables. Using this *adjacency map*, we may derive the appropriate cycles to generate a race-free transition matrix. Let us begin with the requirement that row a be adjacent to rows b, c, and d. This is arranged by assigning a to $(0, 0, 0)$, b to $(0, 0, 1)$, c to $(1, 0, 0)$, and d to $(0, 1, 0)$, as shown in the figure. To make row d adjacent to row c, we can create a cycle through assignment $(1, 1, 0)$, which is referred to as row e in the figure. Similarly, rows b and c can be made adjacent via assignment $(1, 0, 1)$, or f in the figure; and assignment $(0, 1, 1)$, or g in the figure, makes a cycle to connect rows b and d. Figure 6.5.4(a) shows the resulting state table, and Figure 6.5.4(b) shows the final

wx	00	01	11	11
a	(a)	d	(a)	b
b	a	f	g	(b)
c	(c)	(c)	a	a
d	e	(d)	(d)	a
e	c			
f		c		
g				d

(a)

y_2, y_1, y_0 \ wx	00	01	11	10
a 000	(000)	010	(000)	001
b 001	000	101	011	(001)
g 011	—	—	010	—
d 010	110	(010)	(010)	000
e 110	100	—	—	—
111	—	—	—	—
f 101	—	100	—	—
c 100	(100)	(100)	000	000

(b)

Figure 6.5.4 Adding states to create cycles and eliminate races. (a) The state table. (b) The resulting transition table.

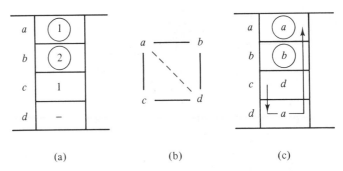

Figure 6.5.5   Using don't cares to create cycles. (a) Flow table fragment. (b) Adjacency diagram. (c) State table fragment.

transition table. Note that all of the entries in the transition table not involved in any of these paths are assigned don't care values (why?).

Before leaving this section, we should note that don't cares in the merged flow table can always be used to create cycles and thus avoid races. Figure 6.5.5 shows a simple illustration of this principle.

6.6 THE ESSENTIAL HAZARD

The essential hazard is quite different from the hazards encountered to this point. This hazard is a function not of the circuit design, but, rather, of the problem statement itself. Thus the essential hazard cannot be removed by simply rearranging the form of the implementing equations: it can be eliminated only by the introduction of delay in the circuit. Fortunately, the essential hazard rarely causes troubles. However, the designer of an asynchronous circuit must be on the lookout for this problem and, when it is encountered, must analyze the resulting circuit to ensure that inherent circuit delays in the design will not cause malfunction. If necessary, the designer will have to introduce physical delay in the design to eliminate problems caused by this hazard.

The essential hazard is fairly easy to identify in the merged flow table. An *essential hazard* occurs whenever a single change in an input variable causes the circuit to end in a different state from the one it would end in if the variable changed three times in succession. Figure 6.6.1(a) shows a flow table having this characteristic, and therefore, an essential hazard. For example if $x = 0$ and we start in state 1, a single change of x to a 1 takes us to state 2. Two more changes of x, from 1 to a 0 and then back to a 1, takes us to state 4 not state 2 and so an essential hazard is indicated. To illustrate how the essential hazard operates, consider the transition table shown in Figure 6.6.1(b). Assume that we start in stable state $(y_1, y_0) = (0, 0)$ and x goes from 0 to 1. Suppose, further, that this change in x is delayed in arriving at secondary y_1 longer than in getting to y_0. Y_0, seeing the change in x, moves over in the table and changes to a 1. This change in y_0 is then seen by secondary y_1, which then begins to change to a 1 also. Suppose, now, that y_1 sees the change in x so that the circuit now moves over to total state $(x, y_1, y_0) = (1, 1, 1)$. Since this is an

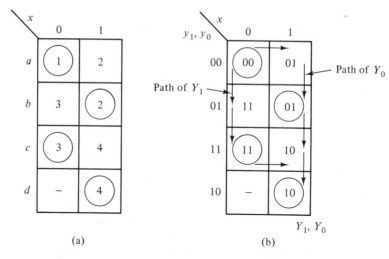

Figure 6.6.1   The essential hazard: (a) flow table with essential hazard; (b) faulty behavior.

unstable state, y_0 changes once again to 0, placing the circuit in the final stable total state (1, 1, 0). Thus, because of the propagation delay differences in the input reaching the two secondary variables, the circuit ends up in state 4 when it should end in state 2.

As mentioned above, any circuit possessing an essential hazard must be examined to determine whether erroneous behavior will result. This analysis basically requires deriving a timing diagram for the conditions associated with the hazard. Problem 6.11 at the end of the chapter will take a closer look at this process.

6.7 SOME DESIGN EXAMPLES

Now that we have studied the process of asynchronous sequential circuit synthesis, let us apply these techniques to the design of some useful circuits. Four designs will be carried out in this section: a switch debouncer, a pulse generator, a double edge-triggered *SR* flip-flop, and the 7474 edge-triggered *D* flip-flop analyzed in Section 6.3.

6.7.1 A Circuit to Debounce Switches

Mechanical switches such as light switches, toggle switches, and the like are usually thought of as devices which, when thrown one way, open a circuit and, when thrown the other, close the circuit. The switch shown in Figure 6.7.1 is called a *double-throw switch*, since it controls two circuits: one when in the up position and one when in the down position. Since switches are mechanical devices, the rocker arm—the portion of the switch that moves from one contact to another—has mass. It also possesses a certain amount of elasticity. Thus, when the rocker arm strikes a contact, it will usually "bounce," perhaps several times, before finally coming to rest. On the other hand, when a contact is broken, if the

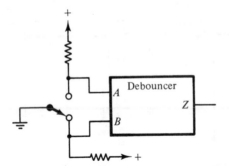

Figure 6.7.1 A switch debouncer circuit.

switch is clean, the break will be "clean." Figure 6.7.2 shows the voltages that might appear at the contacts A and B in Figure 6.7.1 as the switch moves from contact B to contact A and then back again. Note, in this figure, that when this switch is thrown there will be a period of time during which no contact is made. Switches of this type are referred to as *break before make* switches. What we would now like to design is a circuit that will produce a single output change for each single change in the switch position. This circuit is, sometimes, referred to as a *switch debouncer*.

We will begin the design using the example timing diagram shown in Figure 6.7.2. Since this timing diagram represents all of the possible transitions that can occur, any entry not filled in the primitive flow table can be considered a don't care. The primitive flow table derived from this timing diagram and the corresponding merger diagram are shown in Figure 6.7.3.

After merging state 0 with state 1 and state 2 with state 3, we end up with the two-row flow table shown in Figure 6.7.4(a). When we make the assignment on the single state variable y, as indicated, the corresponding excitation table becomes as shown in Figure 6.7.4(b). The resulting excitation equation for the secondary variable Y becomes

$$Y = \bar{B} + yA \qquad (6.7.1)$$

The output Z is clearly equal to the secondary variable y, from these tables. Figure 6.7.5 shows the resulting implementation, which, interestingly, is nothing but the cross-coupled NAND gate SR flip-flop discussed in Chapter 5.

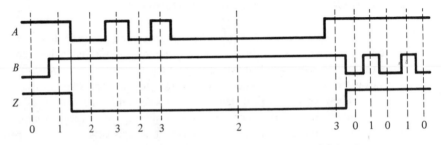

Figure 6.7.2 Typical timing diagram illustrating the switch bounce.

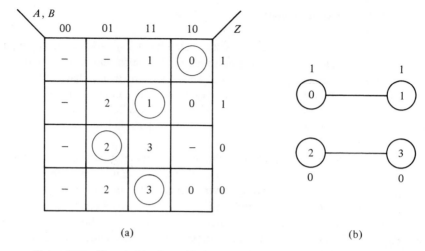

(a)   (b)

Figure 6.7.3   The primitive flow table (a) and merger diagram (b) for the debouncer.

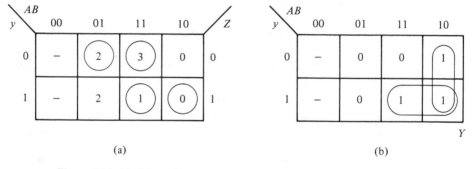

(a)   (b)

Figure 6.7.4   Final flow table (a) and excitation matrix (b) for the switch debouncer.

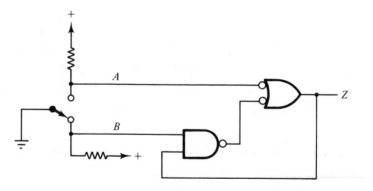

Figure 6.7.5   Final realization of the switch debouncer.

6.7.2 A Pulse Generator

A very useful piece of test equipment is a circuit that can produce an output pulse whenever a switch is pressed. Usually, the pulse required is much shorter than the time during which the switch is pushed. In this example, we will assume that we have a ''debounced'' switch and a regularly occurring string of clock pulses that can be used to create the single pulse required. This circuit is very similar to the gated oscillator described in Section 6.4. The principal difference is that the output is one clock cycle here, rather than several cycles as in the case of the latter.

Figure 6.7.6 shows a typical timing diagram that might occur in the use of this pulse generator. On the basis of this timing diagram we can generate the primitive flow table shown in Figure 6.7.7. There are four entries in this flow table which do not have corresponding occurrences in the timing diagram. These are labeled w, x, y, and z in the figure. A thoughtful analysis of each of these shows how we might logically assign the entries. For example, the entry marked w corresponds to the situation where the circuit is sitting in stable state 2 and G goes low. Since we got to state 2 by starting in state 0 and having G go high, this entry corresponds to a glitch on input G and thus we may ignore it and return to state 0. The remaining three entries corresponding to the situations shown in the figure are assigned values similarly.

Upon merging states 0, 1, and 6, states 4 and 5, and states 7 and 3, in accordance with the merger diagram of Figure 6.7.8(a), we obtain the merged flow table shown in Figure 6.7.8(b). The adjacencies required to avoid race conditions are shown in Figure 6.7.9(a). Using the assignments shown, we may obtain the transition matrix shown in Figure 6.7.9(b). The output matrix, which is shown in Figure 6.7.9(c), may be derived on the basis of the following considerations. First, all outputs corresponding to stable states must be assigned in accordance with the outputs given in the primitive flow table. Second, outputs corresponding to unstable entries in the transition table must be assigned so that no change occurs in an output when the circuit moves between stable states requiring the same output. If, however, the circuit is to move from a stable state having one output value to a stable state having another, the outputs can be assigned as don't cares, since we don't care whether the output changes after arriving at the final state or before. This is the situation with the outputs labeled a, b, and c in Figure 6.7.9(c). The don't care entry in the transition

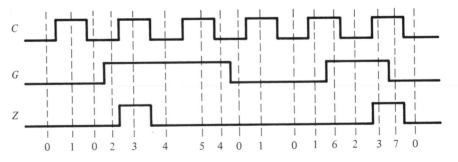

Figure 6.7.6 A typical timing diagram for a pulse generator.

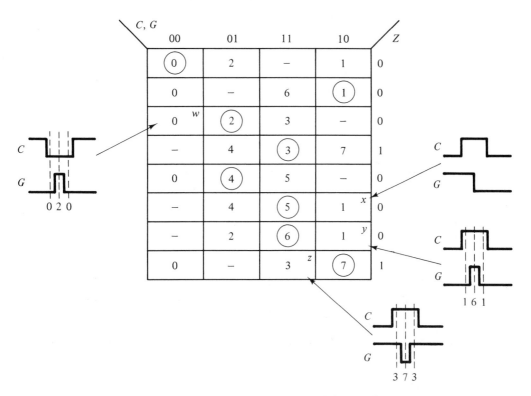

Figure 6.7.7 The primitive flow table for the pulse generator.

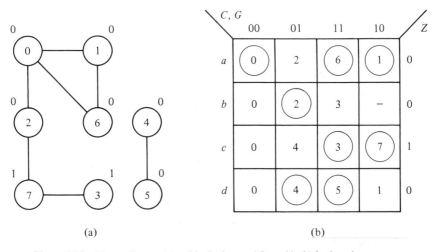

Figure 6.7.8 Merger diagram (a) and the final merged flow table (b) for the pulse generator.

(a)

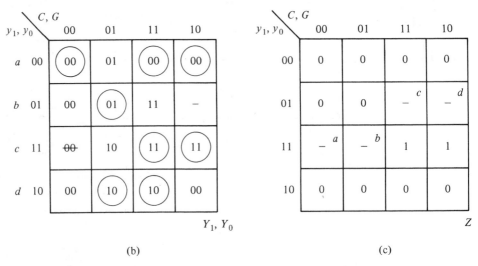

(b)

(c)

Figure 6.7.9 The adjacency diagram (a), transition table (b), and output matrix (c) for the pulse generator.

matrix must, however, be treated a bit more carefully. If the don't care condition can actually be reached from a stable state in the row, then the don't care must be assigned so that the output will not momentarily change. Usually, however, the don't cares cannot actually be reached from a stable state by a single input change (why?). This is the case of entry d in the output matrix. In such a case, the entry may be assigned a don't care value.

The excitation tables generated from the transition table are shown in Figure 6.7.10. By making sure that all adjacent groups of 1s are covered in the excitation tables and using the output matrix shown in Figure 6.7.9(c), we find the resulting implementing equations to be

$$Y_1 = y_1 G + C y_0$$
$$Y_0 = C y_0 + \overline{C} G \overline{y}_1 + \overline{y}_1 y_0 G \qquad (6.7.2)$$
$$Z = C y_0$$

Figure 6.7.11 shows a direct implementation of the equations (6.7.2). Although this circuit will work as shown, we can simplify it somewhat. Note that in the equation for Y_0 we require the generation of the complement of y_1. Although this signal is not directly present in the circuit, we can find a signal that will work. If we assume that the signal at

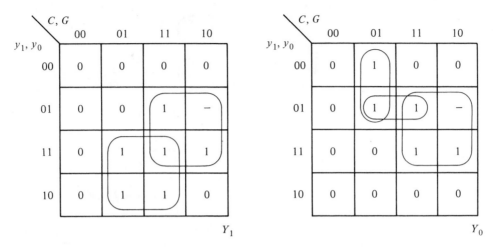

Figure 6.7.10   The excitation matrices for the pulse generator.

point a in Figure 6.7.11 is asserted high, then the function realized is $\overline{G} + \overline{y}_1$. If now we connect point a to point b, removing, of course, the level shifter c, the resulting equation for Y_0 becomes

$$Y_0 = Cy_0 + \overline{C}G(\overline{G} + \overline{y}_1) + y_0G(\overline{G} + \overline{y}_1) \tag{6.7.3}$$

which is logically equivalent to the corresponding equation given in (6.7.2). We can do this, however, only if no hazards are introduced by this factoring. In this case, there are none. The verification of this fact will be explored in Problems 6.8 and 6.9 at the end of the chapter. The resulting realization is shown in Figure 6.7.12.

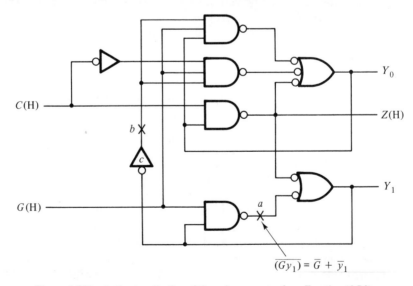

Figure 6.7.11   A direct realization of the pulse generator from Equation (6.7.2).

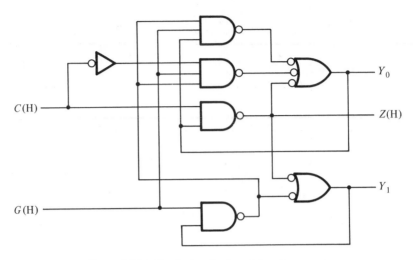

Figure 6.7.12   The final realization of the pulse generator.

6.7.3 A Double Edge-Triggered SR Flip-Flop

The circuit we will design next is one that can be very useful in certain applications but is not currently available as an integrated circuit (IC). This circuit is basically a normal *SR* flip-flop except that the output is set if a low-to-high transition occurs on the set input and is cleared, or reset, if a low-to-high change occurs on the reset input. The output of this flip-flop is unaffected by the input at any other time. From this verbal statement, the primitive flow table can be derived; it is shown in Figure 6.7.13(a). The corresponding merger diagram is shown in Figure 6.7.13(b). Merging the pairs of states 0 and 3, 1 and 2, 5 and 7, and 4 and 6 will produce the merged flow table given in Figure 6.7.13(c).

In order to avoid races, we may assign states as indicated in the adjacency diagram shown in Figure 6.7.14(a) to produce the transition matrix shown in Figure 6.7.14(b). (Note that the rows of the transition table are not in the same order as in the merged flow table.) This transition matrix leads to the following equations for the secondary variables:

$$Y_1 = y_1\bar{y}_0 + \bar{y}_0S + y_1\bar{R}$$
$$Y_0 = y_1\bar{R} + y_0S + y_0y_1 \qquad (6.7.4)$$

Note here that cycles were created, as indicated, in columns $(S, R) = (0, 1)$ and $(1, 0)$ to accommodate the transitions from state 2 to state 3 and state 0 to state 1 without creating a critical race.

The output matrix for this circuit is shown in Figure 6.7.15. As in the last example, all stable states are assigned outputs corresponding to those required in the primitive flow table. Unstable states that are encountered in moving between two stable states with the same output are assigned the common output value. The entries in the table shown in Figure 6.7.15 marked c and d become don't cares, because they are involved in a transition from a stable state with one output value to another stable state with a different output.

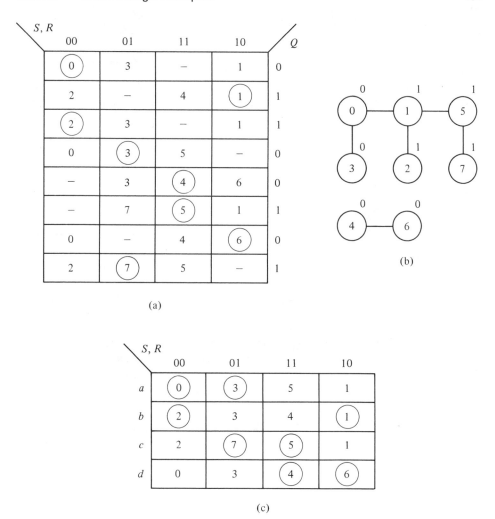

(a)

(b)

(c)

Figure 6.7.13 Derivation of the merged flow table for the dual edge-triggered *SR* flip-flop. (a) Primitive flow table. (b) Merger diagram. (c) Merged flow table.

The entries marked *a* and *b* in the figure are a bit different. These entries are involved in the cycles shown in the transition matrix, Figure 6.7.14(b). We can make the output associated with the first step in each of these cycles a don't care. However, the second step *must* be associated with the final output desired (why?). We thus end with the output matrix shown, from which the equation for the output becomes

$$Z = y_1 \qquad\qquad (6.7.5)$$

a not too surprising result, considering the way in which the flow table was merged.

Implementing Equations (6.7.4) and (6.7.5) directly results in the circuit shown in Figure 6.7.16. As in the last example, note that if the signal at point *a* in the circuit is

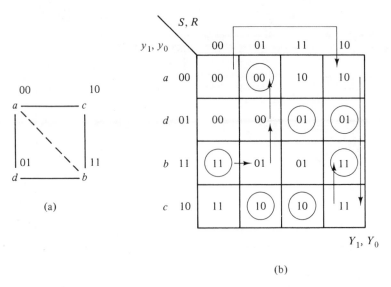

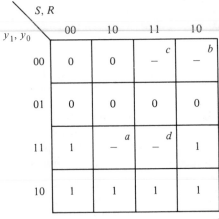

Figure 6.7.14 The resulting transition matrix for the special *SR* flip-flop. (a) Adjacency diagram. (b) Transition table.

assumed to be asserted high, then the logical function implemented at this point is $\bar{S} + \bar{y}_0$. Further, note that if the signal at point b is assumed to be asserted high, then the logical function implemented at this point is $\bar{y}_0 + \bar{y}_1$. Now if we connect point a to point c and point b to point d, removing, once again, the level shifter e, we will end up with a circuit realizing the following equation for y_1:

$$Y_1 = y_1\bar{R} + y_1(\bar{y}_1 + \bar{y}_0) + S(\bar{S} + \bar{y}_0) \tag{6.7.6}$$

Q **Figure 6.7.15** The output matrix for the *SR* flip-f

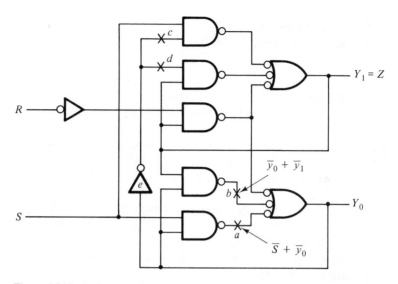

Figure 6.7.16 A direct implementation of the *SR* flip-flop defined by Equations (6.7.4) and (6.7.5).

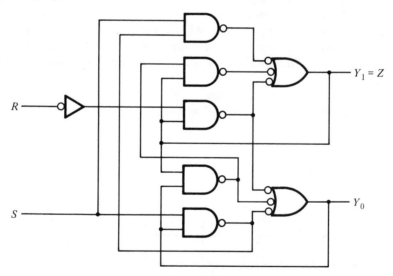

Figure 6.7.17 The final implementation of the dual edge-triggered *SR* flip-flop.

which is logically equivalent to the equation given for Y_1 in equation pair (6.7.4). Before implementing the function in this form, we must, of course, verify that no hazards are created by this factoring. The final realization for this dual edge-triggered *SR* flip-flop is shown in Figure 6.7.17.

6.7.4 Design of the 7474
Edge-Triggered D Flip-Flop

In Section 6.3, we analyzed the 7474 and demonstrated that its behavior is as defined by the manufacturers. Now that we have investigated the process of asynchronous circuit synthesis, it might be useful to design the 7474 and see if our design matches that of the actual device. We will begin this design, as usual, by deriving the primitive flow table. As was done in previous examples, we derive the primitive flow table by accounting for all possible relative changes in the inputs C and D. Figure 6.7.18 shows the completed flow table based on identifying these changes. For example, if we start in state 1, which has a corresponding output of 0, and the clock, C, goes high, the output should stay 0, since input D is 0. This corresponds to moving to state 2 in the figure. If, however, D goes to a 1 before the clock changes, we will go to state 3. In this state, the circuit will be waiting for a 0-to-1 change on C, which will cause the output to take on the value of D by going to state 5. By continuing in this way, we account for all of the possible transitions in the primitive flow table.

We next must reduce the primitive flow table by merging appropriate states. Figure 6.7.19(a) shows the merger diagram corresponding to the primitive flow table of Figure 6.7.18. From this diagram, we can see that states 1, 2, and 4 and states 5, 6, and 8 can each be merged into a single row. We will leave states 3 and 7 as single rows in the flow table. Figure 6.7.19(b) shows the resulting merged flow table. Note that since the rows that were merged had the same outputs, the output becomes associated with rows of the merged table.

The adjacency diagram shown in Figure 6.7.20(a) shows that no problem exists with assigning states to avoid races. The assignment shown in the figure is one of several

C, D				
00	01	11	10	Z
①	3	–	2	0
1	–	4	②	0
1	③	5	–	0
–	3	④	2	0
–	8	⑤	6	1
7	–	5	⑥	1
⑦	8	–	2	1
7	⑧	5	–	1

Figure 6.7.18 Primitive flow table for the 7474 edge-triggered D flip-flop.

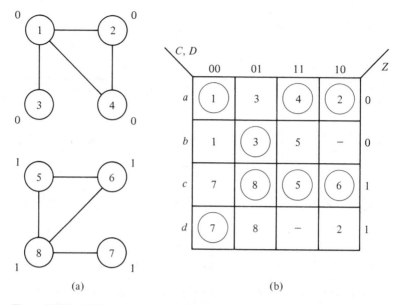

Figure 6.7.19   (a) The merger diagram and (b) the derived merged flow table for the 7474.

possibilities that will work. On the basis of this assignment, we may derive the transition table shown in Figure 6.7.20(b).

If we now look back at the transition table derived during the analysis of the 7474 and shown in Figure 6.3.3, we will observe that our design to this point requires only two state variables, whereas the actual device required three state variables. As we shall see

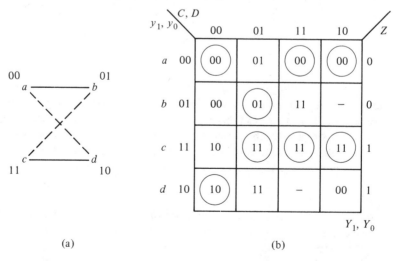

Figure 6.7.20   The transition table for the 7474. (a) Adjacency diagram. (b) The transition matrix.

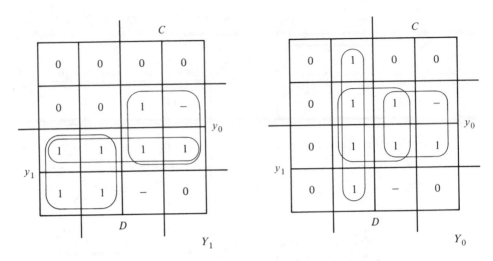

Figure 6.7.21   The excitation matrices for the 7474 D flip-flop.

shortly, the addition of the extra state variable is a product of the desire to reduce the total number of gates in the implementation to a minimum. So, for the moment, let us proceed with the analysis based on the four-state transition table given in Figure 6.7.20. Figure 6.7.21 shows the excitation tables for the two state variables y_1 and y_0, from which the corresponding equations become

$$Y_1 = y_1 y_0 + y_1 \overline{C} + y_0 C \tag{6.7.7}$$
$$= y_0 C + y_1 (y_0 + \overline{C})$$

$$Y_0 = \overline{C} D + y_0 D + y_0 C \tag{6.7.8}$$
$$= y_0 C + D(y_0 + \overline{C})$$

Figure 6.7.22 shows the resulting implementation, in which a total of 7 gates are used to realize Equations (6.7.7) and (6.7.8). Note the marked similarity between this circuit and the one shown in Figure 6.3.2(b).

A level shifter is required in the implementation shown in Figure 6.7.22 to match the output of y_0, which is asserted high, to the input of gate 6, which is asserted low. As we saw in the last two examples, it is quite often possible to obtain the complement of secondary variables at the output of the gate through which the secondary variable is fed back. In this case, this would be the output of gate 4. If we use this output, as shown in the circuit of Figure 6.7.23, we will end up with the following equation for the output Y_1:

$$Y_1 = C y_0 + y_1 (C y_0 + \overline{C}) \tag{6.7.9}$$

which is logically equivalent to Equation (6.7.7) for Y_1. This change, however, produces a static hazard in the circuit in going from state 8 to state 5 or vice versa.[3] Our task now is to remove this static hazard.

[3] This can be verified by the reader by plotting Equation (6.7.9) and observing that two adjacent groups of 1s are no longer connected.

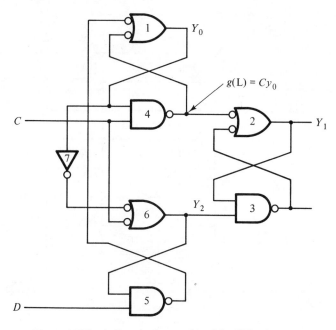

Figure 6.7.22   A direct implementation of the 7474 using 7 gates.

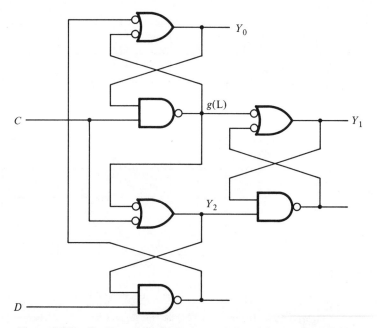

Figure 6.7.23   Circuit resulting from using g to generate the complement of Y_0.

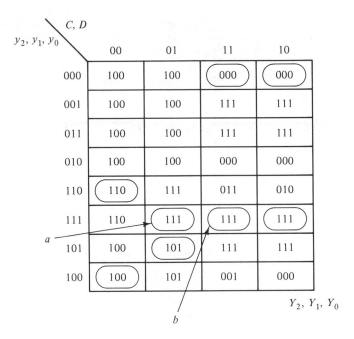

(a) Transition table for the circuit
of figure 6.7.23

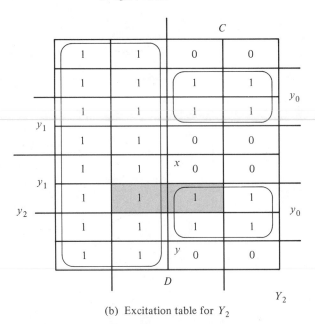

(b) Excitation table for Y_2

Figure 6.7.24   The excitation table showing the presence of a static hazard. (a) Transition table for the circuit of Figure 6.7.23. (b) Excitation table for Y_2.

Before we proceed much further with this example, let us introduce a new secondary variable, Y_2, in the circuit of Figure 6.7.23. This will make the equations simpler and thus, we should hope, the identification of a point in the circuit that may be used to remove the static hazard easier. Doing this produces the equations

$$Y_0 = y_0C + Dy_2 \tag{6.7.10}$$

$$Y_1 = y_0C + y_1y_2 \tag{6.7.11}$$

$$Y_2 = y_0C + \overline{C} \tag{6.7.12}$$

The static hazard clearly shows up in Equation (6.7.12) in the appearance of both the C and $\overline{C}$ terms. The equations for Y_0 and Y_1, on the other hand, do not contain any hazards. Thus, to make the circuit work properly, we must eliminate the hazard in the equation for Y_2. To do this, let us plot these equations to produce a transition table so that we can determine which states are stable, and then we will take a closer look at a plot of Y_2. These plots are shown in Figure 6.7.24(a) and (b), respectively.

The static hazard clearly shows up in the plot of Y_2 shown in Figure 6.2.24(b). As usual, to eliminate this problem, we must connect the two disjoint groups of 1s by placing an additional term into the equation for Y_2. An obvious term to add would be Dy_0. However, this is not really necessary. The static hazard occurs when the circuit moves from one stable state to another in the same row. In this case, the only place where such an event occurs between the two disconnected blocks of 1s is in moving from a to b in Figure 6.7.24(a), corresponding to moving from total state $(C, D, y_2, y_1, y_0) = (0, 1, 1, 1, 1)$ to $(1, 1, 1, 1, 1)$. This is shown by the shaded area in Figure 6.7.24(b). Thus, to eliminate the problem, we need only connect the 1s in the shaded area! A term that will do this and is readily available in the circuit is the output of gate 5, which, if assumed to be asserted low, is Dy_2. If we add this term to the equation for Y_2, two things will happen. First, the static hazard, and its corresponding glitch, will vanish. Second, the entries in the transition table marked x and y, corresponding to total states $(C, D, y_2, y_1, y_0) = (1, 1, 1, 1, 0)$ and $(1, 1, 1, 0, 0)$, will change from $(Y_2, Y_1, Y_0) = (0, 1, 1)$ and $(0, 0, 1)$ to $(1, 1, 1)$ and $(1, 0, 1)$, respectively. Since neither of these states is stable, nor can they be reached by stable states in their respective rows, nor are they involved in any cycles in their column needed to avoid races, this change makes no difference. Thus, we may make this connection and so end with the circuit shown in Figure 6.7.25, which is exactly equivalent to the circuit given for the 7474 by the manufacturer.

6.8 A FINAL COMMENT

The four examples given in Section 6.7 went from fairly easy to quite convoluted. The last example, in particular, illustrates the fact that designing asynchronous sequential circuits, especially if they are to be minimized, requires a great deal of experience and careful analysis. This, along with the existence of the essential hazard and other, more complex hazards that may occur when we allow more than one variable to change at a time, makes designing large circuits of this type very difficult, but not impossible.

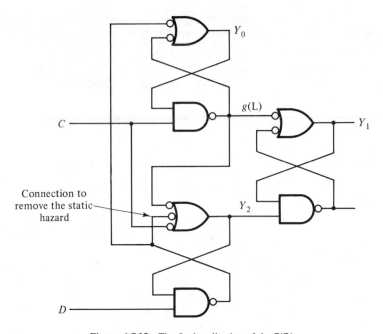

Figure 6.7.25   The final realization of the 7474.

6.9 AN ANNOTATED BIBLIOGRAPHY

A very readable introduction to the ideas of asynchronous sequential circuit analysis and design can be found in the classic text by Maley and Earle. Another classic work, that by McCluskey, gives a slightly different view of the material presented in this chapter and also gives a good presentation of the various hazards encountered in fundamental-mode circuits. (In fact, McCluskey seems to be the originator of the term "fundamental mode.")

MALEY, G. A., and J. EARL, *The Logic Design of Transistor Digital Computers*, Prentice-Hall, Englewood Cliffs, N.J., 1963.

McCLUSKEY, E. J., *Introduction to the Theory of Switching Circuits*, McGraw-Hill, New York, 1965.

Several other, more recent texts also discuss this topic. Among these are the texts by Kohavi, Friedman, and Hill and Peterson.

FRIEDMAN, A. D., *Fundamentals of Logic Design and Switching Theory*, Computer Science Press, Inc., Rockville, Md., 1986.

HILL, F. J., and G. R. PETERSON, *Introduction to Switching Theory and Logical Design*, 3rd ed., Wiley, New York, 1981.

KOHAVI, Z., *Switching and Finite Automata Theory*, 2nd ed., McGraw-Hill, New York, 1978.

Finally, a very fine presentation of asynchronous sequential circuits, with many examples using modern integrated circuit devices, can be found in the book by Fletcher. This presentation contains an analysis of an integrated circuit, the 74120, that was designed to function as either a gated oscillator, as described in Section 6.4, or a pulse generator, as designed in Section 6.7.2.

FLETCHER, W. I., *An Engineering Approach to Digital Design*, Prentice-Hall, Englewood Cliffs, N.J., 1980.

6.10 PROBLEMS

6.1. Find the output Z for the circuit whose flow table and output matrix are given in Figure P6.1(a) if the input signals appear as shown in Figure P6.1(b).

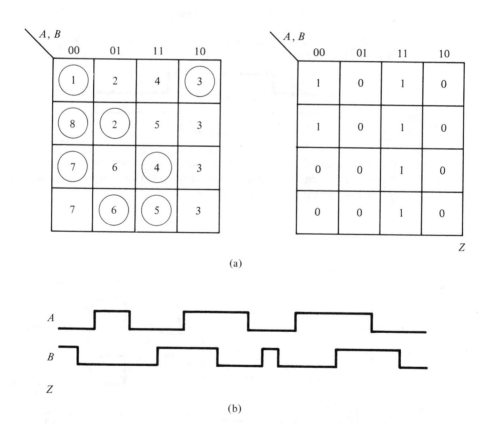

(a)

(b)

Figure P6.1

6.2. Find a race-free assignment for the secondary variables in the flow tables of Figure P6.2.

(1)	(3)	5	7
1	4	(5)	8
1	(4)	(6)	(8)
(2)	4	5	(7)

(a)

1	(3)	(5)	7
(1)	4	(6)	8
2	3	6	(7̇)
(2)	(4)	5	(8)

(b)

Figure P6.2

6.3. Show that the two outputs of the type 7474 flip-flop, $Q(H)$ and $Q(L)$, have the timing relationship shown in Figure P6.3.

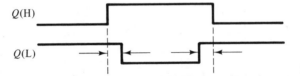

$Q(H)$

$Q(L)$

Figure P6.3

6.4. The circuit shown in Figure P6.4 is claimed by the designer to be free of all glitches and other timing problems that might cause the circuit not to function properly. Analyze this circuit and explain why the designer is far off base.

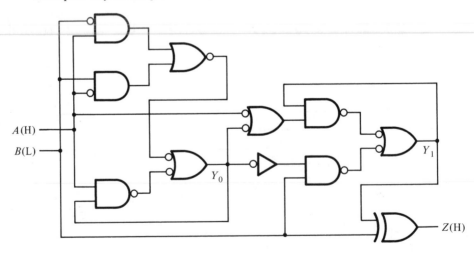

$A(H)$

$B(L)$

Y_0

Y_1

$Z(H)$

Figure P6.4

6.5. Construct a transition table for the circuit shown in Figure P6.5.

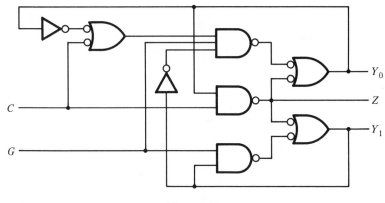

Figure P6.5

6.6. Construct a transition table for the circuit shown in Figure P6.6. Under what conditions will this circuit oscillate? Assuming all gate delays are equal to time interval t, what will be the frequency of oscillation?

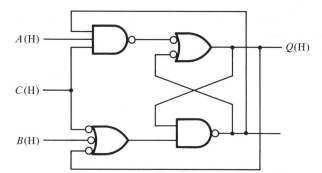

Figure P6.6

6.7. The circuit given in Problem 6.6 operates like a JK flip-flop if a pulse of the correct duration appears on input C. What is the maximum length of this pulse?

6.8. Prove that no static or dynamic hazard is introduced by the use of the term $\overline{G} + \overline{y}_1$ in Equation (6.7.3). (*Hint*: Show that all groups of adjacent 1s are connected.)

6.9. Verify that the implementation of the pulse generator given in Figure 6.7.12, which was derived from Equations (6.7.2) and (6.7.3), is free of hazards and races.

6.10. Consider the merged flow table shown in Figure P6.10. This flow table exhibits an essential hazard. Complete the design and determine, by analyzing the timing of the circuit, whether or not this hazard can cause problems.

(1)	2
3	(2)
(3)	4
1	(4)

Figure P6.10

6.11. Design an edge-triggered T flip-flop as defined by the truth table given in Figure P6.11.

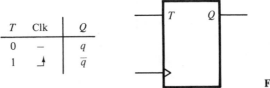

T	Clk	Q
0	$-$	q
1	⌐	$\bar{q}$

Figure P6.11

6.12. Add asynchronous asserted low "set" and "clear" inputs to the edge-triggered T flip-flop of Problem 6.11.

6.13. Does your design of the T flip-flop contain any hazards? If so, show where you might place delays to eliminate them.

6.14. Assuming that the required behavior of the circuit given in Figure P6.4 is given by the transition table shown in Figure P6.14, complete the design of the circuit so as to eliminate all races and hazards.

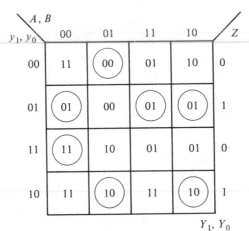

y_1, y_0 \ A, B	00	01	11	10	z
00	11	(00)	01	10	0
01	(01)	00	(01)	(01)	1
11	(11)	10	01	01	0
10	11	(10)	11	(10)	1

Y_1, Y_0 **Figure P6.14**

6.15. Using NOR gates and level shifters only, design a simple SR flip-flop as defined by the truth table of Figure 5.2.3.

6.16. The Motorola 6800 microprocessor requires a two-phase nonoverlapping clock as shown in Figure P6.16(a). Show that the circuit of Figure P6.16(b) produces the required outputs when the input, C, is a regularly occurring clock.

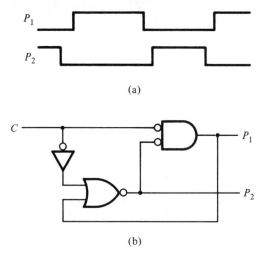

(a)

(b)   **Figure P6.16**

6.17. If the NORs shown in Problem 6.16 are replaced by NAND gates, will the same nonoverlapping two-phase clock be generated? What is the difference in outputs between these two circuits?

6.18. Design a circuit having two inputs, C and G, and two outputs, R and F, in which C is assumed to be a regularly occurring clock and G is a pulse that lasts for at least one complete clock cycle. A pulse equal to a clock pulse is to appear on R when G goes high, and a pulse is to appear on F when G goes low. Figure P6.18 shows a typical timing diagram for this circuit.

6.19. Devise a flip-flop symbol, similar to those given in Chapter 5, for the double edge-triggered SR flip-flop of Figure 6.7.19.

6.20. Add asynchronous set and clear lines to the double edge-triggered flip-flop given in Figure 6.7.17.

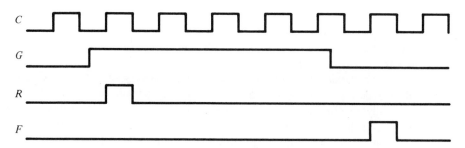

Figure P6.18

Pulse-Mode or Multiply Clocked Sequential Circuits

7

7.1 INTRODUCTION

There are numerous problems that the engineer may encounter in which inputs occur as "pulses" and in which there is no naturally occurring clock that can be synchronized with these pulses to produce the type of clocked sequential circuit discussed in Chapter 5. A couple of examples come to mind immediately: a vending machine in which coins dropped in the slot produce pulses that control the selection and delivery of a canned soft drink or a candy bar; or a demand-access highway intersection in which the arrival of vehicles generates randomly occurring signals to control the sequencing of a traffic light. These examples illustrate a type of sequential circuit in which there is more than one clock signal that can control the output. We will refer to circuits of this type, those having one or more inputs that are "pulses," as *pulse-mode sequential circuits*, or, perhaps better, *multiply clocked sequential circuits*.

One classical form of a pulse-mode circuit is very similar to the clocked sequential circuits described in Chapter 5 except that the state flip-flops are not edge-triggered. Typically, the flip-flops in the feedback paths are the simple *SR* flip-flops shown in Figure 7.1.1. A model for a pulse-mode circuit would then appear as shown in Figure 7.1.2, where the inputs **P** are the "pulsed" inputs and the inputs **X** are "level" inputs. In order for the circuit to operate properly, the flip-flop excitation equations take on the form

$$S_i = P_1 f_{i1}(\mathbf{X}, \mathbf{q}) + \cdots + P_n f_{in}(\mathbf{X}, \mathbf{q}) \tag{7.1.1}$$

$$R_i = P_1 g_{i1}(\mathbf{X}, \mathbf{q}) + \cdots + P_n g_{in}(\mathbf{X}, \mathbf{q}) \tag{7.1.2}$$

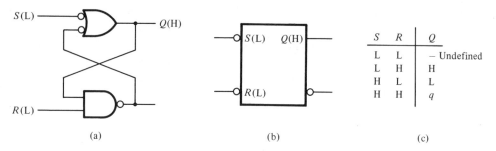

Figure 7.1.1   (a) The siimple cross-coupled NAND gate *SR* flip-flop. (b) *SR* flip-flop symbol. (c) Physical truth table.

for all flip-flops i. A *pulse* input P_j, as used in this model, is then defined as a signal having two characteristics:

1. The pulse is asserted long enough to cause the flip-flop outputs to change.
2. It is shorter than the minimum delay through the combinational logic, so that it is gone, or negated, before the inputs to the flip-flops can change once more.

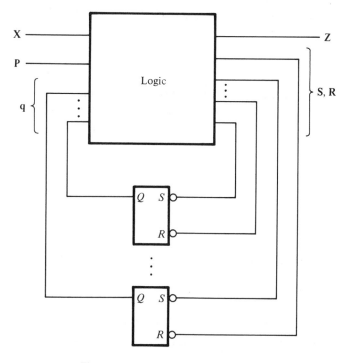

Figure 7.1.2   A general pulse-mode model.

If the first characteristic is not met, the flip-flops may fail to change value when they are supposed to, if they indeed change at all. If the second characteristic is not met, changes in the flip-flop outputs may cause further changes in the flip-flop inputs, which could result in a transition to an incorrect final state. This could happen if the input pulse is still present on the flip-flop inputs when the logic output changes occur. To prevent either case, the width of the input pulses must be very carefully controlled. Such control, especially in light of today's high-speed technology, is chancy, at best.

Because of the many problems arising naturally in engineering that require outputs of a system to be controlled by momentary changes in inputs, the basic concept of a sequential circuit having ''pulsed'' inputs is still important. However, because there are impracticalities in precisely controlling delays, as required by the model suggested above, the view taken in this chapter regarding what constitutes a pulse and how such momentarily occurring signals affect the outputs will be rather different. Our goal here will be to describe a design methodology for multiply clocked sequential circuits that is both physically implementable and reliable.

7.2 THE BASIC PULSE-MODE CIRCUIT MODEL

As suggested above, there is really only one way to make a pulse-mode system function reliably. That is to insert delay in the outputs of the state flip-flops so that input pulses can ''vanish'' long before the changes in their outputs appear at the state inputs to the feedback circuit and can still be long enough to affect the state flip-flops. There generally are two ways to insert such delay. One approach is to use linear circuit elements such as capacitors and resistors to produce the necessary delays. Problems 7.1 through 7.4, at the end of the chapter, will examine this situation more closely. Although this approach can work, it is made unnecessary by a second approach that can be used: delay can be inserted through the use of a second rank of edge-triggered flip-flops that can prevent the state flip-flop output changes from appearing at the logic inputs until the point in time at which the pulsed inputs are negated. This will be the approach taken here.

Figure 7.2.1 shows one form of the general model that we will use for defining a *pulse-mode circuit*. In this model, there are three sets of *inputs*:

1. Level inputs **X**
2. Pulsed inputs **P**
3. Current-state inputs **q**

Combinational logic is used to generate two sets of *outputs*:

1. Normal circuit outputs **Z,** which may be pulsed or level
2. Flip-flop inputs, **S** and **R,** in this model[1]

---

[1] Both the master-rank and the slave-rank flip-flops can, of course, be of any type, as we shall illustrate later.

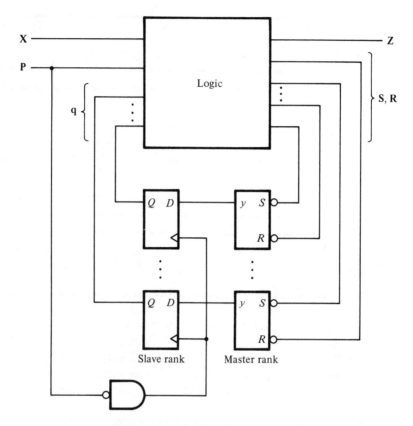

Figure 7.2.1   A multiply clocked sequential circuit model.

The *state flip-flops* are organized into two ranks:

1. A master rank
2. A slave rank

The set and reset inputs to the master rank are controlled by the **S** and **R** outputs of the combinational logic, which are pulsed, as indicated in Equations (7.1.1) and (7.1.2); and the clock inputs of the slave rank are controlled by the point in time at which all pulse inputs are negated. (This model assumes asserted high signals on all inputs, although this is clearly not necessary.)

In this model, a signal will be considered a pulse if it meets two criteria: purpose and form. Specifically, a *pulse* will be defined as a signal

1. Whose purpose is to control the *time* at which a state change is to occur, and
2. Which meets the following physical criteria:

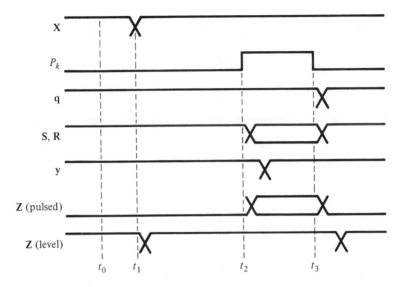

Figure 7.2.2   Timing in the master-slave multiply clocked sequential circuit.

(*a*) A pulse must be long enough to allow the master rank of state flip-flops to change state.

(*b*) No more than one pulse input is to be asserted at any given instant of time.

(*c*) The time between occurrences of input pulses must be long enough that all changes in level inputs will have propagated to the circuit outputs.

On the basis of these specifications, the combinational logic outputs will generally appear as follows:

$$Z_i = h_i(\mathbf{X}, \mathbf{P}, \mathbf{q}) \tag{7.2.1}$$

$$S_j = P_1 f_{j1}(\mathbf{X}, \mathbf{q}) + \cdots + P_n f_{jn}(\mathbf{X}, \mathbf{q}) \tag{7.2.2}$$

$$R_j = P_1 g_{j1}(\mathbf{X}, \mathbf{q}) + \cdots + P_n g_{jn}(\mathbf{X}, \mathbf{q}) \tag{7.2.3}$$

It should be noted that outputs Z_i may be levels or pulses, depending on the function that the output is to perform.[2] It should also be observed that since no two pulses are asserted at the same time, any output that is to appear as a pulse must have a functional form similar to that for the flip-flop S and R inputs as shown in Equations (7.2.2) and (7.2.3).

Before we look at some examples and at the design procedure for circuits having multiple clocks, let us consider some timing aspects of the pulse-mode model shown in Figure 7.2.1. Specifically, let us look at the general form for the timing shown in Figure 7.2.2. Assume that at time t_0 all inputs have been stable for quite some time, with the pulsed inputs being negated. Now suppose that at time t_1 the level inputs change. This will cause the functions f and g in Equations (7.2.2) and (7.2.3) to take on new values,

---

[2] If Z_i, in Equation (7.2.1), is independent of any pulsed inputs, then it will be a level signal.

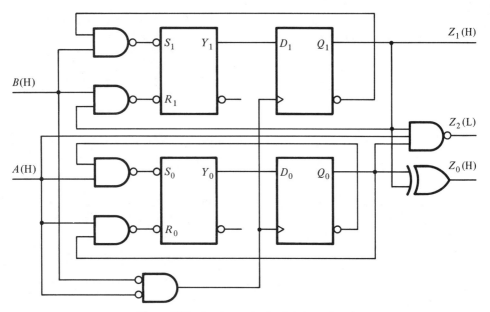

Figure 7.3.1   A pulse-mode circuit to be analyzed.

but since all of the pulsed inputs are 0, all of the S and R flip-flop inputs will be 0 also (a high voltage, in this case). Suppose, now, that at time t_2 pulse input P_k is asserted. (Remember that all other pulsed inputs are still negated, because of the definition of a pulse cited above.) This will also cause the various flip-flop S and R inputs to change, so that the outputs of the master rank, the y_j in the circuit, will change. However, this change cannot be passed on to the input of the combinational circuit, on account of the presence of the slave rank. Therefore, no further change in the outputs or flip-flop inputs can occur. If now the pulse input P_k is negated at time t_3, the master flip-flop outputs will be passed to the outputs of the slave rank, **q,** causing the system state to change and thus setting everything up for the next occurrence of an input pulse.

7.3 AN ANALYSIS EXAMPLE

Let us now examine a typical pulse-mode circuit and see how we can determine its behavior. Figure 7.3.1 shows a circuit having two pulsed inputs, A and B, two level outputs, Z_1 and Z_0,[3] and one pulsed output, Z_2. This particular example has a master rank of simple SR flip-flops and a slave rank of edge-triggered D flip-flops, as shown in the model of Figure 7.2.1. As usual, the objective of analyzing the circuit is to determine how the outputs change with changes on the inputs—pulses, in this case. To begin the analysis of the circuit of

---

[3] These outputs are level because they are dependent on the state variables only and not on any of the pulse inputs.

Figure 7.3.1, let us, as usual, write the flip-flop input equations and the circuit output equations in terms of the circuit inputs and the state variables:

$$S_1 = B\bar{q}_1$$
$$R_1 = Bq_1$$
$$S_0 = A\bar{q}_0 \qquad (7.3.1)$$
$$R_0 = Aq_0$$

$$Z_1 = q_1$$
$$Z_0 = q_0\bar{q}_1 + \bar{q}_0q_1$$
$$Z_2 = Aq_1q_0 \qquad (7.3.2)$$

The equations for the set and reset inputs are plotted in Figure 7.3.2(a). Note here that only two columns, one for each of the pulsed inputs, are shown. We can do this

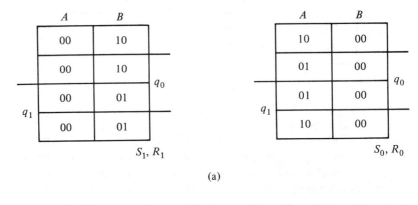

(a)

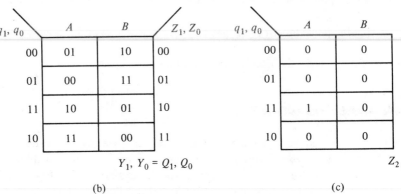

(b)   (c)

Figure 7.3.2 Derivation of the state table for the circuit of Figure 7.3.1. (a) The flip-flop excitation tables. (b) The encoded state table with the level outputs. (c) The pulsed output matrix.

because, by the definition of a pulse given above, no two pulses can be asserted simultaneously and, therefore, there is no need to show all of the possible combinations of pulsed inputs. Using the definition of the simple *SR* flip-flop given in Figure 7.1.1, we obtain the plot for the master-rank flip-flop outputs Y_1 and Y_0 given in Figure 7.3.2(b). Note here that we have also plotted the "level" outputs, Z_1 and Z_0, as being associated with rows in the table. This is done because these outputs are dependent only on the state variables. Since the outputs of the slave rank, Q_1 and Q_0, will be equal to the outputs of the master rank at the time that both input pulses A and B are negated, the table shown in Figure 7.3.2(b) becomes the encoded state table for the pulse-mode circuit. The output table corresponding to the pulsed output Z_2 is given in Figure 7.3.2.(c). Figure 7.3.3 shows a state diagram corresponding to this state table. Since an input pulse is the signal that causes a transition from one state to another, each edge in this figure is labeled with the pulse that causes the transition. To indicate which pulse output is associated with which state transition, the name of the output which occurs on a given state transition will be shown on the transition edge. Thus, the notation A/Z_2 on the edge connecting states 00 and 01 indicates that pulsed input A causes the transition and produces a pulsed output at Z_2. If no output is indicated, then no pulse output is generated on the transition. The value of the level outputs will be shown in the state circles. Thus the state diagram for this example contains elements of both the Moore model, where outputs are associated with the states, and the Mealy model, where outputs are associated with state transitions.

Using the state diagram of Figure 7.3.3, we can now determine how the outputs of the circuit given in Figure 7.3.1 change as pulses occur on the inputs. Suppose we start in state 00. If the input pulses now occur in the sequence A, B, A, B, etc., the outputs will change in the sequence 00, 01, 10, 11, 00, etc., which is basically a normal counting sequence: 0, 1, 2, 3, 0, . . . If, at some point, an A pulse (or a B pulse) occurs twice in succession, the counting sequence will be reversed. For example, the input pulse sequence *ABABBABABAB*, starting in state 00, will produce the output sequence 012303210321. Finally, we note that a pulse will appear on output Z_2 every fourth input pulse only if the circuit is counting in an "up" sequence, that is, 01230123. . . . Thus, Z_2 can be used to indicate whether the circuit is counting up or down.

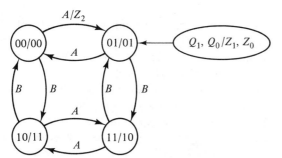

Figure 7.3.3   The derived state diagram for the circuit of Figure 7.3.1.

7.4 THE DESIGN PROBLEM

The synthesis of pulse-mode circuits, as defined here, is not appreciably different from the synthesis of clocked sequential circuits. As we saw in the last section, and, indeed, in the last two chatpers, the first step is to convert the verbal statement of the problem to a state diagram or state table. Once this is done, the rest of the process becomes fairly mechanical. Specifically, for pulse-mode circuits, we must first create a state diagram or state table indicating which inputs and outputs are pulses and which are level signals. We must then derive the master-rank flip-flop input equations along with the output equations. Finally, we implement these equations with the circuit elements available.

Although the process of designing multiply clocked circuits is quite similar to clocked sequential circuit design, there are some variations that will be encountered. In order to illustrate this process more fully, consider the following design problem.

Design Problem

A certain stepping motor has four binary inputs; i.e., each input takes on either a high voltage (12 V) or a low voltage (0 V). If these inputs are changed in a specific sequence, the motor will rotate clockwise 1.8 degrees per step. The reverse sequence will cause the motor to rotate in the counterclockwise direction. Referring to the motor inputs as F_1, F_2, F_3, and F_4, the motor will rotate clockwise one step for each change of these inputs in the sequence $(F_1, F_2, F_3, F_4) = (1010), (1001), (0101), (0110), (1010)$, etc. Reversing this sequence will cause the motor to rotate in the opposite direction. We would like to design a digital circuit that will control this step sequence. The specification for this design requires two pulsed inputs, A and B, and, of course, the four "level" outputs, F_1, F_2, F_3, and F_4. Repetitive pulses on input A are to cause the motor to rotate clockwise, and repetitive pulses on input B are to cause the motor to rotate counterclockwise. The rate at which these pulses occur will determine the rotation rate of the motor. It is, of course, assumed that the input pulses never occur at the same time. It is also assumed that these pulses do not occur at a rate faster than the motor can be stepped.

7.4.1 Setting Up the State Table

To begin the design, we must first develop the state diagram or the state table. In this case, we will begin by creating the state diagram. Figure 7.4.1(a) shows this diagram. Since the outputs in this problem are levels, we reference them to the states, and therefore they appear, as in a Moore model, in the state circles. The edges which connect one state to the next are labeled according to the pulse that causes the state transition. Thus, an edge labeled B means that a pulse on the B input, while the A input is negated, will cause the state transition indicated. Figure 7.4.1(b) shows the corresponding state table.

Encoding the states is generally no different from the case for a clocked sequential circuit. However, as in the case of other sequential circuits, assigning codes to the state variables so that states which are adjacent differ in only one bit position eliminates races and tends to reduce the complexity of the implementing equations. Therefore, the states

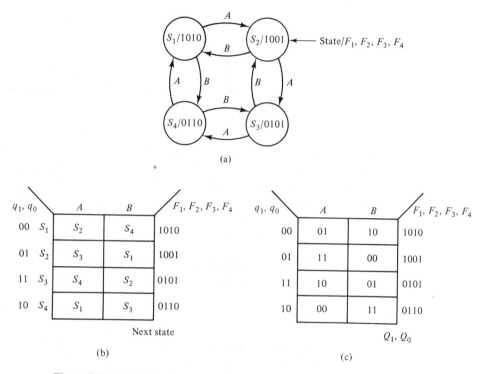

Figure 7.4.1   Derivation of the encoded state table. (a) Stepping motor control state diagram. (b) State table. (c) The encoded state table.

for this motor controller are assigned as shown in Figure 7.4.1(b). The resulting encoded state table is shown in Figure 7.4.1(c). Using either of these tables, (b) or (c), we find the equations for the outputs to be

$$F_1 = \bar{q}_1$$
$$F_2 = q_1$$
$$F_3 = \bar{q}_0 \qquad\qquad (7.4.1)$$
$$F_4 = q_0$$

The reader should try other state encodings to see how the various choices affect the complexity of the output realizations. Obviously, the choice selected here produces output equations which are as simple as we might hope for.

7.4.2 Developing the Master-Rank Flip-Flop Equations

The circuit analyzed in the last section, shown in Figure 7.3.1, had a master rank made up of simple *SR* flip-flops in which the set and reset inputs were controlled to produce the required state transitions. There is, in general, no reason why the master rank of flip-flops

q	Q	S	R
0	0	0	–
0	1	1	0
1	1	–	0
1	0	0	1

Figure 7.4.2   Present-state–next-state table for the simple SR flip-flop of Figure 7.1.1.

could not be made up of some edge-triggered flip-flop such as the D or the JK type. In such a situation, the clock inputs would be controlled specifically by combinations of the pulse and level inputs, while the other flip-flop inputs, D or JK, for example, would be controlled by the level inputs only. To illustrate the use of each, we will design the motor controller using, first, simple SR flip-flops for the master rank, and then edge-triggered D flip-flops, such as the 7474. Problems at the end of the chapter will illustrate the use of other types of edge-triggered flip-flops.

7.4.2.1 Controlling the flip-flop set and reset inputs.

In order to determine the flip-flop input equations, we need to determine what combinations of S and R will cause the flip-flop outputs to change as required by the encoded state table. This was done in Chapter 5 using a present-state–next-state table for the specified flip-flop. Figure 7.4.2 shows this table for the simple SR flip-flop being used here. This is the same table as derived in Chapter 5 and given in Figure 5.2.9(d). The excitation tables for the master-rank state flip-flops can now be derived in exactly the same way as was done in Chapter 5. Specifically, from the current-state table we can determine what the current state is and what the required next state is to be. Using the present-state–next-state table for the SR flip-flop, we can determine the values of (S_1, R_1) and (S_0, R_0) required to produce this transition. For example, we see from the encoded state table that if the circuit is in state $(q_1, q_0) = (0, 0)$ and a pulse occurs on the A input, the circuit is to move to state $(0, 1)$. This requires, by the present-state–next-state table of Figure 7.4.2, that the master-rank SR inputs be $(S_1, R_1) = (0, -)$ and $(S_0, R_0) = (1, 0)$. Figure 7.4.3 shows the excitation tables for the master rank. Using these tables, we can now derive the flip-flop input equations, as follows:

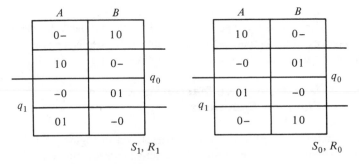

Figure 7.4.3   Master-rank flip-flop excitation tables.

$$S_1 = Aq_0 + B\overline{q}_0$$
$$R_1 = A\overline{q}_0 + Bq_0$$
$$S_0 = A\overline{q}_1 + Bq_1 \qquad (7.4.2)$$
$$R_0 = Aq_1 + B\overline{q}_1$$

The implementation of the motor controller based on Equations (7.4.2) and (7.4.1) and the model shown in Figure 7.2.1 is shown in Figure 7.4.4. This implementation involves a lot of AND-OR logic, which could very easily be created using a PLA device as described in Chapter 4. Problems 7.12 and 7.13, at the end of the chapter, illustrate this point further.

7.4.2.2 Controlling the edge-triggered flip-flop clock input.   As mentioned above, there is no reason why the master rank flip-flops must be of the SR type. We could equally well use an edge-triggered D or JK or T or any other, similar flip-flop. The difference in the design involves only the present-state–next-state tables associated with the chosen flip-flop. In Section 5.2, we derived these tables for the various flip-flops assuming that a single clock was responsible for the change in state. Here, however, there is generally more than one clock that produces state changes. Since the clock inuts of these flip-flops must be controlled by some combination of pulse inputs, the corresponding state

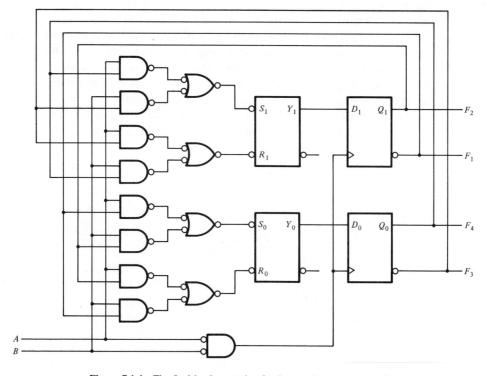

Figure 7.4.4   The final implementation for the stepping motor controller.

q	Q	D	Clk
0	0	0	–
		–	0
0	1	1	1
1	0	0	1
1	1	1	–
		–	0

(a)

q	Q	J	K	Clk
0	0	–	–	0
		0	–	–
0	1	1	–	1
1	0	–	1	1
1	1	–	–	0
		–	0	–

(b)

Figure 7.4.5 Present state/next state tables for the edge-triggered D flip-flop (a) and JK flip-flop (b).

transition tables must be modified to include the flip-flop clock input changes necessary to cause the required state variable changes.

Figure 7.4.5 shows the present-state–next-state tables for the edge-triggerd D and JK flip-flops. Although the clock inputs to these flip-flops are edge-sensitive, their values are shown in these tables as a 0 or a 1. The 1 indicates that the clock is to make an asserted transition, and a 0 indicates it is not. Notice in these tables that some entries show two possible values for the inputs. For example, in the table for the D flip-flop, there are two ways in which a current state of 0 can stay 0. First, if the D input is held at 0, then whether the clock changes or not, the output will not change. Second, if the clock is not changed, the output cannot change and so will remain 0. Similar considerations apply to all of the other table entries shown in Figure 7.4.5(a) and (b).

Let us now implement the master rank of the stepping motor controller using edge-triggered D flip-flops. From the present-state–next-state table for this flip-flop shown in Figure 7.4.5(a) and the encoded state table for the motor controller shown in Figure 7.4.1(c), we can obtain the excitation tables shown in Figure 7.4.6 for the two state flip-flops. There are several places in these tables having two possible values for D and Clk. We must, of course, select one or the other for the final implementation. This can be done by considering the function of each of these inputs. The function of the D input is to take on a stable

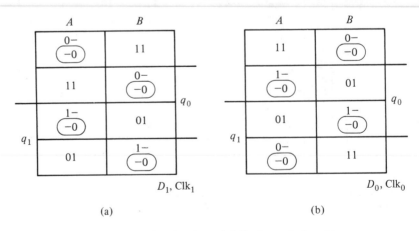

Figure 7.4.6 The edge-triggered D flip-flop excitation tables.

value equivalent to the next output value, while the function of the clock input is to identify the time at which this change is to take place. Thus, the D input must be a signal whose value is stable at the time the clocked input is asserted. In other words, the D inputs *must not* be functions of the pulsed inputs. The Clk inputs, on the other hand, *must* be functions of the pulsed inputs in order to identify when the state transitions are to occur.

In order for the D inputs of the flip-flops to be independent of the pulsed inputs, it is necessary, in any given row of the excitation table, for the value of D to be the same in all of the columns. Thus, for example, in the top row of the excitation table given in Figure 7.4.6(a) (this row corresponds to $(q_1, q_0) = (0, 0)$), we must choose between the two possible entries in the A column so that D_1 is independent of A and B. Since D_1 must be 1 for an input of B, we must choose the circled entry in the A column, since D_1 is a don't care and can thus be made a 1. The remaining circled entries in Figure 7.4.6(a) are selected for the same reason: to make D_1 independent of the pulsed inputs A and B. In exactly the same way, the alternative terms in Figure 7.4.6(b) are selected.

Plots based on these selections are shown in Figure 7.4.7 for each of the flip-flop inputs. The equations derived from these tables are easily seen to be

$$D_1 = \bar{q}_1$$
$$\text{Clk}_1 = A(q_1 \oplus q_0) + B\overline{(q_1 \oplus q_0)} \qquad (7.4.3)$$

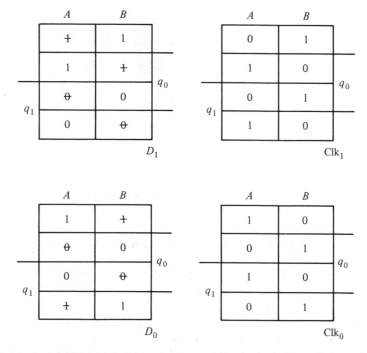

Figure 7.4.7 Individual flip-flop excitation tables for the stepping motor controller implemented using edge triggered D flip-flops.

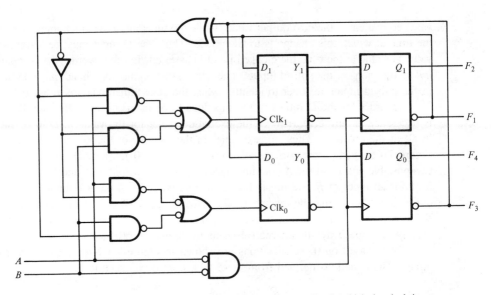

Figure 7.4.8 Implementation of the stepping motor controller in which the clock input of the master rank is controlled.

and

$$D_0 = \bar{q}_0$$
$$\text{Clk}_0 = A\overline{(q_1 \oplus q_0)} + B(q_1 \oplus q_0) \tag{7.4.4}$$

The resulting implementation is shown in Figure 7.4.8.

7.5 SOME FURTHER DESIGN EXAMPLES

The design of multiply clocked or pulse-mode sequential circuits was illustrated by the stepping motor controller designed in Section 7.4. In order to further illustrate this design process and to see how other flip-flop types can be used, two more examples will be given. In the first, we will design a simple combinational lock which requires pushing a set of buttons in a particular sequence to open. In the second example, we will look at a simple vending machine design.

7.5.1 The Design of a Simple Combinational Lock

The problem here is to design a simple combinational lock having two buttons, A and B, by which a person can enter a sequence representing a particular lock "code," and another button, R, used to reset the lock. To open the lock, a person would first push the reset button R and then enter the particular sequence of pushes on the A and B buttons representing the lock's code. If a mistake is made in entering the code, the R button can be pressed and the code sequence reentered. A level output that is used to open the lock is to be generated when the correct input sequence has been entered. The input pulses in this design are to be asserted high, while the level output, Z, is to be asserted low. It will also be assumed that the input pulses, which are derived from the switches A, B, and R, have been "debounced" as described, for example, in Chapter 6. This is necessary so that a single push of a button results in a single input pulse to the circuit.

To begin the design, we must first develop a state diagram. Before we can do this, however, we need to know the code sequence for the lock. In order to keep the problem reasonably simple, let us assume that a three-pulse code is to be used and that for this lock the code sequence is to be ABA. Although such a simple coding is not very practical (there are, after all, only 8 possible such three-pulse codes), it will illustrate the design process very well. Problems 7.17, 7.18, and 7.19, at the end of the chapter, will investigate ways in which such a lock concept might be expanded to make it more practical.

Figure 7.5.1(a) shows the state diagram for this design. This diagram is generated

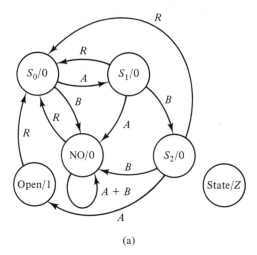

(a)

q_2, q_1, q_0		A	B	R	Z
000	S_0	S_1	NO	S_0	0
001	S_1	NO	S_2	S_0	0
011	S_2	Open	NO	S_0	0
010	NO	NO	NO	S_0	0
110	Open	–	–	S_0	1
111		–	–	–	–
101		–	–	–	–
100		–	–	–	–

(b)

Figure 7.5.1 The state diagram (a) and the state table (b) for the combinational lock.

on the basis of the following considerations. Starting in state S_0, if the correct input sequence is entered, ABA in this case, the system will go first to state S_1, then to state S_2, and finally to the Open state. If an illegal combination occurs at any given time, the system will enter the NO state and stay there, regardless of further pulses on A and B, until the reset button is pushed. Once the system is in the Open state, the output Z will be asserted and the lock will open. The circuit will always return to the initial state S_0 when the reset button R is pushed. Figure 7.5.1(b) shows the corresponding state table. Note here that we have left as don't cares the entries corresponding to the Open state and input pulses on either A or B. This is done because we really don't care where the system goes once the lock has been opened. It is presumed that the operator will press the reset button after closing the lock and before entering the next input sequence.

Our next job in designing this lock is to encode the states. Since there are five states, we will need a total of three state variables for this purpose. We will refer to these as Q_2, Q_1, and Q_0. Assigning these states as shown in Figure 7.5.1(b), we can obtain the assigned-state table shown in Figure 7.5.2. Using the present-state–next-state table for the SR flip-flop as given in Figure 7.4.2, we can derive the excitation tables for the master rank. These are shown in Figure 7.5.3, from which the flip-flop input equations become

$$S_2 = Aq_1q_0$$
$$R_2 = R \tag{7.5.1}$$

$$S_1 = Aq_0 + B$$
$$R_1 = R \tag{7.5.2}$$

$$S_0 = A\bar{q}_1\bar{q}_0$$
$$R_0 = Aq_0 + Bq_1 + R \tag{7.5.3}$$

$$Z = q_2 \tag{7.5.4}$$

The resulting physical implementation is shown in Figure 7.5.4.

q_2, q_1, q_0		A	B	R	Z
S_0	000	001	010	000	0
S_1	001	010	011	000	0
S_2	011	110	010	000	0
NO	010	010	010	000	0
Open	110	–	–	000	1
	111	–	–	–	–
	101	–	–	–	–
	100	–	–	–	–

Q_2, Q_1, Q_0

Figure 7.5.2 The encoded state table for the combinational lock.

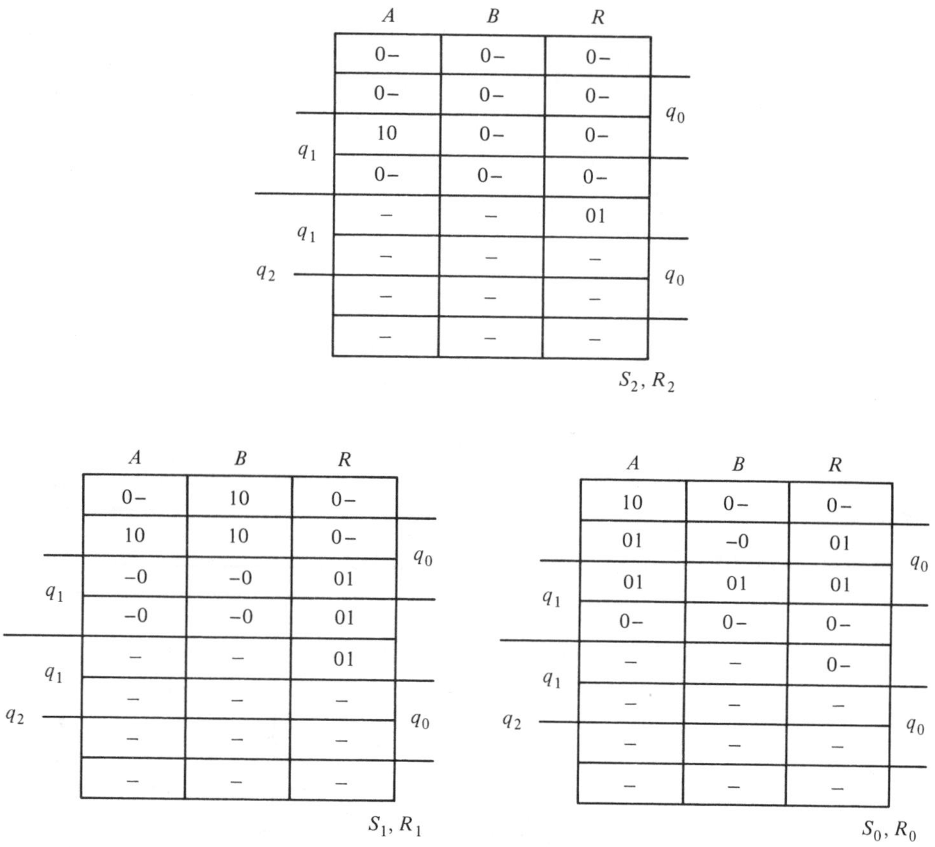

		A	B	R	
		0–	0–	0–	
		0–	0–	0–	q_0
q_1		10	0–	0–	
		0–	0–	0–	
	q_1	–	–	01	
q_2		–	–	–	q_0
		–	–	–	
		–	–	–	

$$S_2, R_2$$

		A	B	R	
		0–	10	0–	
		10	10	0–	q_0
q_1		–0	–0	01	
		–0	–0	01	
	q_1	–	–	01	
q_2		–	–	–	q_0
		–	–	–	
		–	–	–	

$$S_1, R_1$$

		A	B	R	
		10	0–	0–	
		01	–0	01	q_0
q_1		01	01	01	
		0–	0–	0–	
	q_1	–	–	0–	
q_2		–	–	–	q_0
		–	–	–	
		–	–	–	

$$S_0, R_0$$

Figure 7.5.3 Excitation equations for the combinational lock.

7.5.2 A Simple Vending Machine

In this example we will design a controller for a rather primitive vending machine. Although this example is not necessarily realistic by today's vending machine standards, it will help us introduce some additional concepts. (Machines of the type to be described might have been found in gas stations along old U.S. 40 or Route 66 back in the 1940s and 1950s, dispensing cigarettes, soft drinks, chewing gum, or the like.) As shown in Figure 7.5.5, this machine has three pulse inputs, P_{10}, P_X, and R, and three outputs, S, A, and X. The product to be dispensed costs 20 cents, and the machine takes dimes only, although the coin slot is large enough to accommodate any coin. The pulse inputs, therefore, take on the following meaning. P_{10} is a pulse generated when a dime is dropped in the slot. P_X is a pulse which indicates that a coin of some other denomination was inserted in the machine. The pulse R is from the coin return lever, so that you can change your mind and get your money back if you have not yet dropped 20 cents in the machine. It is assumed here that once the 20 cents has been inserted, the product will be delivered. The three outputs are

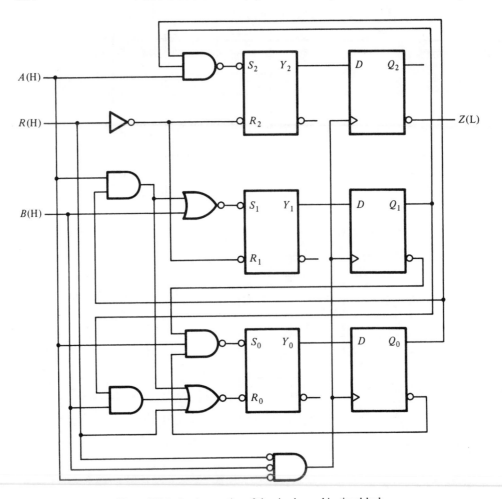

Figure 7.5.4   Implementation of the simple combinational lock.

used to control the following functions. If a coin other than a dime is inserted, it is to be returned to the purchaser immediately. Output pulse X is used for this function. If a dime has been collected and the coin return lever is activated, output pulse A is generated to release the collected dime. Finally, once the 20 cents has been entered, a select pulse, S, is generated to allow the product to be delivered.

Given these specifications, the state diagram shown in Figure 7.5.6(a) can be constructed. In this case, the outputs, since they are pulses, are associated with the state diagram edges and not with the individual states. Notice here that if the coin return lever is pressed before any coins have been inserted, the vending machine generates no output but stays in state S_0. Also note that no output pulses are generated when the first dime is inserted. The machine here needs only to keep track of the fact that the dime has been

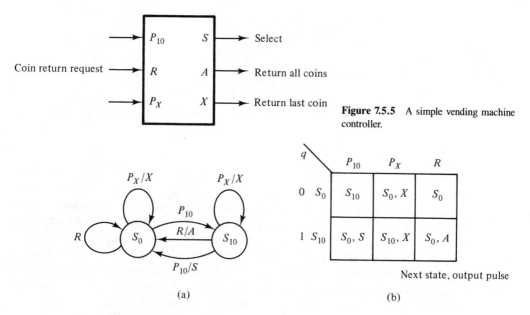

Figure 7.5.5   A simple vending machine controller.

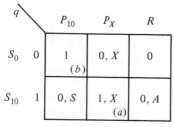

(a)

q	P_{10}	P_X	R
0 S_0	S_{10}	S_0, X	S_0
1 S_{10}	S_0, S	S_{10}, X	S_0, A

Next state, output pulse

(b)

Figure 7.5.6   The state diagram (a) and state table (b) for the simple vending machine.

entered, and this is done by moving to state S_{10}. The state table corresponding to this state diagram is shown in Figure 7.5.6(b).

Since there are only two states, only one state variable need be used. Figure 7.5.7 shows the state assignment and the resulting assigned-state table. In this table, we have shown the output that is associated with a particular state transition and input pulse by its label. For example, the entry 1, X, corresponding to the entry marked (a) in this figure, is used to indicate that the next state is 1 and a pulse is to occur on output X, only. The entry 1, marked (b) in the figure, means that the machine is to go to state 1 and *no* output pulse is to occur, since none is specifically indicated. From this table the output equations can be derived:

$$X = P_X$$
$$S = P_{10}q$$
$$A = Rq$$

(7.5.5)

q	P_{10}	P_X	R
S_0 0	1 (b)	0, X	0
S_{10} 1	0, S	1, X (a)	0, A

Q, output pulse

Figure 7.5.7   The assigned-state table for the vending machine.

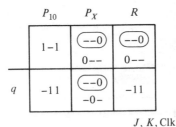

Figure 7.5.8 The initial flip-flop excitation table showing the required choices.

To complete the design, we need to specify the type of flip-flop to be used in the master rank. Suppose, for this example, we select the edge-triggered JK flip-flop whose present-state–next-state table was derived in Section 7.4 and shown in Figure 7.4.5(b). Using this table and the state transition table of Figure 7.5.7, we obtain the flip-flop excitation table shown in Figure 7.5.8. As in the last example, where we used edge-triggered D flip-flops, only the clock input is to be a function of the pulse inputs. Thus, the selection of the J, K, and Clk inputs in the three cells of the figure in which two possible choices are shown must be made so that the J and K inputs are independent of the pulse inputs. This means, as before, that the selection of the alternatives has to be done so that in a given row the value of J is the same for every column. This must also be true for the values of K. The circled entries in the figure show the required selections. These entries are plotted in Figure 7.5.9 for J, K, and Clk.

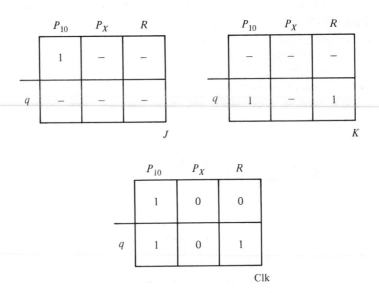

Figure 7.5.9 The final flip-flop input equations.

Using the plots for the JK flip-flop inputs given in Figure 7.5.9, the equations for these inputs become

$$J = 1$$
$$K = 1 \qquad\qquad (7.5.6)$$
$$\text{Clk} = P_{10} + Rq$$

Figure 7.5.10 shows the resulting physical implementation.

In this example, the values for both J and K ended up being a constant 1. It can be shown, as will be done in Problem 7.21, that this is always possible when using JK flip-flops in the master rank. Thus, if the master rank in a multiply clocked sequential circuit is implemented using either JK or T flip-flops (a T is just a JK with the inputs tied together), we need develop equations for the clock inputs only and tie the level flip-flop inputs to a logical 1.

Before leaving this example, one final observation should be made. Note in equations (7.5.6) that the clock input to the master rank is independent of the pulsed input P_X. This means that the master rank cannot change because of input P_X. Thus, there is no need to use P_X to clock the slave rank, since no state change will occur in any case. The implication of this is that the slave rank need be clocked only by the pulse inputs that are actually required to change the state of the master rank, namely, P_{10} and R. This may help simplify a particular circuit implementation.

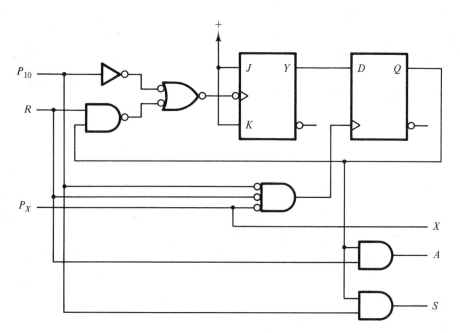

Figure 7.5.10   The final realization for the simple vending machine.

7.6 SOME NOTES ON MIXING LEVEL SIGNALS AND PULSES

Figure 6.3.1, of the previous chapter, repeated here as Figure 7.6.1, shows the definition of two commonly available flip-flops: the 7474 and the 74LS76. These flip-flops have three distinct classes of inputs. The first class are the level inputs associated with the D and the J and K inputs. The second class are the inputs to the clocks, which are edge-sensitive (i.e., the flip-flop changes state on an input edge). The third class of signals are the asynchronous inputs used to set or reset the flip-flop outputs. This latter input set takes precedence over all of the others, as shown in the defining truth tables of Figure 7.6.1. As we have seen in the last few sections of this chapter, pulse-mode circuits can be designed in which either the set and reset inputs of the master-rank flip-flops or the level and clock inputs to these flip-flops are controlled. There is no fundamental reason why we cannot design a system in which we control both, as is possible using the 7474 or 74LS76 type flip-flops. In fact, there are many practical problems in which this approach can be very useful.

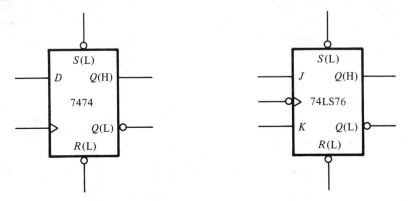

D	Rst	Set	Clk	Q
L	H	H	⬙	L
H·	H	H	⬙	H
–	L	H	–	L
–	H	L	–	H

(a)

J	K	Rst	Set	Clk	Q
L	L	H	H	⬎	q
L	H	H	H	⬎	L
H	L	H	H	⬎	H
H	H	H	H	⬎	$\bar{q}$
–	–	L	H	–	L
–	–	H	L	–	H

(b)

Figure 7.6.1 Flip-flops with asynchronous presets and clears: (a) D flip-flops, type 7474; (b) JK flip-flop, type 74LS76.

To see how we might design a system in which we control the asynchronous inputs as well as the level and the clock inputs, let us first look at the characteristics of the signals used at these three inputs. The *level inputs* are basically signals that stay at some value for realtively long periods of time. Specifically, these signals stay at a fixed value during occurrences of the various input pulses. We may think of the inputs to the flip-flop clock as being rather different from the asynchronous set and reset inputs in that their level at any instant of time is unimportant. The thing that matters is the time at which there is an active level change. Because of this, we will refer to these inputs as *edge inputs*. Since the asynchronous inputs take precedence over all of the others, assertion levels on these inputs must not persist for long periods of time; otherwise, the activity on the other inputs will be masked. Thus, these inputs must be asserted for relatively the shortest periods of time. We will refer to these inputs as being *pulsed inputs*.

Level, edge, and pulsed inputs can be mixed easily under the correct conditions. First, as always, *pulsed inputs cannot overlap*. Second, *no pulse input must be active at the time an edge input occurs*. If this should happen, the edge will have no effect, because the asynchronous inputs take precedence over all other inputs. Finally, *in logical combinations of edge inputs, the level of one edge signal must not mask the occurrence of an asserted transition on another edge input*. This means, assuming an asserted transition on the clock input of low to high, that the clock input must be low when an asserted transition on any edge signal occurs.

To illustrate how we can approach a design in which all of these various input types are used, consider the following problem.

Problem

The Intel 8085 microprocessor has a control line called READY, which can be used to stop the processor at specific times. In particular, if READY is asserted, the processor will run normally. If READY is negated, the processor will stop and remain inactive for as long as READY is negated. A special time of interest to us is during the period of time that the processor is "fetching" an instruction from memory. What we would like to do is design a piece of hardware that will allow us to stop the microprocessor at this point so that we can physically examine various signals in the microcomputer system. We would then like to be able to push a button and have the microprocessor execute the current instruction and then stop at the beginning of the next. This is usually referred to as *single-stepping*. We would also like to have a switch that can be used to select either the single-step mode of operation or a normal mode in which the processor runs continuously without interruption. To implement this function, we clearly need two switches: SS, to *s*ingle-*s*tep the processor, and ST, to select between the run mode and the single-*st*ep mode. We also need some information from the processor identifying the time at which the "fetch cycle" begins so that we can negate the READY input and stop the microprocessor. Using various other outputs on the microprocessor, we can generate two pulsed signals F_1 and F_2 which serve this function. F_1 is a pulse which occurs, once, on entering the fetch cycle. F_2 is a string of pulses that continues for as long as the processor is going

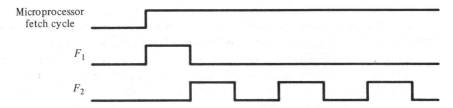

Figure 7.6.2 Timing relationship between pused inputs F_1 and F_2.

through or is stopped in the fetch cycle. There is no overlap between these two pulsed inputs. Our job now is to design the specific hardware to implement this function.[4]

We now have the basic information to design the system. To do the design, we need, first, to classify the various inputs and outputs. The output READY and the input ST are easily classified as level signals, since they are set up for relatively long periods of time. Once the processor is stopped, the effect of pushing switch SS is to cause the processor to execute the current instruction and fetch the next. Since the processor is much faster than the reaction time of whoever has pushed the switch button, the switch will more than likely still be depressed after the processor completes the execution of the current instruction. Thus, input SS is categorized as an edge input, since it is only the time at which it is depressed that is important.[5] The final two inputs, F_1 and F_2, are easily seen to match our idea of pulse inputs. Figure 7.6.2 shows the relative timing of the pulsed inputs F_1 and F_2. Since these signals are generated by the microprocessor, their timing relation is always maintained as shown in this figure.

From these classifications for the various signals, we can now construct the state transition table. This is shown in Figure 7.6.3. In deriving this table, we have assigned state S_0 to the single-step mode of operation and state S_1 to the normal running mode. To begin, suppose the processor is running normally and is, therefore, sitting in state S_1 with the READY output equal to 1. Pulses on inputs F_1 and F_2 and any edge that might occur on the SS input because the single-step button is momentarily depressed should not affect the output, and so the circuit will stay in state S_1.

Suppose now that while the microprocessor is running, the step switch ST is thrown so that input ST = 1. Since we want to negate the READY signal as soon as the microprocessor enters the instruction fetch cycle, we will use F_1 to cause the circuit to switch to the single-step mode. Since we are going to use F_1 to stop the microprocessor, we will make the circuit independent of F_2 by the assignment shown in the F_2 column under ST = 1. Once input pulse F_1 occurs, the circuit switches to state S_0, which causes the microprocessor to stop. As long as the microprocessor is halted, F_1 will not occur again, and thus we have the don't cares shown at (b) in Figure 7.6.3. In order to change the state and cause the processor to run until the next instruction is encountered, the single-step

---

[4] For those readers having some knowledge of the 8085 and wanting to know specifics, F_1 is formed by the signal ALE $S_1 S_0$, and F_2 is generated by CLK $S_1 S_0$, where S_1 and S_0 are the 8085 status signals, CLK is the 8085's clock output signal, and ALE is the 8085's *Address Latch Enable* signal.

[5] We assume here that all switches have been properly "debounced."

q		ST = 0			ST = 1			READY
		F_1	F_2	SS	F_1	F_2	SS	
0	S_0	– (a)	S_1	S_1	– (b)	S_0	S_1	0
1	S_1	S_1	S_1	S_1	S_0	S_1	S_1	1

Figure 7.6.3 State table for the single-step circuit.

switch SS must be asserted. This may happen while the processor is running in the single-step mode (this is highly unlikely, however; why?). We have, therefore, set this entry in the state table equal to state S_1.

Let us now consider what happens if the circuit is in the single-step mode (ST = 1) and the microprocessor has been halted—it is sitting in state S_0—and ST switches to a 0, indicating that the processor should start running normally. This change in ST will cause the circuit to move to the left half of the state table shown in Figure 7.6.3. Since the processor has stopped in a fetch cycle, there will be continuous pulses occurring on the F_2 input. These F_2 pulses will cause the circuit to next move to the run state S_1, as shown in the figure. The don't care shown at (a) in Figure 7.6.3 reflects the fact that once the microprocessor has halted (because READY = 0), pulse F_1 will not occur again until the processor resumes running and encounters the next fetch cycle.

With the state table of Figure 7.6.3 and the state assignment indicated, we arrive at the assigned-state table shown in Figure 7.6.4(a). Since the pulsed inputs F_1 and F_2 control the asynchronous set and reset inputs of the master flip-flop and the edge input SS controls the flip-flop clock, the excitation table becomes as shown in Figure 7.6.4(b). On the basis of this excitation matrix, the flip-flop input equations become

$$S = \overline{ST} \cdot F_2$$
$$R = ST \cdot F_1$$
$$D = \overline{q}(\overline{ST} + ST) + q(\overline{ST} + ST) = 1 \qquad (7.6.1)$$
$$Clk = \overline{ST} \cdot SS + ST \cdot SS = SS$$

We could, of course, implement these equations directly in the master-slave model described in previous sections. However, a little thought can simplify the final realization somewhat. Recall that the reason for the master-slave organization was to prevent changes in those circuit pulse inputs that cause the state flip-flop outputs to change from causing further change in these state flip-flops until all pulse inputs are once again negated. Changes in state flip-flop outputs can affect the state flip-flop inputs only if the inputs are functions of the state flip-flop outputs. If this is not the case, then there is no need for the slave rank of flip-flops. Equations (7.6.1) show that, in this case, none of the flip-flop inputs are functions of their outputs, and so we can eliminate the slave rank. Figure 7.6.5 shows the final implementation for this microprocessor single-step control circuit.

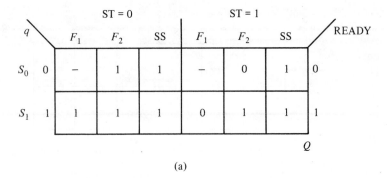

		ST = 0			ST = 1			READY
q		F_1	F_2	SS	F_1	F_2	SS	
S_0	0	–	1	1	–	0	1	0
S_1	1	1	1	1	0	1	1	1

Q

(a)

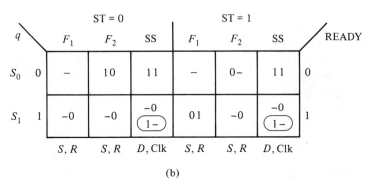

		ST = 0			ST = 1			READY
q		F_1	F_2	SS	F_1	F_2	SS	
S_0	0	–	10	11	–	0–	11	0
S_1	1	–0	–0	$\begin{array}{c}-0\\ \boxed{1-}\end{array}$	01	–0	$\begin{array}{c}-0\\ \boxed{1-}\end{array}$	1
		S, R	S, R	D, Clk	S, R	S, R	D, Clk	

(b)

Figure 7.6.4 The development of the flip-flop excitation table: (a) state table; (b) excitation table.

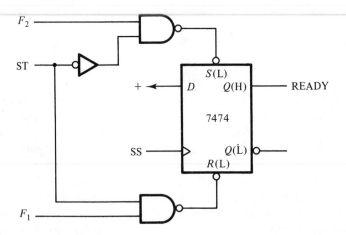

Figure 7.6.5 The final realization of the microprocessor single-step circuit.

7.7 AN ANNOTATED BIBLIOGRAPHY

Although multiply clocked sequential circuits, as discussed in this chapter, are frequently encountered in practical problems, discussion of such circuits is rare in current textbooks. However, the techniques described here are no different, except for the use of the master-slave flip-flop arrangement, from those encountered in the design of pulse-mode circuits as defined by the classical definition given in Section 7.1. References to this model are a bit more readily available. For example, the classic text by McCluskey is an excellent reference for the material covered in this chapter. Hill and Peterson also give a very good presentation of pulse-mode circuits, with many examples.

HILL, J. F., and G. R. PETERSON, *Introduction to Switching Theory and Logical Design*, 3rd ed., Wiley, New York, 1981.

McCLUSKEY, E. J., *Introduction to the Theory of Switching Circuits*, McGraw-Hill, New York, 1965.

Other examples of reference to the classical approach can be found in Kohavi, Nagle, et. al., and Givone.

GIVONE, D. D., *Introduction to Switching Circuit Theory*, McGraw-Hill, New York, 1970.

KOHAVI, Z., *Switching and Finite Automata Theory*, 2nd ed., McGraw-Hill, New York, 1978.

NAGLE, H. T., Jr., B. D. CARROLL, and J. D. IRWIN, *An Introduction to Computer Logic Design*, Prentice-Hall, Englewood Cliffs, N.J., 1975.

Muroga presents a rather different view of pulse-mode circuits in Chapter 8 of his book. Here he refers to these circuits as operating in the *skew mode*.

MUROGA, S., *Logic Design and Switching Theory*, Wiley-Interscience, New York, 1979.

7.8 PROBLEMS

7.1. A buffer with a *Schmitt trigger input* is one in which the output goes from a low to a high voltage when the input goes above a threshold voltage and returns to a low voltage when the input drops below another, quite different, threshold voltage.[6] Such circuits are said to possess *hysteresis*. Figure P7.1(a) shows a plot of input versus output voltage for a buffer having a Schmitt trigger input and the symbol that is used to identify such gates. Gates of this type can be used effectively to produce controlled delays in digital circuits, as we shall explore in this and the next three problems.

Figure P7.1(b) shows a circuit consisting of two buffers, one of which has a Schmitt trigger input, and an *RC* network. Assume that the output of buffer *a* switches between 0 and 5 V and that its output impedance is 0 ohms (Ω). Further assume that the output of the Schmitt trigger buffer goes high when the input goes above 1.7 V and goes low when the input drops below 0.9 V. Draw a timing diagram showing the timing relationship between the input *A* and the signals *X* and *B* in the circuit of Figure P7.1(b). Assume that $R = 10 \text{ k}\Omega$ and $C = 10$ μF.

---

[6] Other types of gates may also have Schmitt trigger inputs.

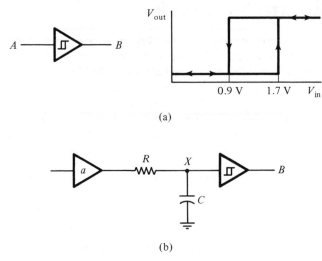

(a)

(b)

Figure P7.1

7.2. What values of R and C in the circuit of Problem 7.1 would be required to produce the following delays? Assume that the propagation delay through each of the buffers is 10 ns (nanoseconds).
  (a) 1 microsecond (μs)
  (b) 1 millisecond (ms)
  (c) 5 s

7.3. Suppose that the minimum propagation delay through the logic in the circuit shown in Figure 7.1.2 is 30 ns and the propagation delay through the flip-flops is 10 ns. What range of input pulse widths will result in a correctly operating circuit? Construct a timing diagram showing the timing relationship between the various signals in the figure.

7.4. How much delay would be required in the outputs of the state flip-flops in the model of Figure 7.1.2 to make the circuit operate properly if the input pulse widths vary between 100 ns and 150 ns?

7.5. Construct a timing diagram for the circuit of Figure 7.3.1, assuming that the circuit starts in state $(Q_1, Q_0) = (0, 0)$ and the inputs occur in the sequence shown in Figure P7.5.

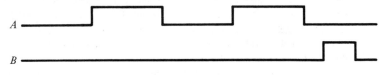

Figure P7.5

7.6. Derive state diagrams and state tables for the multiply clocked sequential circuits shown in Figure P7.6(a) and (b). Assume that A, B, and C are all pulse inputs.

7.7. Construct timing diagrams showing how the circuits of Problem 7.6 respond to the followng input sequences:
  (a) *AABA* applied to the circuit of Figure P7.6(a)
  (b) *ABBCA* applied to the circuit of Figure P7.6(b).

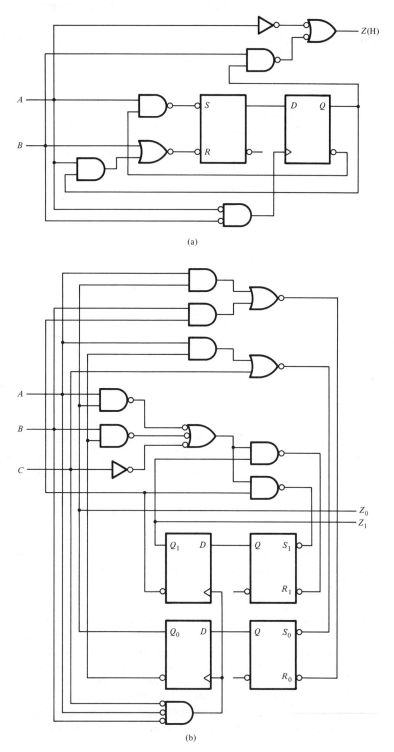

(a)

(b)

Figure P7.6

7.8. Design multiply clocked circuits to implement the machines shown in Figure P7.8. Use SR flip-flops in the master rank.

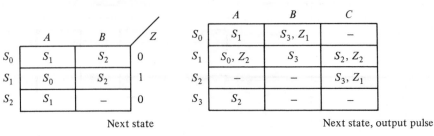

	A	B	Z
S_0	S_1	S_2	0
S_1	S_0	S_2	1
S_2	S_1	–	0

Next state

(a)

	A	B	C
S_0	S_1	S_3, Z_1	–
S_1	S_0, Z_2	S_3	S_2, Z_2
S_2	–	–	S_3, Z_1
S_3	S_2	–	–

Next state, output pulse

(b)

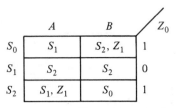

	A	B	Z_0
S_0	S_1	S_2, Z_1	1
S_1	S_2	S_2	0
S_2	S_1, Z_1	S_0	1

Next state, output pulse

(c)

Figure P7.8

7.9. Repeat Problem 7.8 using edge-triggered D flip-flops in the master rank.

7.10. Repeat Problem 7.8 using edge-triggered JK flip-flops in the master rank.

7.11. Design the stepping motor controller of Section 7.4 using edge-triggered JK flip-flops.

7.12. Suppose you are given a programmable logic array having 6 inputs and 8 outputs with 12 product terms. Show a PLA programming diagram similar to that shown in Figure 4.4.5 that would implement the stepping motor controller of Figure 7.4.4.

7.13. Show a PLA programming diagram for the implementation of the master-rank inputs and the slave-rank clock shown in Figure 7.4.8 using the PLA device described in Problem 7.12.

7.14. Derive the present-state–next-state tables, similar to those of Figure 7.4.5, for an edge-triggerd T flip-flop.

7.15. Repeat Problem 7.14 for an edge-triggered SR flip-flop.

7.16. Repeat Problem 7.14 for the XY flip-flop defined in Problem 5.11.

7.17. A circuit, which we will refer to as a pulse identify (PI) circuit, is to be designed having four pulsed inputs, A, B, C, and R, and two outputs, X and Y, which are the outputs of two flip-flops. Design a circuit that will cause the outputs of the flip-flops (X, Y) to set to (00) if pulse R occurs, (01) if pulse A occurs, (10) if pulse B occurs, and (11) if pulse C occurs. Use SR-type flip-flops in the master rank for this problem.

7.18. Suppose you are given n pulse identify circuits as described in Problem 7.17. Design a circuit that will cause the ith PI circuit to set in accordance with the input that produces the ith pulse on the inputs A, B, and C after being reset to 00 by a pulse on input R. For example, suppose that after a pulse occurs on input R, three pulses occur on A, and two pulses occur on B, and one pulse occurs on C. If another pulse occurs on input C, then the output of the seventh PI circuit should set to 11 to reflect the fact that the seventh pulse among A, B, and C has occurred on input C. (*Hint*: Think of a counter that enables one PI after another upon occurrence of pulses on A, B, and C and is reset to enable the first PI on a pulse on R. Once all PIs have been set to their value, a table look-up, using a ROM, could be performed that decodes the pusle sequence to open the lock of Section 7.5.1.)

7.19. How could you apply the ideas presented in Problems 7.17 and 7.18 to the design of a combinational lock whose combination could be easily changed?

7.20. Redesign the vending machine of Section 7.5.2 assuming that both dimes and nickels are accepted.

7.21. Prove that all multiply clocked sequential circuits designed using JK flip-flops in the master rank can always be implemented by a design in which $J = K = 1$. (*Hint*: Consider a row in an excitation table in which one entry goes from a 0 to a 0 and another entry, in the same row, goes from a 0 to a 1. What must always be the choice for J?)

Special Topics in Switching Theory

8

8.1 INTRODUCTION

In this chapter we will examine a number of topics that are of special interest in various areas of digital system design and application. In Section 8.2, for example, we will look at networks of logic elements that can pass information in either direction. Such elements are called *bilateral* elements and are becoming more important in the design of CMOS (complementary metal-oxide-semiconductor[1]) integrated circuits, especially at the VLSI (very large-scale integration) level. In Section 8.3, we will introduce threshold logic. Threshold gates are of special interest because they simulate some types of neuron behavior and so are useful in the simulation of neural networks. Sections 8.4, 8.5, and 8.6 will deal with special methods by which complex switching functions can be implemented. First, we will look at a method for representing a switching function in a form other than the simple SOP or POS forms. These alternative representations can often reduce the amount of hardware required to implement a given function. We will next examine a class of functions called symmetric functions, which occur frequently in real-world designs. The recognition of symmetric functions is important, since they can be implemented in some very economical ways using either bilateral elements or gates. Finally, we will take a closer look at iterative networks, which were first discussed in Chapter 4.

---

[1] Another translation of this acronym is "complementary metal-oxide-silicon."

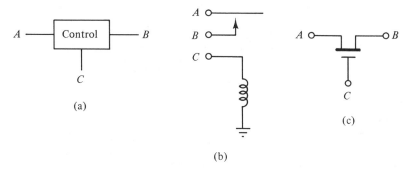

Figure 8.2.1   A model (a), a relay equivalent (b), and an NMOS transistor implementation of a bilateral device.

8.2 *BILATERAL NETWORKS*

In Chapter 4, we introduced the logic gate as a physical device used for the implementation of switching functions. In the gate, information can flow in only one direction: from input to output. In a *bilateral* device, however, information can flow in either direction, although there is generally a preference. The classic bilateral device used for switching function implementation has been the relay. Such circuits have been, and, to a lesser degree, still are, prevalent in telephone switching systems, which, of course, were the systems that prompted the development of switching algebra. Although the use of relays is declining, the use of MOS transistors, which are also bilateral devices, is increasing in the design of VLSI systems. Thus, it is important to understand some of the principles and incumbent problems associated with bilateral devices.

The basic model of a bilateral device is shown in Figure 8.2.1(a). In this model, an information path exists from A to B, or vice versa, which can be enabled by the control signal C.[2] The control line effectively turns the flow of information through the device either on or off. Figure 8.2.1(b) shows these signals as they exist in a relay. In operation, a current supplied to the relay coil at C produces a magnetic field that pulls the contact labeled A into the contact labeled B and thus closes the circuit from A to B. If no current flows in C, a spring pulls contact A up and away from contact B, thus opening the circuit from A to B. In the case of the NMOS transistor shown in Figure 8.2.1(c), a positive voltage applied to the control, or gate, line C will cause current to flow in the circuit AB. If the voltage at C is 0, then the circuit AB will be open and no current can flow. Figure 8.2.2 shows the three types of bilateral elements most commonly encountered. This figure

---

[2] The information referred to here is usually represented by a current flow. Other physical mechanisms exist for carrying information, such as pressure or voltage.

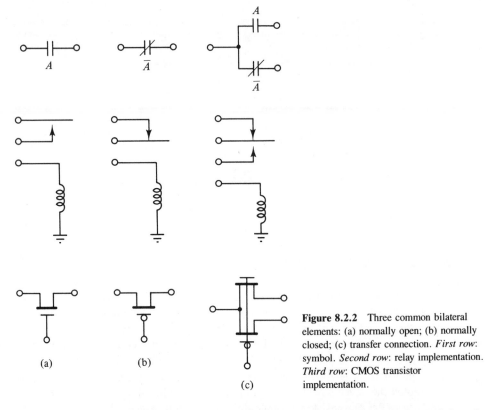

Figure 8.2.2 Three common bilateral elements: (a) normally open; (b) normally closed; (c) transfer connection. *First row*: symbol. *Second row*: relay implementation. *Third row*: CMOS transistor implementation.

(a)   (b)   (c)

shows for each the symbol, the relay arrangement, and the CMOS equivalent[3] of the relay implementation. In what follows, we will discuss bilateral networks in terms of relay contacts simply because they are much easier to understand. In all cases, the CMOS equivalent circuit can be derived using the equivalences shown in this figure.

The basic three types of bilateral elements—normally open, normally closed, and the transfer element—as shown in Figure 8.2.2, can be combined in various ways to produce the necessary logical operations required to implement switching functions. Consider, for example, the relay circuits shown in Figure 8.2.3. In the circuit shown in part (a), we easily observe that the light L will be on if switch A, which is normally open, is closed *and* switch B, also normally open, is closed. Similarly, in Figure 8.2.3(c), the lamp will be lit if either switch C is closed *or* switch D is closed. The *series* connection of switches, therefore, corresponds to the AND function, and the *parallel* connection of switches corresponds to the OR function. The NOT operation can be thought of as a switch that is normally closed and, when thrown, opens up so that the circuit is broken. This is shown

---

[3] CMOS involves the integration of both PMOS and NMOS transistors on the same piece of silicon. Although the behavior of these transistors is not exactly equivalent physically to the corresponding relay circuit shown in Figure 8.2.2, the logical behavior is the same. The bibliography at the end of this chapter gives a number of excellent references to the design of digital circuits using MOS and CMOS technologies.

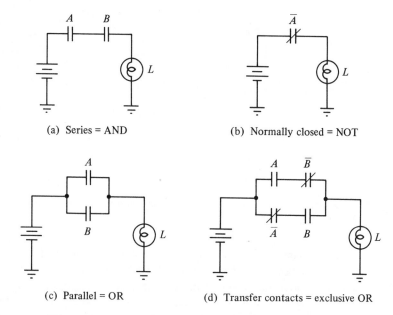

Figure 8.2.3   AND, OR, NOT, and Exclusive OR switch connections.

in Figure 8.2.3(b) and (d). The transfer contact is useful for implementing the Exclusive OR function, as shown in Figure 8.2.3(d). This connection is precisely the one used to wire "two-way" switches in homes, i.e., two switches at different locations which can individually control a single light.

Combinations of series and parallel circuits can be used to implement arbitrary switching functions. For example, Figure 8.2.4(a) shows the *series-parallel* network that implements the function

$$
\begin{aligned}
f(a, b, c, d) &= ab + ac\bar{e} + d\bar{e} + dcb \\
&= a(b + c\bar{e}) + d(\bar{e} + cb)
\end{aligned}
\tag{8.2.1}
$$

This can easily be verified by listing all of the paths that exist from A to B and summing the result to produce Equation (8.2.1).

The implementation of Equation (8.2.1) given in Figure 8.2.4(a) requires that three of the switches, b, c, and e, have two contacts each, since these variables appear in two different parts of the network. Obviously, a switch having two pairs of contacts is going to be more expensive than one having only one pair. Figure 8.2.4(b) shows a non-series-parallel implementation for this function, called a *bridge network*, which requires that each switch have only one pair of contacts. Clearly, the implementation in part (b) is better— simpler—than that of part (a). But how do we find such an implementation?

To investigate the question of how we might reduce the number of contacts in a series-parallel network, let us consider first the implementation of the function

$$
\begin{aligned}
T(w, x, y, z) &= w\bar{x} + wy + xyz \\
&= w(\bar{x} + y) + xyz
\end{aligned}
\tag{8.2.2}
$$

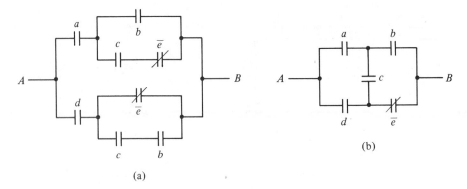

Figure 8.2.4 Series-parallel implementation (a) and reduced network (bridge) realization (b) for the function given by Equation (8.2.1).

A series-parallel realization for this function is shown in Figure 8.2.5(a). In this realization, we can observe that the two occurrences of switch y share a connection at the right side. It would seem reasonable that the bottom switch y could be removed if a connection was made between points 1 and 2 in the figure. The resulting, reduced network is shown in Figure 8.2.5(b). To verify that this reduced network still represents the original function, we need only list the product terms corresponding to all possible paths from left to right. These are the following: $w\bar{x}$, wy, xzy, and $xz\bar{x}$, the latter of which is equal to zero. In this example, since the OR of all of these paths yields the original function, the reduced network realizes the given function. However, if the original equation had been

$$T(w, x, y, z) = wx + wy + xyz$$
$$= w(x + y) + xyz \qquad (8.2.3)$$

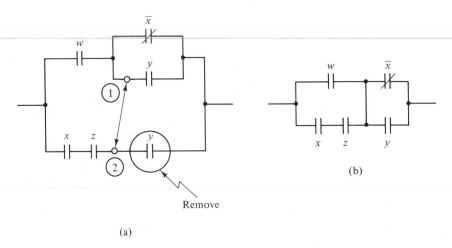

(a)

Figure 8.2.5 Reducing a series-parallel realization by multiple use of contacts. (a) Contacts that can be shared. (b) The reduced network.

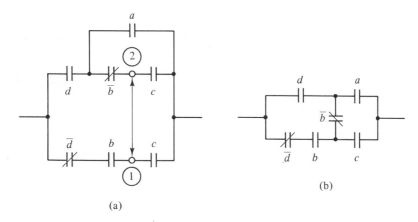

Figure 8.2.6 Implementation for $G(a, b, c, d)$ of Equation (8.2.4). (a) Original series-parallel implementation. (b) Reduced, bridge realization.

a problem would exist with the reduced implementation, since the path $xz\bar{x} = 0$, in the original realization, would now become $xzx = xz$, which is not a product term in Equation (8.2.3). In this case, sharing a contact introduces a *sneak path* which generates a term not in the original expression. Thus, the removal of one of the y contacts would not be possible.

Putting these ideas together, the process of realizing a switching function with relay contacts involves basically the following five steps:

Step 1. Write the expression in a minimal SOP (or POS) form.

Step 2. Factor the expression to reduce the number of literals as much as possible.[4]

Step 3. Implement the factored expression in a series-parallel form.

Step 4. Eliminate the occurrence of multiple contacts that share a connection by removing one of the contacts after the unshared connections are connected.

Step 5. Check for sneak paths by writing the equation for the circuit just formed. If sneak paths exist, remove them by reinserting the appropriate contact removed in step 4.

An example will illustrate this process. Suppose we are required to implement the function

$$G(a, b, c, d) = ad + bc\bar{d} + \bar{b}cd$$
$$= d(a + \bar{b}c) + bc\bar{d} \tag{8.2.4}$$

Equation (8.2.4) gives the original SOP representation required by step 1 and the factored form required by step 2. Figure 8.2.6(a) shows the series-parallel realization for this reduced expression. In this realization, the order of the literals in the series circuit for $bc\bar{d}$ is modified so that the two c contacts can share a connection and so the contacts d and $\bar{d}$ can be

---

[4] We will look at some aspects of this problem in Section 8.4, where we deal with functional decomposition.

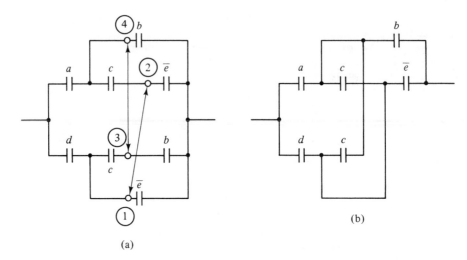

(a)

(b)

Figure 8.2.7   The reduction of the network of Figure 8.2.4. (a) Original series-parallel.
(b) Partially reduced form.

implemented as a transfer contact. After elimination of the bottom c contact, a reduced
network is created as shown in Figure 8.2.6(b). It can be verified, by listing all of the
paths from left to right, that the resulting realization does yield the given function. Note,
also, that this realization requires two simple contacts and two transfer contacts. If the
contacts labeled $\bar{d}$ and b in the bottom series connection in Figure 8.2.6(a) were reversed,
the resulting network would require two separate switches for d and $\bar{d}$ and two switches
for b and $\bar{b}$ (why?).

Let us now go back to the network given in Figure 8.2.4(a) and see how we can get
the reduced version shown in Figure 8.2.4(b). We see in part (a) that both pairs of contacts
b and $\bar{e}$ have a common connection, and so one of each may be eliminated as indicated
in Figure 8.2.7(a) by connecting point 1 to point 2 and point 3 to point 4. After we remove
the two extra contacts, the network of Figure 8.2.7(b) is generated. Note here that the two
occurrences of contact c now are in parallel, and so one can be removed forthwith, yielding
the circuit shown in Figure 8.2.4(b).

This section has only scratched the surface of bilateral switching network design. It
does, however, indicate some of the processes required in the design and, at least, one of
the problems—sneak paths—that can arise in attempting to reduce networks of switches
or any other network of bilateral elements. There is extensive literature in which further
details are given. Some references to it are given at the end of this chapter.

8.3 THRESHOLD LOGIC

In a sense, a threshold gate is a generalized logic gate, for instead of realizing a simple
operation such as AND, OR, or NOT, a single threshold gate can realize fairly complex
switching functions. For example, a single threshold gate can realize the function

$$f(A, B, C, D, E) = AB + AC + ADE + BCDE \qquad (8.3.1)$$

Obviously, if a single gate can realize functions as complex as this, there is certainly some economic advantage in using threshold gates to realize arbitrary switching functions. This, of course, was one of the principal reasons for the early interest in gates of this type. In fact, at least one computer using this technology was built in the 1950s and 1960s. For numerous reasons, some of which will become apparent as we proceed, the anticipated economic advantages of threshold logic were never realized. However, because a threshold gate has properties similar to those of neurons, an increasing interest has developed in the past few years in their potential applications to adaptive control systems, learning automata, and pattern recognition.

Before defining a threshold function, let us first define the following sets of binary n-tuples. Let $f(\mathbf{x}) = f(x_1, x_2, \ldots, x_n)$ be a switching function on n variables. Then define the sets

$$A(f) = \{\mathbf{a} \mid f(\mathbf{a}) = 1\} \qquad (8.3.2)$$

$$B(f) = \{\mathbf{b} \mid f(\mathbf{b}) = 0\} \qquad (8.3.3)$$

$$D(f) = \{\mathbf{d} \mid f(\mathbf{d}) \text{ is unspecified}\} \qquad (8.3.4)$$

where $\mathbf{a}$, $\mathbf{b}$, and $\mathbf{d}$ are binary n-tuples or binary assignments on the variables of f. The set $A(f)$ is simply the set of all *true vectors* of f; the set $B(f)$ represents the set of all *false vectors* of f; and the set $D(f)$ is the set of *don't care vectors*. Using these three sets of binary n-tuples, we can now define a threshold function.

Definition 8.3.1. Let $f(\mathbf{x})$ be a switching function on n variables. Then $f(\mathbf{x})$ is a *threshold function* if there exist a real vector $\mathbf{w} = (w_1, w_2, \ldots, w_n)$, where the w_i's are called the *weights*, and a real number T, called the *threshold*, such that

$$\mathbf{w} \cdot \mathbf{a}(j) \geq T \qquad \text{for all } \mathbf{a}(j) \text{ in the set } A(f) \qquad (8.3.5)$$

and

$$\mathbf{w} \cdot \mathbf{b}(k) \leq T - 1 \qquad \text{for all } \mathbf{b}(k) \text{ in the set } B(f) \qquad (8.3.6)$$

where the notation $\mathbf{w} \cdot \mathbf{x} = w_1 x_1 + w_2 x_2 + \cdots + w_n x_n$. The $(n + 1)$-dimensional vector $[w_1, w_2, \ldots, w_n; T]$ is called the *structure* for $f(\mathbf{x})$.

We can think of a threshold function in a geometric way if we recognize that the set of points which satisfy the equation

$$w_1 x_1 + w_2 x_2 + \cdots + w_n x_n = T \qquad (8.3.7)$$

lie on a plane in Euclidean n-space. Thus, a switching function is a threshold function if an n-dimensional hyperplane can be found that separates the true vectors from the false vectors. For this reason, threshold functions are also often referred to as *linearly separable functions*. This geometric interpretation can easily be illustrated. As an example, Figure 8.3.1 shows a 3-cube in which the vertices are the eight possible binary 3-tuples. Shown

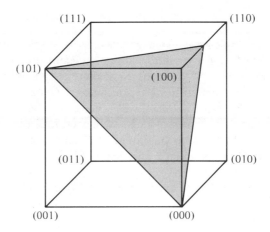

Figure 8.3.1 The hyperplane separating the true and false vectors of the function $x_1\bar{x}_2 + \bar{x}_2\bar{x}_3$.

also in this figure is a plane separating the vectors $A(f) = \{(101), (100), (000)\}$ from the set $B(f) = \{(010), (001), (011), (110), (111)\}$. The function thus defined, namely, $f(x_1, x_2, x_3) = x_1\bar{x}_2 + \bar{x}_2\bar{x}_3$, is a threshold function with a structure, satisfying inequalities (8.3.5) and (8.3.6), of $[1, -2, -1; 0]$.

Before we proceed, we need to make some modification in our notation to avoid confusion later. Since the subject of threshold functions involves both the logical OR operation and the arithmetic sum, we will need some distinctive notation to differentiate between the two. For this purpose, in all that follows *in this section only*, we will use the symbol ∨ to represent the OR operation and + to represent the arithmetic sum. Thus $A ∨ B$ will be taken to be the logical OR of switching variables A and B. This use of ∨ for the OR operation is classic.

Before examining some of the properties of threshold functions, let us take a look at one of the possible physical implementations of a threshold gate. A threshold gate is a very simple device to implement, as is illustrated by the two-input gate shown in Figure 8.3.2. In this example, the output transistor will be turned on, causing the output voltage V_0 to go to zero, if the transistor base voltage, V_B, is greater than zero.[5] The transistor will be off, producing an output voltage of E_2, if V_B is negative. To determine under what conditions this will occur, we first disconnect the transistor and the diode and then write the node equation at the base of the transistor,

$$V_B\left(\frac{1}{R_T} + \frac{1}{R_1} + \frac{1}{R_2}\right) = \frac{V_1}{R_1} + \frac{V_2}{R_2} - \frac{E_1}{R_T} \qquad (8.3.8)$$

from which we see that the base voltage will be positive if

---

[5] Actually, the transistor will turn on if V_B is greater than about 0.6 V. However, this fact does not materially change the following analysis.

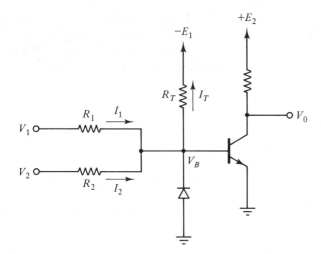

Figure 8.3.2   A simple two-input threshold gate.

$$\frac{V_1}{R_1} + \frac{V_2}{R_2} \geq \frac{E_1}{R_T} \tag{8.3.9}$$

and will be negative otherwise. Inequality (8.3.9) can be interpreted as comparing a weighted sum of input voltages to a scaled output, threshold voltage. If the weighted sum of the inputs is greater than the threshold value, then the transistor turns on; otherwise, it stays off. Comparing inequalities (8.3.9) and (8.3.5), we see that, for this gate, $w_1 = 1/R_1$, $w_2 = 1/R_2$, and the threshold $T = E_1/R_T$.

8.3.1 Unate Functions

Before we investigate some of the properties of threshold functions and methods that we can use to determine whether a given function is linearly separable, let us first take a look at a class of functions that are very important in the study of threshold logic. These functions are referred to as *unate functions* and are defined as follows.

Definition 8.3.2.   Let $f(\mathbf{x})$ be a switching function on n variables. Then $f(\mathbf{x})$ is said to be *positive (negative)* in a variable x_i if there exists an expression for $f(\mathbf{x})$ in which x_i appears everywhere uncomplemented (complemented). $f(\mathbf{x})$ is said to be *unate* if it is positive or negative in all of its variables. Further, $f(\mathbf{x})$ is said to be a *positive function (negative function)* if it is positive (negative) in all of its variables.

From this definition, the following two lemmas naturally arise.

Lemma 8.3.1.   Let $f(\mathbf{x})$ be positive (negative) in the variable x_i. Then $f(\mathbf{x})$ may be factored as

$$f(\mathbf{x}) = x_i f_1 \vee f_2 \qquad (f(\mathbf{x}) = \bar{x}_i f_1 \vee f_2) \tag{8.3.10}$$

where f_1 and f_2 are functions independent of x_i and where f_2 implies f_1, denoted $f_2 \Rightarrow f_1$.

Proof.   If $f(\mathbf{x})$ is positive in the variable x_i, then the factorization is possible, by Definition 8.3.2. Consider now the second part of the theorem. We say that $f_2 \Rightarrow f_1$ if any assignment of the variables making $f_2 = 1$ also makes $f_1 = 1$. Now, if we can find a factorization of the form of Equation (8.3.10) and if f_2 does not imply f_1, then we can obtain the required implication as follows:

$$f(\mathbf{x}) = x_i f_1 \vee f_2 = x_i (f_1 \vee f_2) \vee f_2 = x_i f_3 \vee f_2$$

where $f_3 = f_1 \vee f_2$. Clearly, if $f_2 = 1$, then so also will $f_3 = 1$ and we have found a factorization as specified in the lemma.   QED

An easily proved converse to Lemma 8.3.1 exists and is given as Lemma 8.3.2:

Lemma 8.3.2.   Let $f(\mathbf{x})$ be a switching function on n variables. Then $f(\mathbf{x})$ can be factored as

$$f(\mathbf{x}) = x_i f_1 \vee \bar{x}_i f_2 \tag{8.3.11}$$

where $f_1 = f(\mathbf{x} \mid 1 \rightarrow x_i)$ and $f_2 = f(\mathbf{x} \mid 0 \rightarrow x_i)$. (This principle is usually referred to as the *Shannon decomposition theorem.*) If $f_2 \Rightarrow f_1$ (if $f_1 \Rightarrow f_2$), then $f(\mathbf{x})$ may be written as

$$f(\mathbf{x}) = x_i f_1 \vee f_2 \tag{8.3.12}$$

$$(f(\mathbf{x}) = f_1 \vee \bar{x}_i f_2) \tag{8.3.13}$$

The next theorem and its corollary are also easily proved, and so the proof will not be given here but will be developed in problems at the end of the chapter.

Theorem 8.3.3.   A necessary and sufficient condition for a switching function to be unate is that all of the function's prime implicants share a common minterm.

Corollary 8.3.4.   The minimal SOP expression for a unate function is unique. Another way of saying this is that all of the prime implicants of a unate function are essential.

From these results, we can determine whether a function is unate simply by finding a minimal SOP expression for the function and checking to see that it satisfies the definition, or by finding the prime implicants and looking for a common minterm. Take, for example, the function

$$f(x_1, x_2, x_3, x_4) = \Sigma\, m(3, 6, 7, 11, 13, 14, 15) \tag{8.3.14}$$

The prime implicants for this function are 13, 15(2) and 3, 7, 11, 15(4, 8) and 6, 7, 14, 15(1,8), which clearly have the minterm 15 in common, and the function is, therefore, unate. The minimal SOP expression for this unate function is

$$f(x_1, x_2, x_3, x_4) = x_2 x_3 \vee x_3 x_4 \vee x_1 x_2 x_4 \tag{8.3.15}$$

As we shall see in a moment, all threshold functions are unate. Unfortunately, not all unate functions are linearly separable. In fact, the function just given in Equation (8.3.15) is not a threshold function, even though it is unate.

8.3.2 *Some Basic Threshold Function Properties*

Let us now examine some of the properties of threshold functions. We will begin with two results that show how the signs of the weights and threshold are related to complementation of variables and functions.

Theorem 8.3.5.   Let $f(\mathbf{x})$ be a threshold function with structure $[\mathbf{w}; T]$. Then the function $f(\mathbf{x} \mid \bar{x}_i \rightarrow x_i)$ (the original function with variable x_i replaced by $\bar{x}_i$ everywhere) is a threshold function with structure $[w_1, w_2, \ldots, -w_i, \ldots, w_n; T - w_1]$.

Proof.   The proof follows from the observation that substituting $\bar{x}_i$ for x_i everywhere is equivalent to substituting $1 - x_i$ for x_i in all of the inequalities. Thus we have

$$w_1 x_1 + \cdots + w_i(1 - x_i) + \cdots + w_n x_n \geq T$$

for all $\mathbf{x}$ in $A(f)$, and

$$w_1 x_1 + \cdots + w_i(1 - x_i) + \cdots + w_n x_n \leq T - 1$$

for all $\mathbf{x}$ in $A(f)$; which can be rewritten as

$$w_1 x_1 + \cdots + (-w_i)x_i + \cdots + w_n x_n \geq T - w_i$$

for all $\mathbf{x}$ in $A(f)$, and

$$w_1 x_1 + \cdots + (-w_i)x_i + \cdots + w_n x_n \geq (T - w_i) - 1$$

for all $\mathbf{x}$ in $A(f)$. From these last two inequalities, we see that the resulting function is a threshold function with a structure found by simply replacing w_i by $-w_i$ and T by $T - w_i$.   QED

As an example, consider the function

$$f(\mathbf{x}) = x_1 \vee x_2 x_3 \qquad\qquad (8.3.16)$$

which is a threshold function having a structure $[2, 1, 1; 2]$. By Theorem 8.3.5, the function

$$g(\mathbf{x}) = x_1 \vee \bar{x}_2 x_3 \qquad\qquad (8.3.17)$$

must also be a threshold function, with a structure $[2, -1, 1; 1]$. This can easily be verified by finding all of the assignments on x_1, x_2, and x_3 that make the sum $2x_1 - x_2 + x_3 \geq 1$. The resulting assignments are (100), (110), (101), (111), and (001), from which the function having these as true vectors is just the one given in Equation (8.3.17).

Theorem 8.3.6.   If $f(\mathbf{x})$ is a threshold function on n variables with a structure $[\mathbf{w}; T]$, then $\bar{f}(\mathbf{x})$ is also a threshold function, with structure $[-\mathbf{w}; 1 - T]$.

Proof.   If we multipy the inequalities (8.3.5) and (8.3.6) of Definition 8.3.1 by -1, we obtain

$$-\mathbf{w} \cdot \mathbf{a}(j) \leq -T \qquad \text{for all } \mathbf{a}(j) \text{ in the set } A(f)$$

and

$$-\mathbf{w} \cdot \mathbf{b}(k) \geq 1 - T \qquad \text{for all } \mathbf{b}(k) \text{ in the set } B(f)$$

These inequalities correspond to a new threshold function $g(\mathbf{x})$ having a structure $[-\mathbf{w}; 1 - T]$, in which $A(g) = B(f)$ and $B(g) = A(f)$. But $B(f) = A(\bar{f})$ and $A(f) = B(\bar{f})$, and so $g(\mathbf{x}) = \bar{f}(\mathbf{x})$.   QED

As an example, consider the threshold function given in Equation (8.3.16). The complement of this function is

$$\bar{f}(\mathbf{x}) = \bar{x}_1(\bar{x}_2 \vee \bar{x}_3) \tag{8.3.18}$$

which, by Theorem 8.3.6, must also be a threshold function with structure $[-2, -1, -1; -1]$. The reader should verify this result by setting up the appropriate inequalities and showing that they are all satisfied.

Notice that all of the example threshold functions given to this point were unate. What we will show now is that this is a general property of threshold functions. Since the property of unateness is a very easy property to identify, it can be very useful in identifying which switching functions cannot be threshold functions. This property, unfortunately, cannot be used to determine whether a given function *is* a threshold function, as we shall see shortly.

Lemma 8.3.7.   If $f(\mathbf{x})$ is a threshold function with structure $[\mathbf{w}; T]$, then $f(x \mid 1 \to x_i)$ is a threshold function with the same structure except that $w_i = 0$ and the new threshold equals $T - w_i$. The function $f(\mathbf{x} \mid 0 \to x_i)$ is also a threshold function, with the same structure as $f(\mathbf{x})$ except that $w_i = 0$ (the new threshold is the same as the old, in this case).

Proof.   The proof of this lemma follows directly from the definition of a threshold function (Definition 8.3.1) when we simply plug $x_i = 1$ and $x_i = 0$ into all of the inequalities. Thus, for example, as a result of plugging $x_i = 1$ into the inequalities, the inequality

$$w_1x_1 + \cdots + w_ix_i + \cdots + w_nx_n \geq T$$

becomes

$$w_1x_1 + \cdots + (w_i1) + \cdots + w_nx_n = w_1x_1 + \cdots + (0x_i + w_i) + \cdots + w_nx_n \geq T$$

or, rearranging,

$$w_1x_1 + \cdots + 0x_i + \cdots + w_nx_n \geq T - w_i \qquad \text{QED}$$

An example will help to illustrate this result. Consider the threshold function

$$f(\mathbf{x}) = x_1\bar{x}_2 \vee x_1x_3 \vee x_1\bar{x}_4 \vee \bar{x}_2x_3\bar{x}_4 \tag{8.3.19}$$

whose structure is $[3, -2, 1, -1; 1]$. From Lemma 8.3.7, the function $f(\mathbf{x} \mid 1 \to x_4)$ is a threshold function whose structure must be $[3, -2, 1, 0; 2]$. To verify this, all we need do is to check the appropriate inequalities as follows. First, we note that

$$f(\mathbf{x} \mid 1 \to x_4) = x_1\bar{x}_2 \vee x_1x_3 = g(x_1, x_2, x_3) \tag{8.3.20}$$

which by the lemma is supposed to be a threshold function with the just cited structure. The true vectors of g are (ignoring variable x_4, which have been assigned the permanent value of 1)

$$A(g) = \{(100), (101), (111)\}$$

with the false vectors being the remaining five. Thus, the inequalities required by Definition 8.3.1 become

$$
\begin{array}{lll}
w_1 & \geq T & (100) \\
w_1 + \quad\;\; w_3 \geq T & & (101) \\
w_1 + w_2 + w_3 \geq T & & (111)
\end{array}
$$

and

$$
\begin{array}{lll}
w_1 + w_2 & \leq T - 1 & (110) \\
w_2 + w_3 \leq T - 1 & & (011) \\
w_2 \quad\;\; \leq T - 1 & & (010) \\
w_3 \leq T - 1 & & (001) \\
0 \;\; \leq T - 1 & & (000)
\end{array}
$$

These inequalities are readily verified by substituting $w_1 = 3$, $w_2 = -2$, $w_3 = 1$, and $T = 2$.

We can now present and prove the statement made earlier that all threshold functions are unate.

Theorem 8.3.8. Let $f(\mathbf{x})$ be a threshold function on n variables. Then $f(\mathbf{x})$ is unate.

Proof. Assume that $f(\mathbf{x})$ is a threshold function with structure $[\mathbf{w}; T]$. Without loss of generality, we can factor f as

$$f(\mathbf{x}) = x_1 f_1 \vee \bar{x}_1 f_2$$

where $f_1 = f(\mathbf{x} \mid 1 \rightarrow x_1)$ and $f_2 = f(\mathbf{x} \mid 0 \rightarrow x_1)$ are functions independent of the variable x_1. From Lemma 8.3.7, f_1 is a threshold function with a structure $[0, w_2, \ldots, w_n; T - w_1]$ and f_2 is a threshold function having structure $[0, w_2, \ldots, w_n; T]$. There are now two cases to consider:

Case 1. Assume that $T \geq T - w_1$. Now if $\mathbf{a}$ is a true vector of f_2, then since $w_2 a_2 + w_3 a_3 + \cdots + w_n a_n \geq T \geq T - w_1$, $\mathbf{a}$ must also be a true vector of f_1 and therefore $f_2 \Rightarrow f_1$.

Case 2. Assume that $T < T - w_1$. Then if $\mathbf{a}$ is a true vector of f_1, then since $w_2 a_1 + w_3 a_3 + \cdots + w_n a_n \geq T - w_1 > T$, $\mathbf{a}$ must also be a true vetor of f_2 and so $f_1 \Rightarrow f_2$.

Now, by Lemma 8.3.2, if case 1 holds, then f can be factored as $f(\mathbf{x}) = x_1 f_1 \vee f_2$. If case 2 holds, then f can be factored as $f(\mathbf{x}) = f_1 \vee \bar{x}_1 f_2$. Thus, since, for all variables, f is positive, in case 1, or negative, in case 2, $f(\mathbf{x})$ is unate.   QED

8.3.3 The Determination of Linear Separability

The fact that all threshold functions are unate makes it easier to determine whether a given switching function is a threshold function or not. First, if the given function is not unate, then it cannot be a threshold function, by Theorem 8.3.8. Second, if it is a unate function, we may consider, in our further determination, only the positive version of the given function, where the positive version is just the given function with all variables appearing uncomplemented. We can do this because of Theorem 8.3.5. In fact, as the following lemma, stated here without proof, shows, this may significantly reduce the number of inequalities that need to be solved to determine whether a particular function is a threshold function or not.

Lemma 8.3.9.   Let $f(\mathbf{x})$ be a threshold function with structure $[\mathbf{w}; T]$. Then if $f(\mathbf{x})$ is positive (negative) in variable x_i, then a structure exists in which $w_i > 0$ $(w_i < 0)$.

Basically, this lemma states that given a positive threshold function, a structure exists having all positive weights.

Using the results we have to this point, let us explore how we would determine whether a given switching function is a threshold function or not. Consider, for example, the function

$$f(x_1, x_2, x_3, x_4) = \Sigma\, m(9, 12, 13, 14, 15) \qquad (8.3.21)$$

First we need to derive a minimal sum of products expression for this function. In this case, this is easily done and produces the function

$$f(x_1, x_2, x_3, x_4) = x_1 x_2 \vee x_1 \bar{x}_3 x_4 \qquad (8.3.22)$$

which is clearly unate and so could represent a threshold function, although this is not yet guaranteed. The next step in the determination of linear separability is to make the unate function positive. Thus Equation (8.3.22) becomes

$$f_1(x_1, x_2, x_3, x_4) = x_1 x_2 \vee x_1 x_3 x_4 \qquad (8.3.23)$$

We can now set up and solve the inequalities corresponding to this equation that are stated in Definition 8.3.1. If a solution to the set of inequalities exists, then f_1 is a threshold function and we will have found its structure. Because of Theorem 8.3.5, f must also be a threshold function with the structure determined by that theorem.

Because the function f_1 of Equation (8.3.23) is a completely specified function on four variables, there will be a total of 16 inequalities in the set to be solved:

$$
\begin{array}{llr}
w_1 + w_2 + w_3 + w_4 \ge T & (1111) & \\
w_1 + w_2 + w_3 \phantom{{}+ w_4} \ge T & (1110) & \\
w_1 + w_2 \phantom{{}+ w_3} + w_4 \ge T & (1101) & (8.3.24) \\
w_1 + w_2 \phantom{{}+ w_3 + w_4} \ge T & (1100) \leftarrow & \\
w_1 \phantom{{}+ w_2} + w_3 + w_4 \ge T & (1011) \leftarrow &
\end{array}
$$

which correspond to the true vectors $A(f_1)$; and

$$
\begin{array}{llll}
w_1 & + w_3 & \le T - 1 & (1010) \leftarrow \\
w_1 & \quad + w_4 & \le T - 1 & (1001) \leftarrow \\
w_1 & & \le T - 1 & (1000) \\
& w_2 + w_3 + w_4 & \le T - 1 & (0111) \leftarrow \\
& w_2 + w_3 & \le T - 1 & (0110) \\
& w_2 \quad + w_4 & \le T - 1 & (0101) \\
& w_2 & \le T - 1 & (0100) \\
& w_3 + w_4 & \le T - 1 & (0011) \\
& w_3 & \le T - 1 & (0010) \\
& w_4 & \le T - 1 & (0001) \\
& 0 & \le T - 1 & (0000)
\end{array} \qquad (8.3.25)
$$

which correspond to the false vectors $B(f_1)$.

Although there are 16 of these inequalities, we need not base our solution on all of them. In fact, we can save ourselves a great deal of effort if we make a few simple observations. First, notice that if the inequalities corresponding to assignments (1100) and (1011) (indicated by arrows in inequalities (8.3.24)) are satisfied, then the remaining inequalities of (8.3.24) will be satisfied, since all of the weights are positive, by Theorem 8.3.5. Thus, we need to use only two of the five inequalities corresponding to the true vectors of f_1. Second, notice that if the inequalities corresponding to false vectors (1010), (1001), and (0111) (again, marked by the arrows) are satisfied, then so also will be the remaining inequalities of (8.3.25), again because we know that the weights will be positive. Thus, we end up with a set of 5 inequalities that need to be solved rather than the original 16.

Given the function f_1, it is easy to determine at the outset what this minimal set will be, on the basis of the following definitions.

Definition 8.3.3.   Let $\mathbf{a} = (a_1, a_2, \ldots, a_n)$ and $\mathbf{b} = (b_1, b_2, \ldots, b_n)$ be two binary n-tuples (i.e., n-tuples in which the a_i and b_j all take on values 0 or 1). Then $\mathbf{a}$ is greater than or equal to $\mathbf{b}$ if $a_i \ge b_i$ for all $i = 1, 2, \ldots, n$. Further, $\mathbf{a} > \mathbf{b}$ if there exists at least one a_i such that $a_i > b_i$. If $\mathbf{a}$ is neither greater than, less than, nor equal to $\mathbf{b}$, then $\mathbf{a}$ and $\mathbf{b}$ are said to be *incomparable*.

For example, let $\mathbf{a} = (110101)$, $\mathbf{b} = (100101)$, and $\mathbf{c} = (011001)$. Then, by Definition 8.3.3, $\mathbf{a} > \mathbf{b}$, but $\mathbf{c}$ is not comparable to either $\mathbf{a}$ or $\mathbf{b}$.

Definition 8.3.4.   Let $A(f)$ and $B(f)$ be the sets of true and false vectors, respectively, of switching function f. Let $\mathbf{a}$ be a true vector of f. Then $\mathbf{a}$ is a *minimal true vector of f* if $\mathbf{a} \le \mathbf{x}$ for all $\mathbf{x}$ in the set $A(f)$ to which $\mathbf{a}$ can be compared. Similarly, let $\mathbf{b}$ be a false vector of f. Then $\mathbf{b}$ is a *maximal false vector of f* if $\mathbf{b} \ge \mathbf{x}$ for all $\mathbf{x}$ in the set $B(f)$ to which $\mathbf{b}$ can be compared.

In the above example, the vectors (1100) and (1001) are minimal true vectors and the vectors (0111), (1001), and (0101) are maximal false vectors of f_1.

On the basis of these definitions and the argument given in the above example, it is easy to see that we need to solve only the set of inequalities corresponding to the minimal true vectors and maximal false vectors of a positive function f in order to determine whether it is or is not a threshold function. Thus the inequalities that need to be solved for the above example are the following:

$$
\begin{array}{lll}
w_1 + w_2 & \geq T & (1100) \\
w_1 \quad\ + w_3 + w_4 \geq T & & (1011) \\
w_1 \quad\ + w_3 \quad\ \leq T - 1 & & (1010) \\
w_1 \quad\quad\quad\ + w_4 \leq T - 1 & & (1001) \\
w_2 + w_3 + w_4 \leq T - 1 & & (0111)
\end{array}
\tag{8.3.26}
$$

A conceptually simple approach to solving systems of linear inequalities is by *variable elimination*. For example, we can eliminate the variable T in the inequalities of (8.3.26) by observing that if we use the first and third inequalities, we obtain the inequalities

$$
w_1 + w_2 \geq T \geq w_1 + w_3 + 1
$$

Similarly, we arrive at the set

$$
\begin{array}{l}
w_1 + w_2 \geq T \geq w_1 + w_4 + 1 \\
w_1 + w_2 \geq T \geq w_2 + w_3 + w_4 + 1 \\
w_1 + w_3 + w_4 \geq T \geq w_1 + w_3 + 1 \\
w_1 + w_3 + w_4 \geq T \geq w_1 + w_4 + 1 \\
w_1 + w_3 + w_4 \geq T \geq w_2 + w_3 + w_4 + 1
\end{array}
\tag{8.3.27}
$$

Upon reducing these, we find that

$$
\begin{array}{l}
w_2 \geq w_3 + 1 \\
w_2 \geq w_4 + 1 \\
w_1 \geq w_3 + w_4 + 1 \\
w_4 \geq 1 \\
w_3 \geq 1 \\
w_1 \geq w_2 + 1
\end{array}
\tag{8.3.28}
$$

which can be further reduced to

$$
\begin{array}{l}
w_2 \geq w_3 + 1 \geq 2 \\
w_2 \geq w_4 + 1 \geq 2 \\
w_1 \geq w_3 + w_4 + 1 \geq 3 \\
w_4 \geq 1 \\
w_3 \geq 1 \\
w_1 \geq w_2 + 1 \geq 2 + 1 = 3
\end{array}
\tag{8.3.29}
$$

The inequalities in group (8.3.29) are certainly consistent and can be satisfied by the weights $(w_1, w_2, w_3, w_4) = (3, 2, 1, 1)$. If we plug these values into set (8.3.27), we obtain the result that $5 \geq T \geq 5$. Thus, the function f_2 is a threshold function with a structure of [3, 2, 1, 1: 5], and by Theorem 8.3.5, the original function given in Equation (8.3.21) is a threshold function with structure [3, 2, -1, 1: 4].

8.3.4 Some Final Comments on Threshold Functions

This introduction to threshold logic has been, by necessity, brief. Entire books have been written on the subject, two of which are cited in the bibliography at the end of the chapter. One of the reasons that threshold logic has never been extensively used is seen in the difficulty of determining whether a given function is a threshold function or not. At present, there is only one more or less practical way in which this can be done, and that is by solving the system of linear inequalities associated with each function in question. We have shown here one approach to this process that is more or less suitable for "hand computation." For more complex functions, the use of linear programming makes it possible for a computer to quickly and easily carry out the solution of the required inequalities. Other approaches have been investigated, but none, to this point, are easily carried out and guarantee the identification of all threshold functions. There are some methods which can easily be carried out by hand or computer but which do not guarantee the identification of all threshold functions. Problems 8.11, 8.12, and 8.13, at the end of the chapter, explore one such method. Putting this in another way, there exist no "simple" necessary and sufficient conditions to determine whether a given switching function is a threshold function—this can only be done by solving the given system of linear inequalities.

A second difficulty that arises with threshold logic is that of determining a realization, using the minimal number of threshold gates, for switching functions that are not threshold functions. One approach which works, but requires extensive computation, is to use integer linear programming. The bibliography at the end of the chapter gives some references to the subject of linear programming and integer linear programming formulations that can be used for the general determination of threshold realizability.

For these reasons and others, threshold logic is not extensively used in the design of digital systems today. However, as we mentioned in the introduction to this section, a threshold gate, in some respects, "looks like" a neuron. This analogy comes about because the output of a neuron "fires," or is asserted, if the weighted sum of the inputs is "great enough." Neurons are generally adaptive entities, in that their output behavior can be modified by past experience. This can be done by changing the point at which the weighted sum of inputs causes the output to become asserted, and so a neuron can change its response behavior. Threshold gates are capable of doing exactly the same thing—their threshold, too, can be altered—and, as a consequence, are becoming more and more interesting to investigators involved in adaptive systems research.

8.4 FUNCTIONAL DECOMPOSITION

The simplification procedures presented in Chapter 4 resulted in a sum of products or a product of sums expression which could then be implemented by a "two-level" gate network. For an SOP expression, the first level consists of AND gates to form the product terms and the second level consists of an OR gate to form the sum. This two-level realization is always possible. However, such an approach may require more ICs in the final realization than might be necessary when other devices, such as PLAs or multiplexers, are used. As we saw in Chapter 4, a programmable logic array device (PLA) can be used to implement some very complex functions on a set of n variables with only a single integrated circuit. ROMs and (as indicated in Problem 4.13) multiplexers can also be used in a similar manner. But what happens if the function to be implemented is a function of more variables than the PLA, ROM, or MUX can handle? We will investigate one approach to this problem in this section.

8.4.1 Using a Multiplexer (MUX) to Implement General Functions

Let us begin by taking a brief look at how we can use a multiplexer to implement general functions. Figure 8.4.1 shows the 4-line to 1-line MUX designed in Section 4.3 and its truth table. Consider now how we might implement the function

$$f(x, y, z) = x\bar{y} + y\bar{z} \tag{8.4.1}$$

which is just the function required in Problem 4.13. Recall from Equation (4.3.5) and the truth table of Figure 8.4.1(b) that the output of the MUX is given by

$$Y = \bar{S}_1\bar{S}_0I_0 + \bar{S}_1S_0I_1 + S_1\bar{S}_0I_2 + S_1S_0I_3 \tag{8.4.2}$$

In general, we can write an expression of three variables as

$$g(a, b, c) = \bar{a}\bar{b}(\bar{c}K_0 + cK_1) + \bar{a}b(\bar{c}K_2 + cK_3) \\ + a\bar{b}(\bar{c}K_4 + cK_5) + ab(\bar{c}K_6 + cK_7) \tag{8.4.3}$$

where the K_i are constants determined by the value of the function for the corresponding minterm. Now, comparing Equations (8.4.2) and (8.4.3), we see that to implement a function

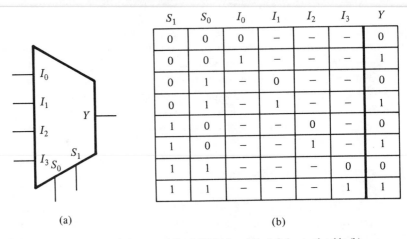

S_1	S_0	I_0	I_1	I_2	I_3	Y
0	0	0	–	–	–	0
0	0	1	–	–	–	1
0	1	–	0	–	–	0
0	1	–	1	–	–	1
1	0	–	–	0	–	0
1	0	–	–	1	–	1
1	1	–	–	–	0	0
1	1	–	–	–	1	1

(a)   (b)

Figure 8.4.1   A 4-line to 1-line MUX (a) and its defining truth table (b).

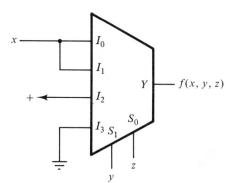

Figure 8.4.2   A MUX implementation of Equation (8.4.1).

of three variables using a 4-line to 1-line MUX, we need to assign two of the variables to the select inputs S_1 and S_0 and associate the third variable with the inputs I_i. We may observe that the terms in parentheses in Equation (8.4.3) take on one of the four values, namely, c, $\bar{c}$, 0, or 1, depending on the associated constants.[6] Thus we can connect the inputs of the MUX either to a 1, to a 0, or to c or $\bar{c}$ depending on the function being implemented.

Let us now go back and look at the function $f(x, y, z)$ given in Equation (8.4.1), which is to be implemented using the MUX of Figure 8.4.1. First, we need to decide which variables are to be associated with the select inputs S_1 and S_0. In this case, let us set $y = S_1$ and $z = S_0$. The reason for this is that x appears in Equation (8.4.1) uncomplemented, whereas both y and z appear complemented. Thus, we will not need the complement of x, and therefore, no extra inverters will be necessary in the final implementation. Next, we need to rewrite this function in the form of Equation (8.4.3). In this case, we obtain

$$f(x, y, z) = \bar{y}\bar{z}(x) + \bar{y}z(x) + y\bar{z}(1) + yz(0) \tag{8.4.4}$$

Using this equation and the associations just mentioned, the implementation of the function using a 4-line to 1-line multiplexer becomes as shown in Figure 8.4.2.

Consider one more example before going on. Suppose we wish to implement the following function using our 4-input MUX:

$$g(A, B, C) = \bar{A} + \bar{B}C + B\bar{C} \tag{8.4.5}$$

Since all of the variables appear in complemented form at least once in this function, there is no particular reason to select one of the variables over the other to be associated with the I inputs. Therefore, just for the sake of discussion, let us associate the variable C with the I inputs and let $A = S_1$ and $B = S_0$. If we now plot this function in the map shown in Figure 8.4.3, we can quickly identify what the functional association is to be with each of the MUX inputs. Specifically, the column $\bar{A}\bar{B}$ must have an associated coefficient of $\bar{C}(1) + C(1) = 1$. Similarly for column $\bar{A}B$. Column AB, on the other hand, has a coefficient

---

[6] A very interesting, and useful, method for plotting the factorization given in Equation (8.4.3) in a two-variable Karnaugh map can be found in Chap. 3 of Fletcher (refer to the annotated bibliography). Using the *variable-entered-map* technique described in this reference, each entry in a two-variable map of a and b contains either 1, 0, c, or $\bar{c}$.

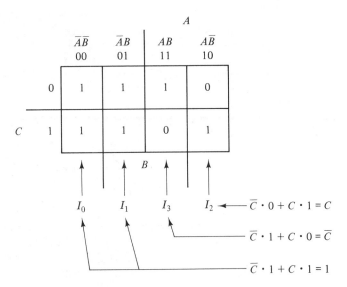

Figure 8.4.3   A plot of the function of Equation (8.4.5) showing the MUX input functions.

of $\overline{C}(1) + C(0) = \overline{C}$, while column $A\overline{B}$ has a coefficient of $\overline{C}(0) + C(1) = C$. Thus $I_0 = I_1 = $, $I_3 = \overline{C}$, and $I_2 = C$. The resulting implementation is shown in Fig. 8.4.4.

These ideas are easily extended to larger MUXs and to functions in which the MUX inputs are associated with more than a single variable. Problems at the end of the chapter investigate this a bit further. As we will next see, we can sometimes cascade MUXs to implement particularly difficult functions, thus reducing the amount of hardware necessary to implement a given function.

8.4.2 The Simple Disjoint Decomposition of Functions

Devices such as MUXs, PLAs, and ROMs can each be used to implement functions on several variables. But what happens if the number of variables required by a given function is greater than can be handled by the given device? Suppose, for example, that we are given a function of five variables and are asked to implement the function using a four-input MUX, if possible. As an example, consider the function

$$f(A, B, C, D, E) = D\overline{E} + CD + ABDE + \overline{ACDE} + \overline{BCDE}$$
$$= \Sigma\, m(1, 2, 6, 7, 9, 10, 14, 15, 17, 18, 22, \qquad (8.4.6)$$
$$23, 26, 27, 30, 31)$$

This function can clearly not be implemented with a single four-input MUX unless we choose to make the inputs to the MUX functions of at least three of the variables. In such a case, the MUX would be realizing the function given in Equation (8.4.2) with the I_i's equal to some function of the remaining three variables. For example, if we were to let

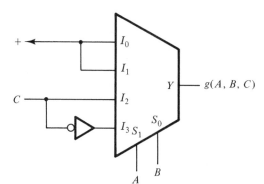

Figure 8.4.4 Final MUX implementation of Equation (8.4.5).

$S_1 = D$ and $S_0 = E$, then the I_i's would be functions of the form $g_i(A, B, C)$. We could, of course, realize each of these functions using a separate four-input MUX. In general, all of these g_i's will be different functions of the remaining variables. However, it may happen that some of the g_i's equal others, or equal the complement of others, or equal a constant 0 or 1. In such a case, the function—in this case, $f(A, B, C, D, E)$—can be factored as

$$f(A, B, C, D, E) = f(g(A, B, C), D, E) \qquad (8.4.7)$$

Such a factorization is termed a *simple disjoint decomposition of f*. We will refer to the function $g(A, B, C)$ as the *independent function*. Without loss of generality, a simple disjoint decomposition is defined as a factorization of the form

$$f(x_1, \ldots , x_n) = f(g(x_1, \ldots , x_p), x_{p+1}, \ldots , x_n) \qquad (8.4.8)$$

where p and $n - p$ are greater than or equal to 2.[7] In the current example, if we can factor f as in Equation (8.4.7), then we can implement $f(A, B, C, D, E)$ using only two MUXs, one to implement $g(A, B, C)$ and the other to implement $f(g, D, E)$. If we plot the function f as shown in Figure 8.4.5, a very striking thing appears. In this plot, each of the rows represents a function of A, B, and C which is either equal to 0 or to 1 or to some function $g(A, B, C)$ or $\bar{g}(A, B, C)$. Thus, we see that $f(A, B, C, D, E)$ does have a simple disjoint decomposition in which

$$g(A, B, C) = AB + C \qquad (8.4.9)$$

The original function can now be factored as

$$f(A, B, C, D, E) = (0)\overline{D}\overline{E} + \overline{(AB + C)}\overline{D}E \\ + (AB + C)DE + (1)D\overline{E} \qquad (8.4.10)$$

A two-MUX realization based on this factorization is shown in Figure 8.4.6.

[7] Every switching function, by the Shannon decomposition theorem, has a simple disjoint decomposition of the form

$$f(x_1, x_2, \ldots , x_n) = x_1 f_1(x_2, \ldots , x_n) + \bar{x}_1 f_0(x_2, \ldots , x_n)$$

Thus, we are interested in simple disjoint decompositions in which the n and $n - p$, in Equation (8.4.8), are 2 or greater.

ABC								
DE	000	001	011	010	110	111	101	100
00	0	0	0	0	0	0	0	0
01	1	0	0	1	0	0	0	1
11	0	1	1	0	1	1	1	0
10	1	1	1	1	1	1	1	1

Row annotations:
- 00 ← 0
- 01 ← $\overline{g}(A, B, C)$
- 11 ← $g(A, B, C)$
- 10 ← 1

Figure 8.4.5   A plot of $f(A, B, C, D, E)$ showing the simple disjunctive decomposition.

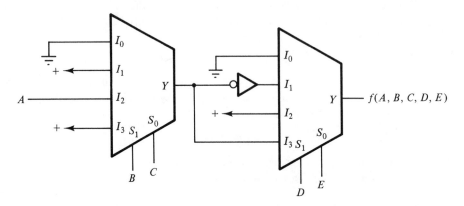

Figure 8.4.6   Final two-MUX realization of $f(A, B, C, D, E)$.

The general problem of determining whether a given switching function has a simple disjoint decomposition is not an easy one. One approach, however, which is very straight-forward, is to use a set of charts which give the location of the minterms with respect to a given partitioning of the variables into two groups of two or more variables each. Such a set of charts is commonly referred to as a *decomposition chart*.[8] For functions of four variables, there are three subcharts in the set. These are shown in Figure 8.4.7, along with the plot of the function

$$f(A, B, C, D) = \Sigma\, m(0, 1, 11, 13, 15) \qquad (8.4.11)$$

In this plot, the 1s of f are shown circled. If a simple disjoint decomposition for f exists, then in at least one of the subcharts, the rows or columns each should be identifiable with

---

[8] In many texts, "decomposition charts" also applies to the charts showing partitions having a single variable, and in at least one book (Kohavi), a zero-variable chart is included in the definition.

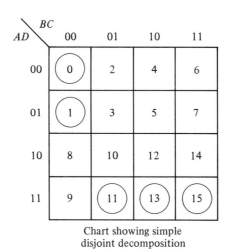

Chart showing simple
disjoint decomposition

Figure 8.4.7   A four-variable decomposition chart showing the plots of Equation (8.4.11).

the function 0, 1, g, or $\overline{g}$, where g is the independent function. From Figure 8.4.7 we see that the plot of f given in the subchart corresponding to the partition AD and BC has rows satisfying this requirement. In particular, the row corresponding to $AD = 00$ can be associated with the independent function

$$g(B, C) = \overline{B}\,\overline{C} \qquad (8.4.12)$$

as can the row corresponding to $AD = 01$. The row $AD = 10$, however, is associated with the function 0. Finally, the row $AD = 11$ corresponds to the function $\overline{g}(A, B) = B + C$. Thus a simple disjoint decomposition exists for f and is given as

$$f(A, B, C, D) = \overline{A}\,\overline{D}g(B, C) + \overline{A}Dg(B, C) + A\overline{D}(0) + AD\overline{g}(B, C)$$
$$= A(\overline{BC}) + AD(B + C) \tag{8.4.13}$$

Let us now return to the five-variable function given in Equation (8.4.6). Figure 8.4.8 shows $f(A, B, C, D, E)$ plotted in two of the ten possible subcharts of a five-variable decomposition chart. Figure 8.4.8(a) shows the plot in terms of the partition ABC and DE. From this chart, we see that the first row corresponds to 0 and the third row to the function 1. The last row can be associated with the function $AB + C$, which is just the independent function given in Equation (8.4.9); and the second row is just the complement of this function. The resulting factorization is that given in Equation (8.4.10).

Consider now the chart shown in Figure 8.4.8(b). Note that although the rows of

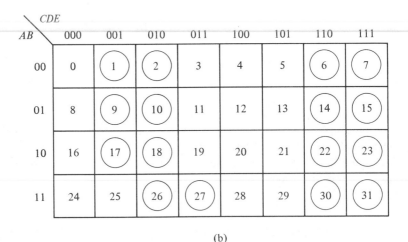

(a)

(b)

Figure 8.4.8   Two possible plots of the function of Equation (8.4.6).

this chart cannot be associated with an independent function, its complement, or the constant 0 or 1, the columns can. In this particular case, we end up with a simple disjunctive decomposition of the form

$$f(A, B, C, D, E) = 1(\overline{CD\overline{E}} + C\overline{DE} + \underline{CDE}) + \underline{AB(\overline{CDE})} + \overline{(AB)}(\overline{CDE})$$
$$= D\overline{E} + CD + (AB)\overline{C}DE + \overline{(AB)}(\overline{CDE}) \tag{8.4.14}$$

where the independent function is $g(A, B) = AB$.

There are, of course, many other forms of decomposition possible. Some of these are explored in the references given at the end of the chapter.

8.5 SYMMETRIC FUNCTIONS

As we saw in Section 8.4, it is sometimes possible to simplify the implementation of a function by using more than two levels of gates. One class of functions which generally has simpler realizations if multiple levels of logic are used is the symmetric function class, to be defined below. The implementation of symmetric functions is especially simple using bilateral devices in a non-series-parallel, or bridged, network configuration. Further, such bilateral implementations can easily be accomplished without introducing sneak paths. The purpose of this section, then, is to introduce the concept of a symmetric function. We will first introduce some basic definitions and properties of these functions. We will then show how we can determine whether a given function is a symmetric function. And finally, we will show a non-series-parallel network of bilateral elements that can be used to implement all symmetric functions on n variables.

8.5.1 Some Basic Properties of Symmetric Functions

Before introducing the concept of a symmetric function, let us first define some notation that will be useful in what follows.

Definition 8.5.1.   Define x^a as follows:

$$x^a = \overline{x} \text{ if } a = 0 \qquad \text{and} \qquad x^a = x \text{ if } a = 1 \tag{8.5.1}[9]$$

The purpose of this notation is to make it possible to refer to literals rather than variables in defining a symmetric function.

Definition 8.5.2.   A switching function $f(x_1, x_2, \ldots, x_n)$ on n variables is said to be a *symmetric function of the literals* $x_1^{j1}, x_2^{j2}, \ldots, x_n^{jn}$, or, simply, a *symmetric function*, if and only if it is invariant under any permutation of the n literals.

Take, for example, the function

---

[9] Some authors reverse this definition. See, for example, the Hill and Peterson text cited in the bibliography. The reason we choose the notation used here is that a vector of the form (100), corresponding to the superscripts on the product term $x_1^1 x_2^0 x_3^0$, produce a minterm $x_1\overline{x_2}\overline{x_3}$ which corresponds to the superscript vector.

$$f(x_1, x_2, x_3) = x_1 x_2 \bar{x}_3 + x_1 \bar{x}_2 x_3 + \bar{x}_1 x_2 x_3 \tag{8.5.2}$$

This function remains exactly the same function if we interchange, for example, the variables x_1 and x_2, as follows:

$$f(x_2, x_1, x_3) = x_2 x_1 \bar{x}_3 + x_2 \bar{x}_1 x_3 + \bar{x}_2 x_1 x_3 \tag{8.5.3}$$

Any other permutation of literals x_1, x_2, and x_3 results in precisely the same function as given in Equation (8.5.2), and thus $f(x_1, x_2, x_3)$ is a symmetric function of these literals.

Theorem 8.5.1. A necessary and sufficient condition that a switching function $f(x_1, x_2, \ldots, x_n)$ on n variables be symmetric on the literals $x_1^{j_1}, x_2^{j_2}, \ldots, x_n^{j_n}$ is that it be definable by a set of integers $M = \{a_1, a_2, \ldots, a_m\}$ such that the function takes on the value 1 when and only when a_i, $i = 1, 2, \ldots, m$, of the literals take on the value 1.

Proof. First, assume that the function f is 1 if any a_k of the literals are 1. Clearly, any permutation of a_k literals will not change the fact that the function is 1. Thus, a function is symmetric if it can be represented by the set M.

Next, assume that f is symmetric. Let $\mathbf{b} = (b_1, b_2, \ldots, b_n)$ be an assignment on the n variables for which f is 1, that is, for which $f(b_1, b_2, \ldots, b_n) = 1$. Assume that k of the n bits in this assignment are 1. Now a permutation of the literals of symmetry amounts to a permutation of the bits in the assignment $\mathbf{b}$. Since such a permutation of bits in this assignment cannot change the number of 1s, and since f is symmetric, k must be an element of the set M. A similar argument holds for all other assignments which make $f = 1$ and for which the number of 1s in the assignment differs from k. Thus, a symmetric function generates the set M.   QED

We will denote a symmetric function as $S_M(x_1^{j_1}, \ldots, x_n^{j_n})$. For example, the function of Equation (8.5.2) is a symmetric function which is 1 if any two of its variables are 1 and so will be denoted as

$$f(x_1, x_2, x_3) = S_2(x_1, x_2, x_3) \tag{8.5.4}$$

Another example will further illustrate the notation and the concept. Consider the function

$$\begin{aligned} g(A, B, C) &= S_{2,3}(A, B, \bar{C}) \\ &= ABC + A\bar{B}\bar{C} + \bar{A}B\bar{C} + AB\bar{C} \end{aligned} \tag{8.5.5}$$

This function is symmetric with the set $M = \{2, 3\}$. To verify the symmetric property, we would need to check the functions resulting from all possible permutations of the literals A, B, and $\bar{C}$; for example, the permutation $(\bar{C}, A, B) \rightarrow (A, B\,\bar{C})$. Upon making the substitutions indicated in Equation (8.5.5), i.e., replacing A by $\bar{C}$, B by A, and $\bar{C}$ by B, the resulting function would be

$$g(A, B, C) = \bar{C}A\bar{B} + \bar{C}\bar{A}B + CAB + \bar{C}AB \tag{8.5.6}$$

which is the same as given in Equation (8.5.5).

A number of properties of symmetric functions can be derived from the definition and Theorem 8.5.1. Some of these, along with some simple examples, are given below without proof. Problems at the end of the chapter will explore the validity of these results and will introduce some other interesting results.

Theorem 8.5.2.   Let S_M and S_N be two symmetric functions on the same set of n literals. Then

(i) $S_M + S_N = S_P$ where $P = M \cup N$

(ii) $S_M S_N = S_Q$ where $Q = M \cap N$

where the symbols $\cup$ and $\cap$ represent the set union and the set intersection, respectively.

A simple example can be shown by considering the following functions $S_{0,1}$ and $S_{1,4}$, which are symmetric functions on the same set of four literals:

$$S_{0,1} + S_{1,4} = S_{0,1,4}$$
$$S_{0,1} S_{1,4} = S_1$$

The reader should verify these identities by expanding both functions into SOP expressions and then ANDing and ORing these expressions are indicated.

Theorem 8.5.3.   Let S_M be a symmetric function on some set of n literals. Then

$$\overline{(S_M)} = S_{\overline{M}}$$

where $\overline{M}$ is simply the set of numbers in the range 0 to n that are not in the set M.

For example, consider the symmetric function on four variables given above, namely, $S_{0,1}$. By Theorem 8.5.3, the complement of this function becomes

$$\overline{(S_{0,1})} = S_{2,3,4}$$

Theorem 8.5.4.   Let $S_M(x_1^{j1}, x_2^{j2}, \ldots, x_n^{jn})$ be a symmetric function on the n literals $x_1^{j1}, x_2^{j2}, \ldots, x_n^{jn}$. Let $M = \{a_1, a_2, \ldots, a_k\}$. Then, without loss of generality,

$$S_M(x_1^{j1}, x_2^{j2}, \ldots, x_n^{jn}) = x_1^{j1} S_N(1 \cdot x_2^{j2}, \ldots, x_n^{jn}) + \overline{(x_1^{j1})} S_M(0 \cdot x_2^{j2}, \ldots, x_n^{jn})$$

where $N = \{a_1 - 1, a_2 - 1, \ldots, a_k - 1\}$, in which $0 - 1$ is ignored.

As an example, the function

$$
\begin{aligned}
f(A, B, C) &= S_{0,2}(A, B, C) \\
&= \overline{AB}\overline{C} + AB\overline{C} + A\overline{B}C + \overline{A}BC \\
&= A(B\overline{C} + \overline{B}C) + \overline{A}(\overline{B}\overline{C} + BC) \\
&= AS_1(B, C) + \overline{A}S_{0,2}(B, C)
\end{aligned}
\qquad (8.5.7)
$$

8.5.2 The Identification of Symmetric Functions

The identification of symmetric functions is not always easy. A simple, brute-force approach would be to simply try all permutations of the variables and observe the resulting functions. A far less exhaustive approach can be found if we first understand some of the characteristics of these functions.

Let us begin by assuming that we are dealing with functions which are *symmetric in uncomplemented variables only*. Actually, we can do this without loss of generality, because any symmetric function which involves complemented variables can be transformed to a symmetric function of uncomplemented variables only by simply substituting new, uncomplemented variables for the complemented variables. For example, the function $g(A, B, C)$ of Equation (8.5.5), which is symmetric in the variables A, B, and $\overline{C}$, can be converted to a function $h(A, B, D)$ that is symmetric in the uncomplemented variables A, B, and D simply by replacing every $\overline{C}$ in Equation (8.5.5) by D. The resulting function becomes

$$\begin{aligned} h(A, B, D) &= g(A, B, C\,|\,D \to \overline{C}) \\ &= AB\overline{D} + A\overline{B}D + \overline{A}BD + ABD \\ &= S_{2,3}(A, B, D) \end{aligned} \qquad (8.5.8)$$

Now, let $S_M(x_1, x_2, \ldots, x_n)$ be a symmetric function and let k be an element of the set M. From the proof of Theorem 8.5.1, there must be $\binom{n}{k}$ minterms which have exactly k 1s, where the notation $\binom{n}{k}$ indicates the number of combinations of n things taken k at a time, which is defined as follows:

$$\binom{n}{k} = \frac{n!}{(n-k)!k!} \qquad (8.5.9)$$

For example, the minterms for the function $S_{1,3}(A, B, C)$ are just 001, 010, 100, and 111, which are the three minterms having one of three variables set to 1 and the one minterm having three of three variables set to 1. Note also that the number of minterms for which $A = 1$ is 2, which is the number of minterms for which $B = 1$ and is also the number for which $C = 1$. In general, the number of minterms in which a variable equals 1 must be the same for all variables, since the function is symmetric. We can summarize these facts by listing the minterms, in binary form, in a table as follows:

A	B	C	Row sum
0	0	1	1
0	1	0	1
1	0	0	1
1	1	1	3
2	2	2	Column sum

The column sum in this table indicates the number of minterms for which the respective variables (or, more generally, literals) are 1. The row sums are used to count the number of occurrences of minterms having a specific number of literals. In this case there are three minterms having a single variable equal to 1 and one minterm having all three variables equal to 1. This is, of course, precisely what is required for the function $S_{1,3}(A, B, C)$.

Let us now apply these ideas to see whether the following function is symmetric in its uncomplemented variables:

$$f(A, B, C, D) = \Sigma\, m(3, 5, 6, 7, 9, 10, 11, 12, 13, 14) \qquad (8.5.10)$$

We will begin our investigation by making up a table, as was done for the function $S_{1,3}(A, B, C)$ given above. This is shown in Table 8.5.1. This table shows that the column sums for all of the variables are the same and so the function may be symmetric. We need to check the row sums next to see that they occur the correct number of times. In this table there are two types of minterms: those having two 1s and those having three 1s. If this function is symmetric, then there must be $\binom{4}{2} = 6$ rows having a row sum of 2, and there must be $\binom{4}{3} = 4$ rows having a row sum of 3. We can see from Table 8.5.1 that both of these conditions are satisfied and so this function must be symmetric and equal to

$$f(A, B, C, D) = S_{2,3}(A, B, C, D) \qquad (8.5.11)$$

The reader can easily verify this by writing out the function in minimal SOP form and then permuting the variables in various ways.

Let us next consider the function $g(A, B, C, D)$ shown in Table 8.5.2(a). In this table, we see that all of the column sums are not the same, and thus we know that if this function is symmetric, it is not symmetric in all of its uncomplemented variables. We do not, however, at this stage have any evidence to indicate that g is not symmetric in some

TABLE 8.5.1 TESTING
THE FUNCTION $f(A, B, C, D)$.

A	B	C	D	Row sum
0	0	1	1	2
0	1	0	1	2
0	1	1	0	2
0	1	1	1	3
1	0	0	1	2
1	0	1	0	2
1	0	1	1	3
1	1	0	0	2
1	1	0	1	3
1	1	1	0	3
6	6	6	6	Column sum

TABLE 8.5.2 TESTING TABLE FOR $g(A, B, C, D) = \Sigma\, m(1, 2, 4, 7, 8, 13, 14)$

A	B	C	D	Row sum		A	$\overline{B}$	C	D	Row sum
0	0	0	1	1		0	1	0	1	2
0	0	1	0	1		0	1	1	0	2
0	1	0	0	1		0	0	0	0	0
0	1	1	1	3		0	0	1	1	2
1	0	0	0	3		1	1	0	0	2
1	1	0	1	3		1	0	0	1	2
1	1	1	0	3		1	0	1	0	2
3	4	3	3	Column sum		3	3	3	3	Column sum
			(a)						(b)	

set of literals. We may note that this function consists of seven minterms. Thus a column sum of 4 in this table can be made a column sum of $7 - 4 = 3$ if the column is complemented. By complementing the B column to produce the table shown in Table 8.5.2(b), we obtain column sums of 3 for all of the variables. Now if the row sums occur the requisite number of times, the function g will be a symmetric function on the variables A, $\overline{B}, C$, and D. In this case, we must have $\begin{pmatrix} 4 \\ 2 \end{pmatrix} = 6$ rows having a row sum of 2 and one

row having a row sum of 0, which is, in fact, the case here. Thus the function $g(A, B, C, D)$ is symmetric and equal to

$$g(A, B, C, D) = S_{0,2}(A, \overline{B}, C, D) \tag{8.5.12}$$

The preceding example illustrates, in an indirect manner, one mechanism for determining whether or not a given function is symmetric. Specifically, we have the following theorem:

Theorem 8.5.5.   A switching function on n variables cannot be symmetric in any set of literals if either (a) more than two column sums occur or (b) exactly two column sums occur whose sum does not equal the number of minterms of the function.

The validity of this result follows from the fact that in either case, complementation of any set of columns cannot result in a column sum that is the same for all of the columns. With this theorem, we now have a very powerful method to determine whether a function is *not* symmetric. This, unfortunately, does not immediately simplify the job of determining whether the function *is* symmetric.

Let us next consider the example shown in Table 8.5.3. We see in this case that although the column sums are all the same, the row sums do not match the requirements. In this case, that amounts to having one row with a sum of 0, one row with a sum of 4, and six rows with a sum of 2. Although the row sums are not correct, the function may still be symmetric in some set of literals. Unfortunately, in this case, since the column sums are all the same, we have no clue as to which variables need to be

TABLE 8.5.3 A SET OF TESTING TABLES FOR THE FUNCTION $h(A, B, C, D) = \Sigma\, m(0, 5, 6, 9, 10, 15)$

A	B	C	D	Row sum
0	0	0	0	0
0	1	0	1	2
0	1	1	0	2
1	0	0	1	2
1	0	1	0	2
1	1	1	1	4
3	3	3	3	Column sum

(a)

$A = 0$

B	C	D	Row sum
0	0	0	0
1	0	1	2
1	1	0	2
2	1	1	Column sum

$A = 0$

B	C	D	Row sum
0	0	1	1
0	1	0	1
1	1	1	3
1	2	2	Column sum

(b)

$A = 0$

$\overline{B}$	C	D	Row sum
1	0	0	1
0	0	1	1
0	1	0	1
1	1	1	Column sum

$A = 1$

$\overline{B}$	C	D	Row sum
1	0	1	2
1	1	0	2
0	1	1	2
2	2	2	Column sum

(c)

complemented. However, Theorem 8.5.4 may be of some help here. First we will select an arbitrary variable of h, say A, and then factor the function using the Shannon decomposition theorem to obtain

$$h(A, B, C, D) = \overline{A}h(0, B, C, D) + Ah(1, B, C, D) \qquad (8.5.13)$$

Now if the two functions $h(0, B, C, D)$ and $h(1, B, C, D)$ are symmetric in the same set of literals, then by Theorem 8.5.4 the original function $h(A, B, C, D)$ is also symmetric. Table 8.5.3(b) shows the resulting plots. We see from this plot that if we complement either column B or columns C and D, symmetric functions will result. Complementing column B produces the plot shown in Table 8.5.3(c), from which we see that from Equation (8.5.13) and Theorem 8.5.4 the function $h(A, B, C, D)$ becomes

$$\begin{aligned}
h(A, B, C, D) &= \overline{A}S_1(\overline{B}, C, D) + AS_2(\overline{B}, C, D) \\
&= S_2(\overline{A}, \overline{B}, C, D)
\end{aligned} \qquad (8.5.14)$$

This example illustrates, again in an indirect way, another condition that will indicate that a given switching function cannot be symmetric in any set of literals. In this example, the column sums were all the same, namely, 3. Further, this column sum was exactly half of the number of rows. Thus, any column could be complemented without changing this column sum. Because of this we were able to complement a subset of the columns, A and B in this case, to produce a table showing that the original function was symmetric. If the column sums had been the same but not equal to half of the number of rows, then complementing any subset of the columns, except all of them, would have resulted in a table in which the column sums differed, and thus the function could not have been symmetric.

Complementing all of the columns cannot change the fact that the row sums are not the required values (why?). Thus we have the following theorem.

Theorem 8.5.6. Any switching function that has a testing table in which the column sums are equal but not equal to exactly half of the number of minterms and that does not have the requisite number of row sums *cannot* be symmetric in any set of literals.

Theorems 8.5.4, 8.5.5, and 8.5.6 give us the basic tools necessary to determine whether or not a given function is symmetric. The determination process can be summarized in the following steps:

Step 1. Prepare a testing table and obtain the column and row sums. Next, check the table to determine on the basis of Theorems 8.5.5 and 8.5.6 whether the function cannot be a symmetric function, as follows:
- (a) If there are more than two column sums or if there are exactly two column sums the sum of which is not equal to the number of rows, *stop—the function is not symmetric*.
- (b) If there is exactly one column sum and it is not equal to exactly half of the number of rows in the table and the row sums do not occur the requisite number of times, then *stop—the function is not symmetric*.

Go to step 2.

Step 2. If the column sums are all the same, then proceed to step 3. Otherwise, complement the necessary columns to make the column sums the same. If the resulting row sums occur the required number of times, then the function is symmetric. Otherwise, it is not.

Step 3. If the column sums are all the same and the row sums occur the required number of times, the function is symmetric. If the row sums do not occur the necessary number of times and the column sum is equal to half the number of rows, then expand the function about any one of its variables to produce a function of the form

$$f(x_1, x_2, \ldots, x_n) = \bar{x}_1 h(x_2, \ldots, x_n) + x_1 g(x_2, \ldots, x_n)$$

Test the functions h and g to determine whether they are symmetric in the same set of literals. If either or both is not symmetric or if both are symmetric but not on the same set of literals, then f is not symmetric. If both h and g are symmetric in the same set of literals, then f is symmetric, by Theorem 8.5.4.

8.5.3 Bilateral-Network Realizations of Symmetric Functions

The implementation of a symmetric function in a bilateral network is particularly easy. Figure 8.5.1 shows a general network that realizes all of the symmetric functions that are symmetric on the literals a, b, and c (i.e., the uncomplemented variables). The extension

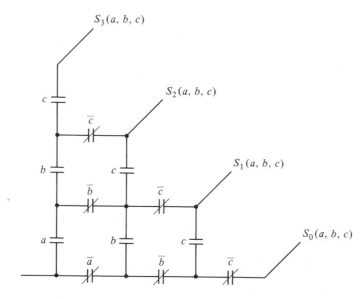

Figure 8.5.1   A three-variable bilateral network that realizes all of the symmetric functions on the three literals a, b, and c.

of this pattern to four and more variables is obvious. To obtain a function such as $S_{0,1}$, we simply tie the given outputs together, as shown in Figure 8.5.2(a). After removing the unused contacts, we obtain the network shown in Figure 8.5.2(b). Noting that the contacts c and $\bar{c}$ now are in parallel, the circuit can be simplified further as shown in part (c) of the figure.

The implementation of functions which are symmetric in one or more complemented variables follows in exactly the same manner except that the contacts corresponding to the complemented variables are complemented. For example, the implementation of the function $S_2(a, \bar{b}, c)$ is shown in Figure 8.5.3.

8.6 ITERATIVE NETWORKS

In Section 4.3, we introduced the idea of an iterative network in the design of the adder and comparator circuits. In these examples we designed a circuit to implement the specified function for a single bit and then cascaded the result to implement the function for n bits. In order to do this we had to create our designs in such a way that information could be passed not only from input to output but from one bit position to the next. In the case of the adder, information was passed from right to left, i.e., from low-order bit to high-order bit. In the case of the comparator, this bit-to-bit transfer of information went from left to right. In general, of course, this bit-to-bit information could be required to flow in both

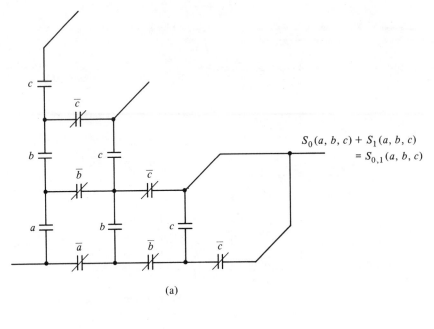

$$S_0(a, b, c) + S_1(a, b, c)$$
$$= S_{0,1}(a, b, c)$$

(a)

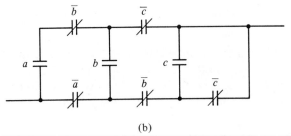

(b)

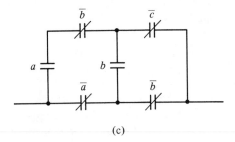

(c)

Figure 8.5.2   The implementation of $S_{0,1}(a, b, c)$: (a) initial implementation; (b) implementation after removing the unused contacts; (c) final, simplified implementation.

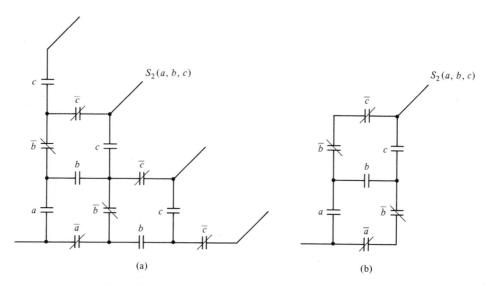

Figure 8.5.3 Implementation of $S_2(a, b, c)$: (a) initial; (b) final.

directions.[10] Figure 8.6.1 shows a model for a network of this type in which information flows from left to right. Such a network, one made up of a cascade of identical *cells*, is called an *iterative network*. The cell in the model of Figure 8.6.1 consists of some switching network which has cell inputs $\mathbf{X}_i$, cell outputs $\mathbf{F}_i$, and what we will refer to here as *secondary* variables consisting of the secondary inputs $\mathbf{C}_{i+1}$ and secondary outputs $\mathbf{C}_i$, where the boldface notation implies that, generally, there is more than one variable involved. The secondary variables carry the intercell information. Although the switching network making up each cell is usually combinational, there is no reason why it could not be a sequential circuit. As we shall see in a moment, the secondary variables serve a function remarkably similar to that of the state variables in a sequential circuit.

The synthesis process is most easily explained by using an example. Suppose we are required to determine whether an n-bit number contains four or more 1s. As was the case in the adder problem of Chapter 4, writing and implementing an equation that realizes this function for even reasonably small values of n, such as, for example, 8, would be quite difficult. We will therefore use the idea of an iterative network to carry out the design. To begin the design, consider cell i. In this case, the cell input is just the ith bit of the input number. The cell output is to be 1 if there are four or more bits that are 1 in the set from the ith bit to the highest-order bit. This cell output is 0 otherwise. The cell secondary inputs must then carry the information about the number of 1s to the left of the ith bit, and the cell secondary outputs must pass on the number of 1s to the left of the $(i - 1)$st bit. Figure 8.6.2 shows a *cell table* for this iterative network cell. Each row in this cell table, corresponding to the secondary inputs to cell i, indicates the particular number of

---

[10] Circuits of this type, requiring information flow in both directions, possess feedback and thus have the characteristics of sequential circuits. Because of this we will not discuss such circuits here. However, we will show in one of the problems at the end of the chapter a technique that can be used for handling this situation.

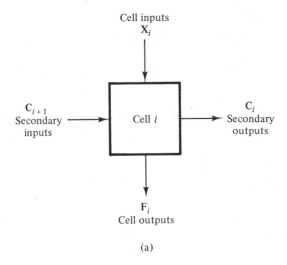

Cell inputs
X_i

C_{i+1}
Secondary
inputs

Cell i

C_i
Secondary
outputs

F_i
Cell outputs

(a)

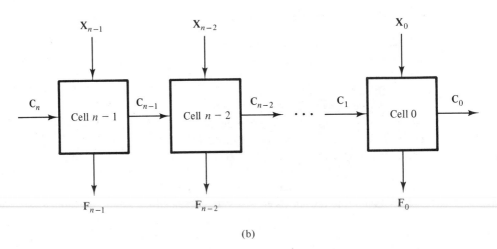

X_{n-1} X_{n-2} X_0

C_n Cell $n-1$ C_{n-1} Cell $n-2$ C_{n-2} ... C_1 Cell 0 C_0

F_{n-1} F_{n-2} F_0

(b)

Figure 8.6.1 A model for an iterative network: (a) model for the cell; (b) A cascade of cells.

1s seen to the left of the current cell. The entries in each row show that the secondary outputs, which are the secondary inputs to cell $i-1$, should be. These secondary outputs then represent the number of 1s to the left of the $(i-1)$st cell.

Since there are five rows in the cell table of Figure 8.6.2, we will need three secondary variables to distinguish these rows. Figure 8.6.3 shows one of the many possible encoded cell tables for this function. By plotting the secondary outputs and the cell outputs in a Karnaugh map, we see that the resulting functions become

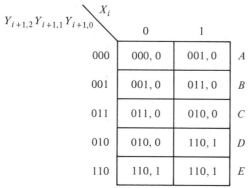

Figure 8.6.2 The cell table for an iterative network which checks for four or more 1s.

$Y_{i+1,2}Y_{i+1,1}Y_{i+1,0}$ \ X_i	0	1	
000	000, 0	001, 0	A
001	001, 0	011, 0	B
011	011, 0	010, 0	C
010	010, 0	110, 1	D
110	110, 1	110, 1	E

$Y_{i,2}, Y_{i,1}, Y_{i,0}, F_i$

Figure 8.6.3 The encoded cell table for the four 1s checker.

$$F_i = Y_{i,2} = Y_{i+1,2} + Y_{i+1,1}\overline{Y}_{i+1,0}X_i$$
$$Y_{i,1} = Y_{i+1,1} + Y_{i+1,0}X_i \qquad (8.6.1)$$
$$Y_{i,0} = Y_{i+1,0}\overline{X}_i + \overline{Y}_{i+1,1}X_i$$

The resulting implementation is shown in Figure 8.6.4(a). Concatenating these cells produces the final iterative network shown in Figure 8.6.4(b). Note here that the secondary inputs to the leftmost cell are all 0, corresponding to the situation of no 1s to the left. Finally, we note that the circuit output, the one that shows that the input number **X** contains four or more 1s, is just the output of the rightmost cell, F_0.

If we look at the cell table of Figure 8.6.2, we notice a remarkable similarity to the state table of a sequential machine. In fact, there is an interesting correspondence between these two switching circuit types. If we were given the problem of developing a clocked sequential circuit having a single input and a single output which was to be 1 whenever the number of 1s in the string of input bits was 4 or greater, we would derive a state table

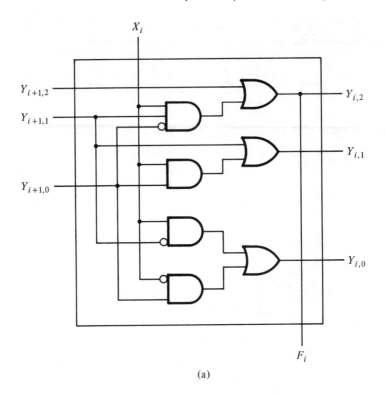

(a)

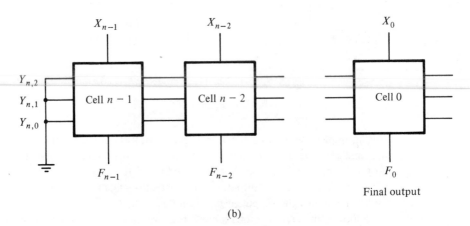

(b)

Figure 8.6.4 A final realization for the four 1s checker: (a) the cell implementation; (b) the iterated set of cells.

that was identical to the cell table of Figure 8.6.2. In the case of the sequential circuit, we have a time sequence of inputs X; and in the iterative network we present all of the inputs at one time. Both circuits perform the same function, but one does it in *time* and the other in *space*. This space-time trade-off can sometimes be used to advantage.

8.7 AN ANNOTATED BIBLIOGRAPHY

A classic text by Caldwell presents an extensive treatment of contact networks. Good discussions of this topic can also be found in Kohavi and the book by Hill and Peterson.

CALDWELL, S. H., *Switching Circuits and Logic Design*, Wiley, New York, 1958.

HILL, J. F., and G. R. PETERSON, *Introduction to Switching Theory and Logical Design*, 3rd ed., Wiley, New York, 1981.

KOHAVI, Z., *Switching and Finite Automata Theory*, 2nd ed., McGraw-Hill, New York, 1978.

A very readable book dealing with VLSI design is the text by Weste and Eshraghian, in which the authors use contact networks to describe the functioning of certain CMOS transistor circuits. For the reader who would like to get a bit more information on VLSI design, the classic book by Mead and Conway is a good choice, while the book by Muroga covers a tremendous number of related topics and presents a very complete list of references.

MEAD, C., and L. CONWAY, *Introduction to VLSI Systems*, Addison-Wesley, Reading, Mass., 1980.

MUROGA, S., *VLSI Systems Design*, Wiley, New York, 1982.

WESTE, N., and K. ESHRAGIAN, *Principles of CMOS VLSI Design: A Systems Approach*, Addison-Wesley, Reading, Mass., 1985.

The principal classic texts that deal with threshold logic are those by Muroga and by Lewis and Coates. Threshold logic is discussed as well in chapters of Hill and Peterson and of Kohavi.

LEWIS, P. M. II, and C. L. COATES, *Threshold Logic*, Wiley, New York, 1967.

MUROGA, S., *Threshold Logic and Its Applications*, Wiley, New York, 1971.

Two excellent books that deal with linear programming are those by Gass and by Cooper and Steinberg. Both are quite readable. In addition to linear programming, Cooper and Steinberg discuss other optimization problems and methods of solution. The threshold logic text by Muroga also shows the appropriate formulation of the linear programming problem for determining whether a given function is a threshold function.

COOPER, L., and D. STEINBERG, *Introduction to Methods of Optimization*, Saunders, Philadelphia, 1970.

GASS, S. I., *Linear Programming; Methods and Applications*, 2nd ed., McGraw-Hill, New York, 1964.

There are many books that describe the use of the multiplexer in the implementation of switching functions. These include Hill and Peterson, Mano, and Givone. In addition, Chapter 4 of Ercegovac and Lang goes into great detail, with many examples, on the use not only of multiplexers but of demultiplexers as well.

ERCEGOVAC, M. D., and T. LANG, *Digital Systems and Hardware/Firmware Algorithms*, Wiley, New York, 1985.

GIVONE, D. D., *Introduction to Switching Circuit Theory*, McGraw-Hill, New York, 1970.

MANO, M. M., *Digital Logic and Computer Design*, Prentice-Hall, Englewood Cliffs, N.J., 1978.

An extensive treatment of functional decomposition may be found in either Hill and Peterson or Kohavi. These books also do a very good job of discussing symmetric functions.

Iterative network design is discussed by numerous authors. Notable texts are those by Hill and Peterson, Kohavi, Ercegovac and Lang, Friedman, Givone, and Roth.

FRIEDMAN, A. D., *Fundamentals of Logic Design and Switching Theory*, Computer Science Press, Inc., Rockville, Md., 1986.

ROTH, C. H., *Fundamentals of Logic Design*, 2nd ed., West Publishing, St. Paul, 1978.

8.8 PROBLEMS

8.1. Using contact networks, implement the following functions. Reduce your implementation as much as possible.
   (a) $f(a, b, c, d) = ab\bar{c} + a\bar{b}d + \bar{b}c + \bar{a}d$
   (b) $h(A, B, C, D) = AB\bar{C} + \bar{A}D + \overline{ADBC}$
   (c) $g(w, x, y, z) = \Sigma\, m(7, 11, 13, 14)$

8.2. Design a contact network that can turn a light on or off independently from three different locations (a three-way switch).

8.3. How many possible paths are there in the contact network shown in Figure P8.3? Under what conditions will sneak paths occur in this circuit?

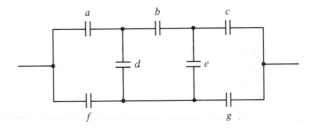

Figure P8.3

8.4. The contact network shown in Figure P8.4 implements the function

$$f(A, B, C) = AB + AC + BC$$

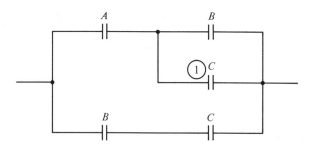

Figure P8.4

Show that the top C contact (marked 1 in the figure) can be eliminated if a transfer contact is introduced.

8.5. Generally, a transfer contact is cheaper than two pairs of contacts. With this in mind, find a minimal contact realization for the function given in Equation (8.2.3).

8.6. The function given in Equation (8.3.1) is unate. Show that all of its prime implicants are essential.

8.7. The function specified by the hyperplane shown in Figure 8.3.1 is a threshold function and is unate. Show that the two prime implicants of this function share a minterm. What is this minterm?

8.8. Prove Theorem 8.3.3.

8.9. Determine which of the following functions are threshold functions by setting up and solving the system of linear inequalities associated with the minimal true and maximal false vectors. You will want to make the functions positive first.
  (a) ABC
  (b) $A + \overline{B}\,\overline{C}$
  (c) $\overline{A}(B + C)$
  (d) $AB + AC + \overline{B}\,\overline{C}$
  (e) $AB + BC$
  (f) $AB + CD$

8.10. Prove that the function of Equation (8.3.15) is not a threshold function by showing that no structure can be found that satisfies Definition 8.3.1.

8.11. *Prove*: Let $g(\mathbf{x})$ be a positive threshold function on n variables with structure $[\mathbf{w}; T]$. Then
  (a) $yg(\mathbf{x})$ is a threshold function with structure $[w_y, \mathbf{w}; T_y]$, where

$$w_y \geq w_i + 1 - T \qquad \text{and} \qquad T_y = T + w_y$$

  (b) $y + g(\mathbf{x})$ is a threshold function with structure $[w_y, \mathbf{w}; T_y]$, where

$$w_y \geq T \qquad \text{and} \qquad T_y = T$$

8.12. *Prove*: Let $f_1(\mathbf{x})$ and $f_2(\mathbf{x})$ be two positive threshold functions on the same set of n variables with structures $[\mathbf{w}; T_1]$ and $[\mathbf{w}; T_2]$, respectively. Assume that $T_2 > T_1$. Then

$$yf_1(\mathbf{x}) + f_2(\mathbf{x})$$

is also a threshold function, with structure $[w_y, \mathbf{w}; T_y]$, where

$$w_y = T_2 - T_1 \qquad \text{and} \qquad T_y = T_2$$

8.13. Using the results of Problems 8.11 and 8.12, determine structures for each of the following threshold functions:

(a) $AB + AC + AD + BCD$

(b) $ABC + ABD$

(c) $A\overline{B} + A\overline{C}\overline{D} + \overline{B}\overline{C}\overline{D}$

(d) $AB + ACD + ACE + ACF + A\overline{D}E + \overline{B}\overline{C}\overline{D} + \overline{B}CE$

8.14. Implement the following functions, using one 4-line to 1-line MUX:

(a) $A\overline{B} + A\overline{C} + \overline{B}C$

(b) $A(B + \overline{C})$

(c) $A\overline{B}\overline{C} + \overline{A}B\overline{C} + \overline{A}\overline{B}C + ABC$

(d) $\overline{A}B\overline{C} + A\overline{C}D + BCD$

8.15. Determine which of the following functions have simple disjoint decompositions, and identify the independent function.

(a) $f(w, x, y, z) = \Sigma\, m(0, 2, 4, 6, 10, 11, 14, 15)$

(b) $g(A, B, C, D) = \overline{A}B + \overline{A}C + \overline{A}D + BCD$

(c) $h(a, b, c, d) = b\overline{c} + a\overline{c}d + \overline{a}\overline{c}d + \overline{a}bcd + ab\overline{c}d$

(d) $F(W, X, Y, Z) = \Pi\, M(1, 2, 3, 5, 6, 9, 11, 12, 14)$

8.16. Derive a complete five-variable decomposition chart. Use this chart to determine any and all simple disjoint decompositions for the following functions:

(a) $ABD\overline{E} + AB\overline{D}E + CD\overline{E} + C\overline{D}E + \overline{A}CD + \overline{B}CD$

(b) $\Sigma\, m(0, 1, 4, 7, 8, 9, 10, 11, 12, 13, 14, 15, 18, 19, 21, 22)$

(c) $\Sigma\, m(3, 6, 7, 8, 9, 10, 12, 13, 18, 19, 23, 24, 25, 28, 29, 30)$

(d) $\overline{A}BC\overline{E} + \overline{A}\overline{B}CD + \overline{A}\overline{B}C + ABE + \overline{B}E + \overline{C}DE$

8.17. Prove Theorem 8.5.2. (*Hint:* Consider the set of true vectors for each function as defined in Equations (8.3.2) through (8.3.4).)

8.18. Prove Theorem 8.5.3.

8.19. *Prove:* If S_M is a symmetric function, then

$$\overline{(S_m)} = S_N$$

where $N = \{n - a_1, n - a_2, \ldots, n - a_m\}$.

8.20. Is the dual of a symmetric function symmetric?

8.21. Design an iterative network cell having a single input, X_i, and a single output, Z_i, such that $Z_i = 1$ if the number of 1s on the inputs to the right of cell i and including cell i is odd.

8.22. Design an iterative network cell having a single iniput, X_i, and a single output, Z_i, such that $Z_i = 1$ if $X_{i+2} = X_i$ and zero otherwise.

8.23. An iterative network cell is to be designed having one input, X_i, and one output, Z_i, such that $Z_i = 1$ if $(X_{i+2}, X_{i+1}, X_i) = (1, 0, 1)$ and 0 otherwise. Design the cell and show the necessary boundary values for the secondary variables.

8.24. Design an iterative network cell having one input, X_i, and one output, Z_i, such that $Z_i = 1$ if there are an even number of 1s to the left of cell i *and* an even number of 1s to the right of cell i. (*Hint:* It is probably easiest to break this design into two parts; one part to check for an even number of 1s to the left and one part to check for an even number of 1s to the right.)

Large-Scale System Design

9

9.1 INTRODUCTION

Up to this point we have dealt with the analysis and design of rather small digital systems. The real world is generally a bit more complex. The small-scale systems we have been dealing with are characterized by having a few states and a very small number of inputs and outputs. In the case of a computer, it is easily seen that the design would involve an extremely large number of inputs and outputs and an equally large number of states to implement all of the necessary instructions. Trying to carry out a design of such complexity with the tools discussed in the last few chapters would be very difficult indeed. The usual practice in engineering, when confronted with very complex design problems, is to break the design up into smaller, more manageable pieces and then, by connecting the pieces, implement the desired system.

Figure 9.1.1 shows a model for large-scale systems that can be used to simplify the design process.[1] Basically, we can think of a large digital system as consisting of two main parts: a processing section and a control section. The processing section, or processing unit, takes information on its inputs, processes this information in some well-defined manner, and then produces outputs which represent the desired system response. It is the principal responsibility of the control section to ensure that this process is carried out in the correct manner. This is done by sending information to the processing unit to tell it what is to be done at each step of the process. The control unit also gets information back from the processing unit telling the controller how the process is going. This information is then

[1] This model can also be used for a computer by adding a block for memory. Such a computer model is generally referred to as the von Neumann model.

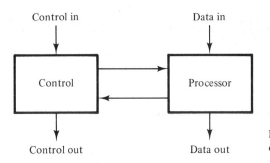

Figure 9.1.1   A model of a large-scale digital system.

used by the controller to set up the next step in the required process. The controller may also receive information from outside the system that can affect the process, and it can also generate control information to be used by other, external systems.

In this chapter we will examine the design process and the components involved in the processing unit and then take a look at some general methods for specifying and designing the controller. We will then go through two complete design examples to illustrate the processes described.

9.2 REGISTERS

The processing unit consists of, among other things, combinational logic for carrying out arithmetic and logical operations. This logic is designed using methods already described. Since information coming in is processed over some finite period of time, the processing unit must also have hardware to temporarily store both incoming information and intermediate results. This is the function of a *register*. A register is a collection of *binary cells*, each of which stores one bit of information. Usually, a binary cell is made up of a flip-flop, and so a register is an ordered collection of flip-flops.

A particularly useful type of flip-flop is one that, in addition to its normal characteristics, has preset and clear capability, i.e, a flip-flop that can be set to 1 or cleared to 0 without being clocked. The 7474 and 74LS76, which are edge-triggered D and JK flip-flops, respectively, are good examples of flip-flops possessing this characteristic. These devices were discussed in the last two chapters and will be used extensively in what follows. The reader should review the defining truth tables for those flip-flops given in Figure 7.6.1.

9.2.1 Storage Registers

There are many types of registers useful in the design of large-scale systems. Perhaps the most common is used to temporarily store information, usually in the form of one or two bytes.[2] Figure 9.2.1 shows a schematic of a *storage register* using edge-triggered D flip-flops. Latch-mode D flip-flops, or transparent latches, such as the one shown in Figure

---

[2] Two bytes of information is quite often referred to as a *word*, and four bytes may be referred to as a *double word*.

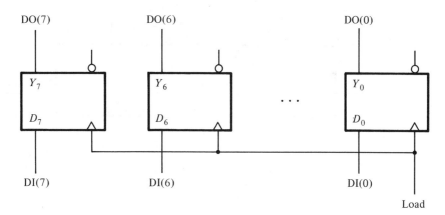

Figure 9.2.1   An 8-bit storage register.

5.2.5, are also quite often used for this type of register. In Figure 9.2.1, information that is present on the input lines DI(i) will appear on the output lines DO(i) when the clock is asserted. This information will not change until the next clock pulse occurs, and during the intervening time, the stored data is available for use by other processing elements in the system.

Every moderately complex system may use several registers of varying size and function. It would clearly be inconvenient, and perhaps a bit confusing, to show all of these various registers in the detail given in Figure 9.2.1. What we will do to avoid this complexity is to replace the detailed schematic diagram with a block diagram showing the various inputs, outputs, and control lines associated with the register. Figure 9.2.2(a) shows such a block diagram symbol for the storage register of Figure 9.2.1. As systems grow in complexity and size to what one might encounter in the design of a computer, even this simple symbol may be too detailed for clear presentation of the entire system—and this, after all, is what a schematic drawing is supposed to do. Figure 9.2.2(b) shows a more concise symbol for the storage register. In this symbol the hash mark (/) on a line indicates that this line actually represents not just one wire, but several wires. The number of wires represented is given by the number adjacent to the hash mark. The labels DO⟨7:0⟩ and DI⟨7:0⟩ are used to indicate that the eight lines going out are given the names DO(7), DO(6), etc., and those coming in are given corresponding names. The convention taken here is that the left number of the 7:0 is the index given to the most significant bit, while the right number, 0 in this case, is the index given to the least significant bit. The remaining indices are assigned in successive order.[3]

---

[3] A new logic symbology standard, IEEE Std. 91–1984, has recently been proposed which basically uses a uniform symbol to represent all register functions. Although it has some features that would strongly recommend it, and many more that would not, this standard will not be used here. The Appendix gives an introduction to this standard and shows the standard symbols used for many of the registers, counters, and similar devices, designed in this chapter.

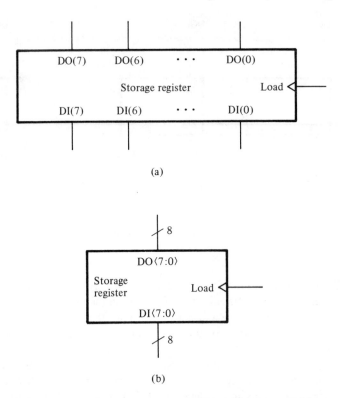

(a)

(b)

Figure 9.2.2   Symbols for a storage register. (a) Block diagram symbol for a storage register. (b) A simplified symbol for a storage register.

9.2.2 Shift Registers

Another very useful register is the *shift register*. The shift register is used for a number of chores, including converting parallel information to serial and vice versa. Figure 9.2.3 shows a shift register that shifts information in the register one bit position left each time the clock line is asserted. If one bit of information appears on the right_in line at each clock assertion, a complete byte of information will be assembled in the register and appear at the outputs after eight clock pulses. Of course, a right-shifting register looks similar except that each flip-flop output is fed to the flip-flop on its right.

Obviously, the design of a simple shift register need not be accompanied by the general sequential circuit design procedure discussed in Chapter 5, although the end result would be the same. In fact, it might be instructive, at this point, to review this process. To simplify our task of designing, directly, a shift register that shifts left, using methods of Chapter 5, we will assume that we are designing a 3-bit shift register. We begin by constructing a state diagram as shown in Figure 9.2.4. Although this state diagram is not overly complex, one can certainly imagine how large it would become for an 8-bit shift register.

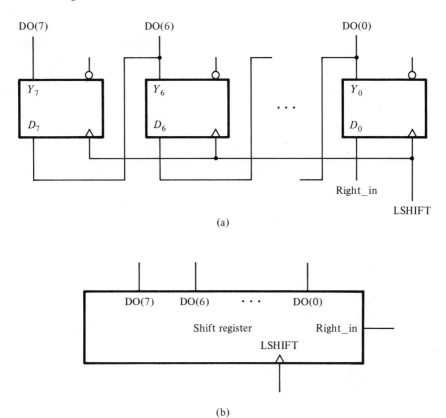

(a)

(b)

Figure 9.2.3 (a) An 8-bit shift left register. (b) Block diagram symbol for an 8-bit shift register.

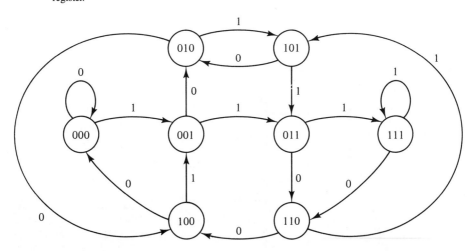

Figure 9.2.4 State diagram of a 3-bit left-shifting shift register.

The next step in the design process is to develop the flip-flop input equations. Assuming D flip-flops, this task is fairly simple. Figure 9.2.5 shows the flip-flop excitation table from which the flip-flop input equations can be derived. The resulting equations are

$$D_2 = y_1$$
$$D_1 = y_0 \qquad (9.2.1)$$
$$D_0 = x$$

Letting $Y_i = \text{DO}(i)$ and $x = \text{Right\_in}$, we end up with a 3-bit version of Figure 9.2.3. This is certainly what we expected from the statement of the original problem.

What we should observe from this example is the following. There is no need to take complex steps to solve simply stated problems: in this case, that a register is to be designed that shifts a bit coming in from the right, one bit position left on each clock pulse. Even registers having very complex functions may be designed without the use of the formal procedures if each function is designed separately, with all functions being integrated into a whole at the end. This process is most readily shown by the next example.

A more general type of shift register, called a *universal shift register*, is one that can shift right or left and can be loaded in parallel, if so desired, thus giving it the capability of a storage register as well. Using the philosophy just espoused, this register can be designed as follows. If we assume the use of D flip-flops in our register, then the D input to the ith flip-flop must come from one of three sources. For a left shift, this input must come from the $(i - 1)$st flip-flop; for a right shift, it must come from the $(i + 1)$st flip-flop; and for a parallel load, it must come from the ith input. By using a multiplexer, as, for example, the 4-line to 1-line MUX of Figure 4.3.8, we may select one of these three lines as an input to each flip-flop. In fact, if we use this multiplexer, we can add a "do nothing" operation, in which the output does not change when a clock occurs. Figure 9.2.6 shows a block diagram symbol for such a register and a schematic diagram showing how the internal flip-flops are interconnected. The select lines for the MUX are coded to function as shown in Table 9.2.1. The notation $L/\overline{R}$ used in this table and in Figure 9.2.6

$y_2 y_1$ ╲ $y_0 x$	00	01	11	10
00	000	001	011	010
01	100	101	111	110
11	100	101	111	110
10	000	001	011	010

$D_2 D_1 D_0$

Figure 9.2.5 Excitation table for the left-shifting shift register. In terms of Figure 9.2.4, the $Y_i = \text{DO}(i)$; $x = \text{Right\_in}$; and the clock which drives this circuit is equal to LSHIFT.

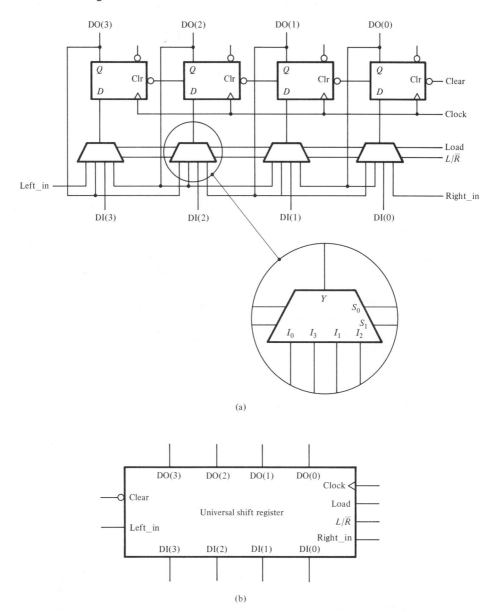

Figure 9.2.6 A universal (left-right) presettable shift register: (a) schematic diagram; (b) block diagram.

indicates that the register is to shift left when the line is high and is to shift right when the line is low.

In this coding, the combination 11 is used to cause the contents of the shift register to remain unchanged after each clock pulse. It does this by reloading each flip-flop with

TABLE 9.2.1 UNIVERSAL SHIFT
REGISTER MULTIPLEXER CONTROL

$L/\overline{R}$	Load	Function
0	0	shift right
1	0	shift left
0	1	load input information into register
1	1	do nothing

what it already contains. Large systems generally have a common clock that is distributed to all of the registers in the system. Thus, this 11 combination plays the role of disabling the register during certain periods of time when the information in the register is being used by other portions of the system. Another approach to this register disable would be to AND the clock line with a separate enable signal so that the clock can effectively be turned off. This alternative, however, has some disadvantages, which we shall look at a bit later.

Finally, we note that in this shift register, D flip-flops having an asynchronous clear input are used to add a clear capability to the register without increasing the complexity of the design. The use of this added capability will be further illustrated in the examples of Section 9.5.

9.2.3 Counters Revisited

Counters are used quite often in large-scale systems to keep track of the number of times a certain process is repeated. Sometimes a single counter may be used at different times for different count values. A counter which is quite useful for such functions is one which can be initially cleared to 0 and also preset to some given value, like a storage register. What we would like to have is a counter which can be cleared, enabled to count, and preset to some value, and which supplies a signal out indicating when the counter has reached its maximum count value, usually $11 \cdots 1$. Figure 9.2.7 shows a binary 4-bit look-ahead carry counter that meets these requirements. The design of this counter is based on the following considerations. First, the counting function is most easily carried out by the use of T flip-flops, as was discussed in Chapter 5. Second, the storage register function would appear to be carried out most easily by the use of D flip-flops. A JK flip-flop can easily be made into either a D or a T flip-flop by appropriate connections to its inputs. Specifically, for a T flip-flop, the J and K lines must be equal, and for a D flip-flop these inputs must be complements of each other. Thus, if an input L is used to select either the load function, when $L = 1$, or the count function, when $L = 0$, and if $DI(i)$ is the data input to the ith flip-flop, then the appropriate input equations become

$$J_i = L \, DI(i) + \overline{L} \, DO(i - 1) \cdots DO(0)$$
$$K_i = L \, \overline{DI(i)} + \overline{L} \, DO(i - 1) \cdots DO(0)$$

$$(9.2.2)$$

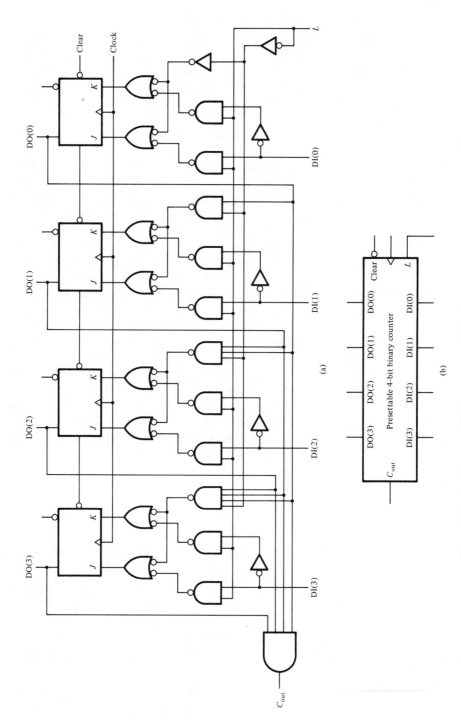

Figure 9.2.7 A 4-bit presettable binary counter: (a) implementation; (b) block diagram.

As an example of the use of this type of counter, consider the following problem. A circuit is to be designed having a 1-MHz clock as an input. A single pulse is to be produced at the output at a rate which is some fractional value of the input frequency. In particular, this rate is to be in the range of from one-sixteenth of the input frequency to half the input frequency. Further, this output rate is to be *programmable*; i.e., the frequency should be changeable by an external digital system. A circuit to perform this function will be referred to as a *programmable frequency divider*. Such a device is commonly used in serial communications to match the receiver data rate to the transmitter data rate and vice versa.

To begin the solution of this problem, we must first come up with an *algorithm*, or plan of attack, which will satisfy the problem's requirements. Consider the presettable counter of Figure 9.2.7. If this counter is preset to, say, 12 and is clocked by the 1-MHz input clock, then the C_{out} line will be asserted on the third input clock pulse when the count value equals 15. Now consider what happens if C_{out} is fed back to the L input and 12, that is, 1100, is held on the counter inputs, say, by connecting them to a high or a low voltage as necessary. Since we have connected C_{out} to the L line, the counter will be loaded with 12 on the next, or fourth, clock pulse, and C_{out} will go low. After three more clock pulses, C_{out} will again go high and the process will repeat. We can see that the output, C_{out}, is produced on every fourth clock pulse. Obviously, if the counter inputs were held at 9, C_{out} would be produced on every seventh clock pulse, and so on. Figure 9.2.8 shows the timing involved at the point where C_{out} is asserted. The important thing to note here is that, because of propagation delays in the flip-flops, C_{out} is asserted after the clock goes high and will be asserted at the time the next low-to-high transition of the clock occurs.

In order to be able to program the counter input value, we need to place a 4-bit storage register on the input to the counter so that an external system can load the register with the desired count. The storage register output will then serve as the input to the counter, so that the load-count-load sequence can continue repeatedly. Figure 9.2.9 shows the final design that meets the requirements of the problem. In this design, the pulse that is to be generated can be obtained from the C_{out} line. Note that the frequency division is obtained by subtracting the division factor n from 16 and loading this number into the storage register. Thus, the counter will be loaded with numbers in the range of 0, for a division factor of 16, to 14, for a division factor of 2. One might ask the question, what happens if a division factor of 1 is required? Will the frequency divider work in this case?

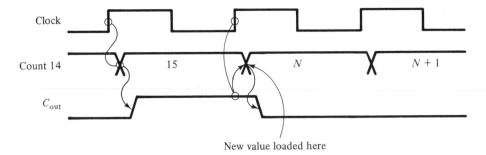

New value loaded here

Figure 9.2.8 C_{out} timing with respect to the clock and count value.

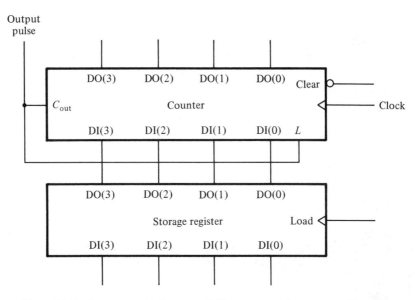

Figure 9.2.9 A programmable frequency divider constructed from a presettable counter and a storage register.

The answer to this question is no, and the reason is simple. A division factor of 1 would require loading the counter with 15 each time the counter reaches 15! Since C_{out} is the signal that generates the output pulse, this would result in C_{out} always staying high, resulting in a DC, or steady, signal out.

The solution to the problem of designing a programmable frequency divider required the generation of an *algorithm* to perform the task required. Basically, an algorithm is nothing more than a specific plan to solve a specific problem. Algorithms may be defined in many ways, although the most common is to somehow list the steps needed to perform a given task. In the case of the frequency divider, the steps are:

1. Load the counter with $16 - n$, where n is the division factor.
2. Count successive clock pulses until the count reaches 15.
3. Then go back to step 1.

In Section 9.4, we shall look at methods for defining complex algorithms that can easily be converted into hardware realizations. We shall first, however, introduce a special notation that can make the design process and algorithm implementation easier.

9.3 A REGISTER TRANSFER NOTATION

The systems we are concerned with here are digital systems made up of combinational logic that is designed to perform some processing task; registers for temporarily storing information, counting events, and the like; and other sequential circuits for controlling the processes. In order to define algorithms for such systems it is necessary to define a notation which shows how information in the system is to be processed. We will refer to such a notation as a *register transfer notation*,[4] since it is used to show how information in one register is processed and passed on to another register for further processing. In the frequency divider designed at the end of Section 9.2.3, for example, the register involved was a counter, which was first loaded with a constant and then incremented. After each increment, the count value was tested to see whether it was equal to 15. If not, the counter was incremented again. If the value was 15, then the counter was loaded with the constant once more and the process continued. All of the elements making up this process will be defined in what follows.

9.3.1 The Basic Notation

In Section 9.2, we defined a *register* as an ordered set of binary cells, each storing one bit of information. In order to refer to a given register, it must have a name. Since we may also wish to refer to individual bits, or collections of bits, we must also indicate how the bits of the register are numbered. Thus, in general, a register will be indicated by the notation

$$\text{Register\_name}\langle i{:}j \rangle \qquad\qquad (9.3.1)$$

where i is the leftmost, or most significant, bit of the register and j is the rightmost, or least significant, bit. The intervening bits are numbered successively from i to j. For an n-bit register, it will usually be the case that $i = n - 1$ and $j = 0$. Thus, the 4-bit register whose name is CAT will be denoted as CAT$\langle 3{:}0 \rangle$, which is equivalent to the ordered set of bits (CAT(3), CAT(2), CAT(1), CAT(0)). Once a register is defined, we may refer to it by its name only, if such a reference can be made without confusion. Thus, register CAT$\langle 3{:}0 \rangle$ may also be referred to simply as register CAT.

It is quite often necessary to refer to some subset of the bits of a register. Such a subset is referred to as a subregister. For example, the left half of register DOG$\langle 15{:}0 \rangle$ would be DOG$\langle 15{:}8 \rangle$, which we might wish to refer to as register LEFT__DOG. Noncontiguous bits in a register may also be referred to by use of the notation.

$$\text{Register\_name}\langle a{:}b,\ c{:}d,\ .\ .\ . \rangle \qquad\qquad (9.3.2)$$

where the indices a, b, c, d, etc., are all in the range of i to j of expression (9.3.1). For example, the subregister RAT$\langle 12{:}10,\ 1{:}0 \rangle$ of register RAT$\langle 15{:}0 \rangle$ would be the ordered set of bits (RAT(12), RAT(11), RAT(10), RAT(1), RAT(0)).

In a computer, information flows from one register to another, usually after some intermediate process occurs. By the transfer of information from register A to register B, we mean that after the transfer, register B contains a copy of the contents of register A and register A is unchanged. We will denote a simple *register transfer* by the notation

---

[4] Although there are currently no standards for a register transfer notation, the notation given here is typical of what may be found in the literature.

$$\text{Register\_1}\langle a{:}b\rangle \rightarrow \text{Register\_2}\langle c{:}d\rangle \qquad (9.3.3)$$

where the leftmost bit of register 1 is copied into the leftmost bit of register 2, the next bit of register 1 is copied into the next bit of register 2, and so on. Obviously, the register transfer makes sense only if *both registers contain the same number of bits*.

As an example of the application of this notation, let us define the 8-bit register $A\langle 7{:}0\rangle$ and require that the contents of this register be shifted right one bit position, with the leftmost bit being unaffected. This transfer would be denoted as

$$A\langle 7{:}1\rangle \rightarrow A\langle 6{:}0\rangle \qquad (9.3.4)$$

Thus, if A contained (01101011) before this transfer, then A will contain (00110101) after the transfer.

Many times it is necessary to preset a register to some constant value. This was done in Section 9.2.3 with the presettable counter used in the frequency divider example. We will indicate the presetting of a register by the notation

$$n \rightarrow \text{Register}\langle i{:}j\rangle \qquad (9.3.5)$$

where n is the number to be loaded into the register—in binary, of course. Thus, $145 \rightarrow A\langle 7{:}0\rangle$ would mean that register A would contain (10010001) after the transfer is completed. The use of a constant is also essential for counting. For example, the notation $A + 1 \rightarrow A$ would mean that the contents of register A are incremented by 1 so that the number in the register is 1 greater after the transfer than before.

Functions of registers are also easily indicated. The general form for functions of two registers would be

$$f(A\langle a{:}b\rangle, B\langle c{:}d\rangle) \rightarrow C\langle i{:}j\rangle$$

or simply

$$f(A, B) \rightarrow C \qquad (9.3.6)$$

For example, we might write $A + B \rightarrow C$ to mean that register C is to be loaded with the arithmetic sum of the contents of registers A and B. Thus, if A and B are 8-bit registers and A contains (00010111) and B contains (00100100), then after the transfer, C would contain (00111011).

It very often happens that one of several possible transfers is to be executed, depending on some condition such as the value of the number held in a register. This was the case for the frequency divider of Section 9.2.3. In order to represent such transfers, we need to introduce a pair of register functions. The first is called the *value function*, denoted

$$\text{val}\,(\text{Register}\langle i{:}j\rangle) \qquad (9.3.7)$$

and defined as the numeric value of the number held in the register. For example, if register $A\langle 7{:}0\rangle$ contains (00111011), then val $(A) = 59$ (base 10). The second register function we will need is called the *characteristic function*, denoted

$$\text{ch}\,(\text{Register}\langle i{:}j\rangle, k) \qquad (9.3.8)$$

and defined as being 1 if val (Register$\langle i{:}j \rangle$) $= k$ and 0 otherwise. Thus, if val $(A) = 59$, then ch $(A, 59) = 1$, whereas ch $(A, 60) = 0$. As an example, consider the following transfer:

$$\text{ch } (R, 0)\, A + \text{ch } (R, 1)\, B + \text{ch } (R, 2)\, (A \text{ plus } B)$$
$$+ \text{ch } (R, 3)\, (A \text{ minus } B) \rightarrow C \qquad (9.3.9)$$

where the registers are defined as $R\langle 1{:}0\rangle$, $A\langle 7{:}0\rangle$, $B\langle 7{:}0\rangle$, and $C\langle 7{:}0\rangle$. The result of this transfer is that register C will be loaded with the contents of register A if val $(R) = 0$, the contents of register B if val $(R) = 1$, the arithmetic sum of registers A and B if val $(R) = 2$, or the arithmetic difference between registers A and B if val $(R) = 3$.

If the register being used to select the function is only 1 bit long, then the logical value of this bit may be used to control the transfer. Thus, for example, the notation

$$xA\langle 7{:}0\rangle + \bar{x}B\langle 7{:}0\rangle \rightarrow C\langle 7{:}0\rangle$$

would mean that C is loaded with the contents of register A if $x = 1$ or the contents of register B if $x = 0$.

Using the notation just developed, we can now describe the frequency divider circuit by the following register transfer:

$$C_{\text{out}}(\text{Storage\_register}\langle 3{:}0\rangle) + \overline{C}_{\text{out}}(\text{Counter}\langle 3{:}0\rangle \text{ plus } 1) \rightarrow \text{Counter}\langle 3{:}0\rangle$$

where $C_{\text{out}} = \text{ch } (\text{Counter}\langle 3{:}0\rangle, 15)$.[5]

9.3.2 Hardware Considerations

A typical storage register was shown in figure 9.2.2. An examination of this figure shows that there are three types of signals associated with the register: *input*, *output*, and *control* (the clock signal used to load the register). Using registers of this type, the transfer

$$A\langle 7{:}0\rangle \rightarrow B\langle 7{:}0\rangle$$

is carried out by connecting the outputs of register A to the inputs of register B and clocking the Load signal of register B. Figure 9.3.1 shows this interconnection.

More complex registers, such as the universal shift register of Figure 9.2.7, require more than just the clocking signal for control. In this case, in addition to the clock signal, two other control lines, Load and $L/\overline{R}$, are required to control the specific function of the register. Thus, we may observe that the control signals associated with a given register are made up of two types of signals: *timing* (the clock) and *function* (Load and $L/\overline{R}$). In general, all registers will have associated with them four classes of signals, namely,

1. Inputs
2. Outputs

---

[5] The use of the plus $(+)$ can often be misinterpreted—does it mean arithmetic addition or logical OR? In what follows, we will spell out the arithmetic "plus" whenever confusion can occur.

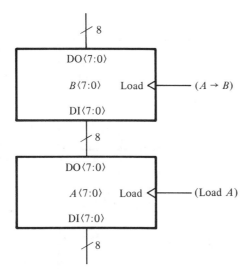

Figure 9.3.1 A pair of registers used to implement the transfer $A <7:0> \rightarrow B<7:0>$.

3. Function select
4. Timing

Since these signals are common to all registers, we need not give each a separate name or specific identity in the register transfer notation described above. The connections required to implement a specific transfer can be inferred from the transfer notation itself. In particular, outputs associated with the register or function of registers on the left of the transfer arrow will be connected to inputs of the register on the right of the transfer arrow. The required connections for function control may be inferred from the register function described. The transfer is finally carried out by the timing or clock signal associated with the receiving register.

For example, suppose we are given the register transfers

$$XL(A) + \overline{X}R(B) \rightarrow R(A) \qquad (9.3.10)$$

$$XB(0) + \overline{X}B(3) \rightarrow A(3) \qquad (9.3.11)$$

where the registers and subregisters are defined as $A\langle 3{:}0\rangle$, $B\langle 3{:}0\rangle$, $L(A) = A\langle 3{:}1\rangle$, $R(A) = A\langle 2{:}0\rangle$, and $R(B) = B\langle 2{:}0\rangle$. This transfer causes A to shift right, with the low-order bit of B going into the high-order bit of A if $X = 1$. If $X = 0$, A is simply loaded with the contents of B. Using the universal shift register shown in Figure 9.2.7 and a 4-bit version of the general storage register shown in Figure 9.2.3, the interconnections required to implement the transfers of expressions (9.3.10) and (9.3.11) are as shown in Figure 9.3.2. These connections are made on the basis of the following considerations. Since we are never shifting right, Right_in need not be connected to anything. The derivation of the control equations is based on the MUX control defined in Table 9.2.1 of Section 9.2.2 and goes as follows. To shift right, $X = 1$, $L/\overline{R} = 0$, and Load = 0. To load register

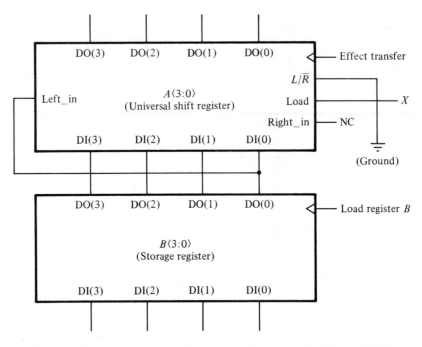

Figure 9.3.2   An implementation of the transfers of expressions (9.3.10) and (9.3.11).

A, $X = 0$, $L/\overline{R} = 0$, and Load $= 1$. Shifting left and doing nothing are not required by this problem. Thus we have the following truth table:

X	$L/\overline{R}$	Load	
0	0	0	shift right
1	0	1	load

and therefore

$$L/\overline{R} = 0 \qquad \text{and} \qquad \text{Load} = X \qquad\qquad (9.3.12)$$

From these equations we see that $L/\overline{R}$ needs to be grounded and Load needs to be connected to input X. Transfer (9.3.11) further indicates that $B(0)$ must be connected to Left_in of register A.

    In large-scale systems, there will generally be many registers, each having several control signals. In order to write the specific set of control equations for a given system, we must associate a control signal with a specific register. We will do this using the notation

$$\text{Register\_name[control\_signal]}$$

where the square brackets are used to indicate that the name enclosed is the name of a control signal. In the above case, we will write equation pair (9.3.12) as

$$A[L/\overline{R}] = 0 \qquad \text{and} \qquad A[\text{Load}] = X \qquad (9.3.13)$$

The register transfers (9.3.10) and (9.3.11) do not indicate specific connections between registers A and B. For example, these transfers do not indicate, specifically, that DO(0) of register $B\langle 3:0 \rangle$, or $B[\text{DO}(0)]$, is to be connected to input DI(0) of register A and Left_in of register $A\langle 3:0 \rangle$, or $A[\text{Left\_in}]$ and $A[\text{DI}(0)]$. These connections are dependent on the specific registers used to implement the required transfers. The point of this is that the register transfer notation developed in Section 9.3.1 is used to describe *what transfers must occur to implement an algorithm but not how the transfers are actually implemented in hardware*. The hardware implementation must be inferred from the transfer itself and the specific choice of registers used to implement the transfer. We shall see further examples of this in what follows.

9.4 FLOW CHARTS AND STATE DIAGRAMS

The specification of a control algorithm for a digital system requires writing a specific sequence of register transfers. In computer programming, flow charts are used extensively to define algorithms. Since we are dealing here with control processes that involve the manipulation of information in particular ways and in particular sequences, it would seem that a flow chart would be a convenient way of describing the required algorithm. As described earlier, the control unit of large-scale digital systems is nothing but a simple clocked sequential circuit having inputs, outputs, and states. As we shall see in a moment, flow charts and state diagrams are equivalent, in a limited sense. Thus a flow chart, with its graphic representation of information and process flow, is an ideal way of representing a sequential circuit making up the control mechanism in a large-scale system.

9.4.1 The Flow Chart

Although flow charts used for programming purposes have many different elements, only four will be needed in what follows. These are shown in Figure 9.4.1 and are defined in the following paragraphs.

The *entry point* flow chart element is used to indicate one of two things: the starting point of the algorithm being implemented, or a continuation point in an algorithm when the flow chart becomes too large to be included on one page. This second use for the entry point requires a corresponding *exit point*. Control algorithms that are useful never stop (except, possibly for the control of a bomb!).[6] Thus, there is actually no *terminal point* in an algorithm. The exit point element is used only to indicate the label of the entry point

---

[6] The AT&T computer 3B2 has a shutdown mechanism that represents a terminal point in the algorithm. When the computer's power switch is manually turned off, the computer updates all disk information necessary and then the computer's control mechanism, not the operator, finally shuts the power off.

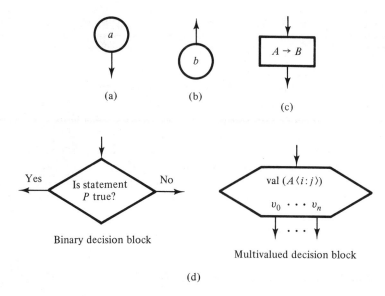

Figure 9.4.1 The basic flow chart elements: (a) entry point; (b) exit point; (c) transfer block; (d) the decision block.

element for continuation of the algorithm. Figure 9.4.2 shows an example of how the entry point and exit point flow chart elements are used. In this case, the portion of the algorithm on page 1 is continued on page 2 at the page 2 entry point A_2. When the page 2 portion of the algorithm is completed, it returns to page 1 via the exit and entry point pair labeled A_1.

The *transfer block* is used to indicate, explicitly, what transfer is required at a particular stage in the algorithm. This transfer is indicated within the block by the use of the register transfer notation described in Section 9.3.

The *decision block* is used to identify which path in an algorithm is to be followed next. This is done by indicating, in the block, what condition is required to continue on a given path. Figure 9.4.1(d) shows two common ways in which the condition can be specified. A decision block always has one entry point but will have two or more exit

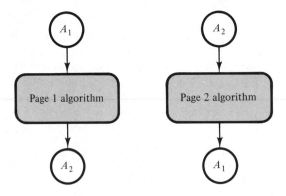

Figure 9.4.2 Using entry and exit points to continue algorithms on multiple pages.

points, depending on the indicated condition. For example, if the condition is val $(A\langle 1{:}0 \rangle)$, then there will be four possible ways to leave the decision block, one for each value of register $A\langle 1{:}0 \rangle$.

9.4.2 Flow Chart–State Diagram Equivalence

Figure 9.4.3(a) shows a typical state diagram having one input, X, one output Z, and three states. We may make a correspondence between these elements and those of a flow chart in the following way. First, the way in which we leave a decision block in a control algorithm is dependent on information coming into the control unit from outside (refer to Figure 9.1.1). Thus, the information in a decision block constitutes *input* information to the controller. Second, the transfers indicated in the transfer blocks are brought about by outputs from the controller and thus transfer blocks represent *outputs* from the system. Finally, at any given instant of time, we will be somewhere in the flow chart, either waiting for the next input to be read or simply waiting for the next clock pulse. Thus, the state of the system corresponds to edges in the flow chart. The position of the states is indicated in the flow chart by the hash marks (/) on edges corresponding to the state, with the state label adjacent to the hash mark. Figure 9.4.3(b) shows the flow chart equivalent to the state diagram of Figure 9.4.3(a) based on these equivalences.

Given a flow chart with the states indicated, we may easily derive the corresponding state diagram. To see how we might do this, consider the flow chart shown in Figure 9.4.4(a), which represents some arbitrary algorithm. Let us begin at state S_0. From the flow chart, we see that we will stay in state S_0 if $Q_1 = 0$, or if $\overline{Q}_1$ is 1. Thus there will be a self-loop on state S_0 with input $\overline{Q}_1$. Further, since there are no transfers in this path in the flow chart, there are no outputs generated. The resulting state diagram elements are shown in Figure 9.4.4(b) as the self-loop on state S_0. Note that this path is dependent only on input Q_1 and so the other inputs, Q_2 and Q_3, become don't cares and are therefore not shown. In the state diagram, a dash (–) on the output side of any slash (/) is used to indicate that no transfers are to occur. All of this amounts to a shorthand notation for Q_1, Q_2, Q_3/T_1, T_2, T_3, T_4, T_5, $T_6 = 0{-} {-}/00000$, where the dashes here are don't cares, as usual. In a similar way, we go from state S_0 to state S_1 if $Q_1 = 1$, and since there are no transfers in this path, the edge from state S_0 to state S_1 in the state diagram is labeled $Q_1/{-}$, which corresponds to Q_1, Q_2, Q_3/T_1, T_2, T_3, T_4, T_5, $T_6 = 1{-} {-}/000000$.

Before proceeding, let us define what we mean by a path. As used here, a *path is a sequence of flow chart elements which takes us from one state to another*. Consider now the possible paths from state S_1. An examination of the flow chart of Figure 9.4.4(a) shows that there are three paths from state S_1: two going to S_2 and one going to S_3. The conditions for traversing these paths are dependent on inputs (decisions) Q_2 and Q_3 only. In particular, we will go from S_1 to S_2 if either $Q_2 = 1$ or $Q_2 = 0$ and $Q_3 = 1$. Otherwise, we will go to state S_3. The transfers required in each path are easily found from the flow chart. The corresponding state diagram edges are shown labeled in Figure 9.4.4(b) using our shorthand notation. For example, the edge labeled $\overline{Q}_2 Q_3/T_1$, T_2, T_4 is equivalent to Q_1, Q_2, Q_3/T_1, T_2, T_3, T_4, T_5, $T_6 = {-}01/110100$. The rest of the control state diagram is easily determined by continuing to list all paths along with their associated conditions and transfers.

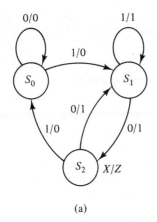

(a)

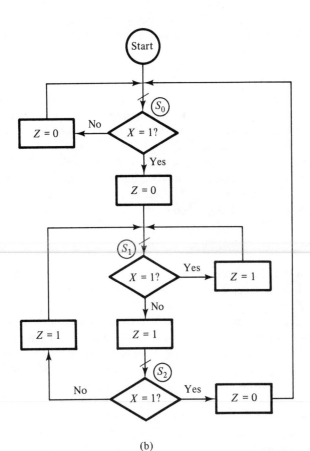

(b)

Figure 9.4.3 State diagram (a) and flow chart equivalent (b).

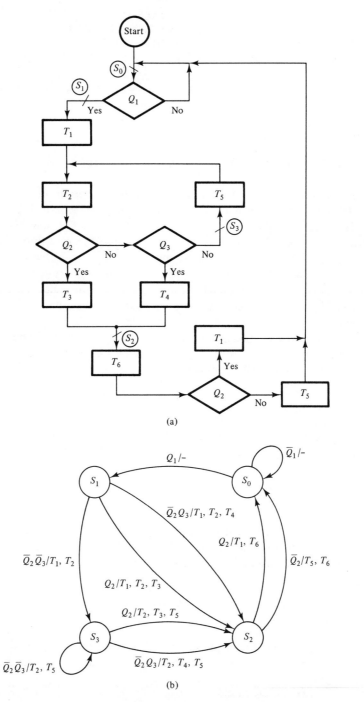

(a)

(b)

Figure 9.4.4 Derivation of a state diagram from a flow chart: (a) a control algorithm flow chart; (b) the equivalent state diagram.

TABLE 9.4.1
STATE ASSIGNMENT
FOR FIGURE 9.4.3

State	Y_1	Y_2
S_0	0	0
S_1	0	1
S_2	1	1

9.4.3 Derivation of the Control Equations

To see how we may derive the control equations, specifically, the outputs and the next-state equations, let us refer, for the moment, back to the state diagram of Figure 9.4.3(a). The design procedure described in Chapter 5 started by assigning values to the state variables needed to encode the various states. In this case, we have three states, so we will need two state variables, Y_1 and Y_0. We may arbitrarily assign states as shown in Table 9.4.1. We are now ready to write the appropriate equations. Let us begin with the equation for the output, Z. The sequential circuit represented by the state diagram of Figure 9.4.3(a) is a Mealy machine, and therefore we know that the output Z is a function of both the input X and the current state. Without constructing the assigned-state table, and ignoring the don't care conditions,[7] we may write an equation for Z by observing that $Z = 1$ if we are in state S_1 and $X = 1$ or $X = 0$ or, on the other hand, if we are in state S_2 and $X = 0$. The resulting equation is

$$Z = s_1(X + \overline{X}) + s_2\overline{X}$$
$$= s_1 + s_2\overline{X} \tag{9.4.1}$$

where the lowercase s_i's represent the current state. As was done in Chapter 5, we will represent the next state using capital S_i's. Equation (9.4.1) can be written in terms of the state variables by replacing the s_i's by the assignment given in Table 9.4.1. The resulting equation becomes

$$Z = \overline{y}_1y_0 + y_1y_0\overline{X} = \overline{y}_1y_0 + y_0\overline{X} \tag{9.4.2}$$

The next-state equations can be derived in a similar manner. In particular, we may write, directly from the state diagram, the following three equations relating the next state to the current state and input:

$$S_0 = s_0\overline{X} + s_2X$$
$$S_1 = s_0X + s_1X + s_2\overline{X} \tag{9.4.3}$$
$$S_2 = s_1\overline{X}$$

The first of these three equations says, for example, that we will be in state S_0 if either we started in state S_0 and $X = 0$ or we started in state S_2 and $X = 1$. To develop the

---

[7] Since two state variables can encode four states and we have only three states, don't cares will be associated with the output and the state variable equations.

equations for the state variables, we can construct a "compound" truth table based on the state assignment given in Table 9.4.1 that shows the assignment for the next state and the equation, from equation set (9.4.3), required to reach this next-state assignment. This is shown in Table 9.4.2. From this table we may write the equations for the state variables by writing the sum of the expressions for which each state variable is 1. Thus, we have

$$Y_1 = S_2 = (s_1\overline{X})$$
$$Y_2 = S_1 + S_2 = (s_0X + s_1X + s_2\overline{X}) + (s_1\overline{X}) \tag{9.4.4}$$

or, upon substituting the assignments for the states given in Tables 9.4.1 and 9.4.2,

$$Y_1 = (\overline{y}_1y_0\overline{X})$$
$$Y_2 = (\overline{y}_1\overline{y}_0X + \overline{y}_1y_0X + y_1y_0\overline{X}) + (\overline{y}_1y_0\overline{X}) \tag{9.4.5}$$
$$= \overline{y}_1X + y_0\overline{X}$$

It can be shown that Equations (9.4.2) and (9.4.5) are the equations that would arise from the assigned-state table with the don't cares set to 0. (This should be verified by the reader.)

Equations (9.4.1) and (9.4.4), from which Equations (9.4.2) and (9.4.5) are derived, may be obtained directly from the flow chart shown in Figure 9.4.3(b) by observing what input conditions are required for each path in the flow chart and which transfers occur when traversing these paths. For example, $Z = 1$ whenever we go over path S_1 to S_1, S_1 to S_2, or S_2 to S_1. The first and last paths require that $X = 1$, and the second path requires that $X = 0$. Since we can obtain the design equations directly from the flow chart, there is, therefore, no need to derive an equivalent state diagram, although this is always possible.

Let us now return to the flow chart of Figure 9.4.4(a) for a bit more complex example of the process used for deriving the design equations from the flow chart. From this figure, we can see that there are conditions associated with each path and each transfer and these conditions need not be the same. For example, the condition for taking the path S_1-S_2 is that $Q_2 = 1$ or $Q_3 = 1$ or both. However, transfer T_3 occurs only if $Q_2 = 1$ whether we take path S_1-S_2 or path S_3-S_2. Note also that the path S_2-S_0 has no conditions on it at all, although the transfers T_1 and T_5 do have associated conditions.

In order to summarize all of these possibilities in a systematic way, we may construct a *path-transfer table* which identifies each path and its associated condition and all transfers that are required in traversing the path along with their associated conditions. The path-transfer table corresponding to the flow-charted algorithm of Figure 9.4.4(a) is shown in Figure 9.4.5. We may now write the design equations based on information in this table.

TABLE 9.4.2 TABLE OF NEXT-STATE CONDITIONS

Next state		Condition
Y_1	Y_2	
$S_0 = 0$	0	$s_0\overline{X} + s_2X$
$S_1 = 0$	1	$s_0X + s_1X + s_2\overline{X}$
$S_2 = 1$	1	$s_1\overline{X}$

Path	Path condition	Transfer	Transfer condition
S_0-S_0	$\overline{Q}_1$	–	–
S_0-S_1	Q_1	–	–
S_1-S_2	$Q_2 + Q_3$	T_1, T_2	–
		T_3	Q_2
		T_4	$\overline{Q}_2\overline{Q}_3$
S_1-S_3	$\overline{Q}_2\overline{Q}_3$	T_1, T_2	–
S_3-S_2	$Q_2 + Q_3$	T_2, T_5	–
		T_3	Q_2
		T_4	$\overline{Q}_2\overline{Q}_3$
S_3-S_3	$\overline{Q}_2\overline{Q}_3$	T_2, T_5	–
S_2-S_0	–	T_6	–
		T_1	Q_2
		T_5	$\overline{Q}_2$

Figure 9.4.5   Path-transfer table for the flow chart of Figure 9.4.4.

Let us begin by writing the next-state equations. Consider, first, the ways in which we can end up in state S_0. There are two paths for which S_0 is the terminal state: S_0-S_0 and S_2-S_0. By ANDing the initial states and their corresponding path conditions and then ORing the results, the next-state equations for S_0 can be written. Thus we obtain

$$S_0 = s_0\overline{Q}_1 + s_2 \tag{9.4.6}$$

The remaining equations are derived in a similar fashion and are found to be as follows:

$$S_1 = s_0 Q_1 \tag{9.4.7}$$

$$S_2 = s_1(Q_2 + Q_3) + s_3(Q_2 + Q_3) \tag{9.4.8}$$

$$S_3 = s_1\overline{Q}_2\overline{Q}_3 + s_3\overline{Q}_2\overline{Q}_3 \tag{9.4.9}$$

Derivation of the transfer equations proceeds in a similar manner. In the case of the transfers, however, transfer conditions must be ANDed with the path conditions *and* current state. Let us consider transfer T_1, for example. T_1 is associated with paths S_1-S_2, S_1-S_3, and S_2-S_0. By ORing the conditions required for the transfer T_1 for each of these paths, we obtain the equation

$$\begin{aligned} T_1 &= [s_1(Q_2 + Q_3)] + [s_1\overline{Q}_2\overline{Q}_3] + [s_2 Q_2] \\ &= s_1 + s_2 Q_2 \end{aligned} \tag{9.4.10}$$

The remaining equations are derived similarly:

$$T_2 = s_1 + s_3 \tag{9.4.11}$$

$$T_3 = (s_1 + s_3)Q_2 \tag{9.4.12}$$

$$T_4 = (s_1 + s_3)\overline{Q}_2\overline{Q}_3 \tag{9.4.13}$$

$$T_5 = s_3 + s_2\overline{Q}_2 \tag{9.4.14}$$

$$T_6 = s_2 \tag{9.4.15}$$

To write all of the equations in terms of the state variables requires that we make a state assignment. Since there are four states, we will, of course, need two state variables, y_1 and y_0. When we make the assignment as shown in Table 9.4.3, the transfer equations become

$$T_1 = \overline{y}_1 y_0 + y_1 \overline{y}_0 Q_2 \tag{9.4.16}$$

$$T_2 = \overline{y}_1 y_0 + y_1 y_0 = y_0 \tag{9.4.17}$$

$$T_3 = y_0 Q_2 \tag{9.4.18}$$

$$T_4 = y_0 \overline{Q}_2 \overline{Q}_3 \tag{9.4.19}$$

$$T_5 = y_1 y_0 + y_1 \overline{y}_0 \overline{Q}_2 = y_1 y_0 + y_1 \overline{Q}_2 \tag{9.4.20}$$

$$T_6 = y_1 \overline{y}_0 \tag{9.4.21}$$

On the basis of the state assignment of Table 9.4.3, we may determine the state variable equations as follows:

$$\begin{aligned} Y_1 &= S_2 + S_3 \\ &= (s_1 + s_3)(Q_2 + Q_3) \\ &= y_1(Q_2 + Q_3) \end{aligned} \tag{9.4.22}$$

$$\begin{aligned} Y_0 &= S_1 + S_3 \\ &= s_0 Q_1 + (s_1 + s_3)\overline{Q}_2\overline{Q}_3 \\ &= \overline{y}_1 \overline{y}_0 Q_1 + y_1 \overline{Q}_2\overline{Q}_3 \end{aligned} \tag{9.4.23}$$

TABLE 9.4.3
STATE ASSIGNMENT
FOR FIGURE 9.4.4

State	Y_1	Y_0
S_0	0	0
S_1	0	1
S_2	1	0
S_3	1	1

Equations (9.4.16) through (9.4.23) represent the final design equations. Once the state-variable flip-flops are specified, the flip-flop input equations can be generated, as was done in Chapter 5; the basic design process is then complete.

9.4.4 The Placement of States for Register Control and Timing

Before we look at some design examples, we need to examine carefully the derivation of the timing and control equations which are specific to the registers used to implement the given algorithm. As we observed in Section 9.3.2, there are two classes of signals associated with the control of a register: *timing* and *function*. We may think of the function signals as level signals generated by combinational logic which implement the transfer equations. These level signals are functions only of the state variables and the various system inputs. The timing signals, on the other hand, represent quite a different situation.

Let us examine the timing problem by looking at the 4-bit presettable binary counter of Figure 9.2.7. We will refer to this counter as register C. There are three possible transfers that can be associated with this counter: load, clear, and increment the register. Suppose, for a given application, that the following two transfer equations are derived from the control flow chart:

$$(n \rightarrow C) = LDC = s_i f(\mathbf{x})$$
$$(C + 1 \rightarrow C) = INCC = s_j g(\mathbf{x}) \tag{9.4.24}$$

where $f(\mathbf{x})$ and $g(\mathbf{x})$ are functions of the application inputs $\mathbf{x}$ and where n is a constant. From a comparison with Figure 9.2.8, we see that the function control signal L is simply equal, in this case, to the transfer LDC. We will denote this as

$$C[L] = LDC = s_i f(\mathbf{x}) \tag{9.4.25}$$

where the square brackets identify the argument as a control line associated with register C, as was done in section 9.3. In order to carry out the transfers of expression (9.4.24), we must clock the register if either transfer is to occur. Thus we need to determine C[clock]. The simplest approach, and the one to be taken here, is to simply AND the system clock, SYSCLK, with the conditions. Thus

$$C[\text{clock}] = (LDC + INCC) \cdot SYSCLK \tag{9.4.26}$$

With this arrangement, the transfer will actually occur on the rising edge of the system clock, since the counter was designed using rising edge–triggered flip-flops. The question that next arises is, When do we change the state of the control system? There are two simple approaches we might take here: we can cause the state change to occur at the same time as the transfer or after the transfer. In what follows we will take the latter approach, in which the transfer will occur first, on the low-to-high transition of SYSCLK, and then the state change will occur, on the high-to-low transition of the clock, as shown in Figure 9.4.6. This two-phase type of operation is fairly common. Thus, the clocking of register C will happen before any change in state actually occurs.

This two-phase clocking scheme does present a potential difficulty that must be

Change state   Perform transfer   Change state

SYSCLK

Figure 9.4.6 Assumed timing for control algorithm implementation.

avoided. Since the transfers associated with a path occur before the state changes, we must ensure that no transfer in a path can cause the path conditions to change in such a way as to cause the system to end up in a state other than the one it would have ended up in before the transfer or in such a way as to affect other transfers in the path. An example is shown in Figure 9.4.7(a). To see what happens in this case, assume that val $(A) = n$ when the system reaches state i. According to the flow chart, we should then increment A and go to state j. However, what will actually happen is that A will be incremented, which will cause the answer to the question "Is $A = n$?" to change from yes to no. Two things will then occur. First, it is clear from the flow chart that we will end up not in state j, but, rather, in state k. Second, transfer T_1 will also be executed. This can be seen from Figure 9.4.7(b) by observing that

$$\text{clock for transfer } T_1 = \overline{\text{ch}}\,(A, n)\, s_i\, \text{YSCLK} \qquad (9.4.27)$$

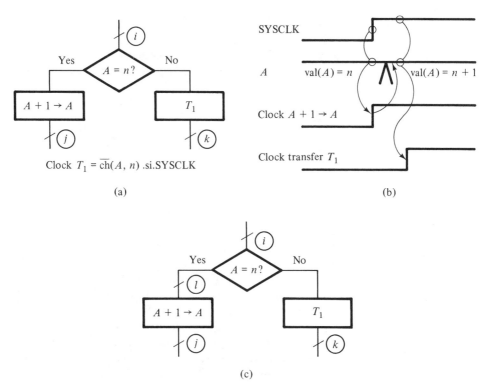

Clock $T_1 = \overline{\text{ch}}(A, n)$ .si.SYSCLK

(a)

(b)

(c)

Figure 9.4.7 State placement that can cause undesired transfers and state changes. (a) Faulty state placement. (b) Unwanted transfer caused by states of part (a). (c) Position of state to correct the problem of part (a).

Now, when we enter state i, this clock signal is low, because $\overline{ch}\,(A, n) = 0$. However, once A is incremented, $\overline{ch}\,(A, n) = 1$ and Equation (9.4.27) becomes 1, so that the clocking of transfer T_1 occurs. Thus, we not only end up in the wrong state, but we perform a transfer that is not intended. This problem is easily solved by simply placing a state between a transfer and any decision point that can be affected by that transfer. Thus, in Figure 9.4.7(a), we can eliminate the problem by placing an extra state between the decision block and the transfer $A + 1 \rightarrow A$, as shown in Figure 9.4.7(c).

From this example, we see one important factor in the placement of states in a control flow chart:

Rule 1.   *A state must separate any transfer from any decision which is affected by the transfer.*

This is not the only criterion for state placement, though. In general, we may place as many transfers between states as can be physically carried out simultaneously. On the other hand, we could not, for example, simultaneously perform the transfers $0 \rightarrow C$ and $C + 1 \rightarrow C$.[8] Thus, a second rule for state placement is:

Rule 2.   *A state must separate transfers which cannot be performed simultaneously.*

We need to start somewhere the process of state placement. In all that follows, we will start by placing a state at the entry point to an algorithm. Thus the third rule for state placement is:

Rule 3.   *Place a state at the entry point to an algorithm.*

Starting with the entry point state, the next state should be placed as far down in the algorithm as is allowed by the other two rules.

These rules should be taken as guides to the placement of states in a control flow chart and not as absolute rules never to be broken. In fact, as we shall see in the next section, any of these guides may be broken if no undesired behavior results.

Before leaving the subject of register control, let us look at an alternative to the two-phase timing scheme described above. It would seem that we could eliminate the problems of wrong states and unwanted transfers if we caused the state to change at exactly the same time that a transfer is made. In fact, as long as all of the flip-flops—registers as well as state flip-flops—are edge-triggered, this can be done. However, there is a serious problem associated with this approach, namely, clock skew. When a clock signal is distributed throughout a large system, it may happen that the clock arrives at one register before it arrives at another, thus causing the registers to change at slightly different times. This situation is referred to as *clock skew* and is caused by different propagation delays from one register to another. In the example cited above, the clocking of the register involved

---

[8] This is, of course, equivalent to the single transfer $1 \rightarrow C$. However, in the control algorithm it may be essential that the clearing of C and the incrementation of C be separate events.

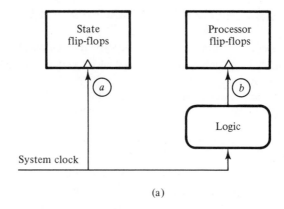

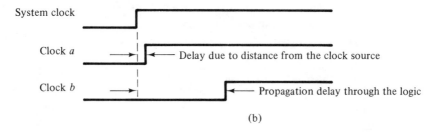

Figure 9.4.8   An illustration of clock skew: (a) implementation of control and processor clocks; (b) resulting timing and clock skew.

in the transfer was derived using combinational logic, whereas the clock controlling the state change comes directly from the clock generator or the system clock. Thus there will be a significant time delay between the arrival of the clocking signals for the register and that of the signals for the state flip-flops, as shown in Figure 9.4.8. The resulting clock skew can produce the same unwanted system behavior as the two-phase clocking scheme described above.

One way of eliminating, or at least reducing, clock skew is to connect the clock generator directly to the clock input of *all* of the flip-flops without going through any combinational logic. Unfortunately, this requires that all of the registers be designed to have a "do nothing" mode, as shown for the universal shift register of Figure 9.2.6, whose control functions were given in Table 9.2.1. Obviously, such a requirement will increase the complexity of the register design. Furthermore, even if this is done, clock skew can still occur on account of differences in the physical distances between registers. In this case, registers close to the clock generator will receive the clock before registers farther away.[9] For these reasons, we will use the two-phase clocking system in all that follows.

---

[9] The propagation delay of signals on wires is of the order of 1 nanosecond per foot (30 cm). Thus, in very high-speed systems, where propagation delays through gates and flip-flops may be of the order of tens of picoseconds, even very short differences in path length can lead to incorrect system operation.

9.5 THE DESIGN PROCESS AND SOME EXAMPLES

The process of designing a large-scale system is not much different from the design process followed in Chapter 5. However, since such systems are generally characterized by having two components, the control unit and the processor unit, as shown in Figure 9.1.1, we must not only specify the controller, on the basis of a given algorithm, but the processing unit as well. We may outline the design process as follows:

1. *Define the problem.* Identify exactly what the system is supposed to do, in a global sense.
2. *Identify the registers and other elements in the processor.* What hardware is required to perform the required task?
3. *Develop a control algorithm.* On the basis of the problem and the assumed processor hardware, develop a control algorithm flow chart. This step and step 2 generally must be done together, or, at least, iteratively.
4. *Develop the transfer and state variable equations.* This step proceeds as was described in Section 9.4.
5. *Write the specific register and other control equations.* On the basis of the specific registers required to implement the algorithm and the control equations of part 4, derive the necessary equations to make all of the processor components function as required.

Perhaps the best way to illustrate this process is by giving some examples. Two will be given here. The first example is a hardware system that is used to multiply two unsigned positive 8-bit numbers to form a 16-bit result. The second example, a digital speedometer for a bicycle, illustrates an alternative timing scheme, as well as some interesting asynchronous timing situations.

9.5.1 A Serial Hardware Multiplier System

In this example, we will design a hardware system that multiplies two 8-bit unsigned (positive) numbers together to produce a 16-bit result. The algorithm we shall use is based on the usual pencil-and-paper method, which is best illustrated with an example. Suppose the problem is to multiply 1001 by 1011. The work is carried out as follows:

$$
\begin{array}{r}
1001 \\
\times\ 1011 \\
\hline
1001 \\
1001 \\
0000 \\
1001 \\
\hline
1100011
\end{array}
$$

This process begins by multiplying the multiplicand by the rightmost bit of the multiplier to form a partial product. The multiplicand is next multiplied by the next-least significant bit of the multiplier, with the result being added to this partial product, after shifting one bit position, to form the next partial product. This process is then repeated for each of the remaining bits of the multiplier.

Figure 9.5.1(a) shows the general organization of the hardware needed to carry out

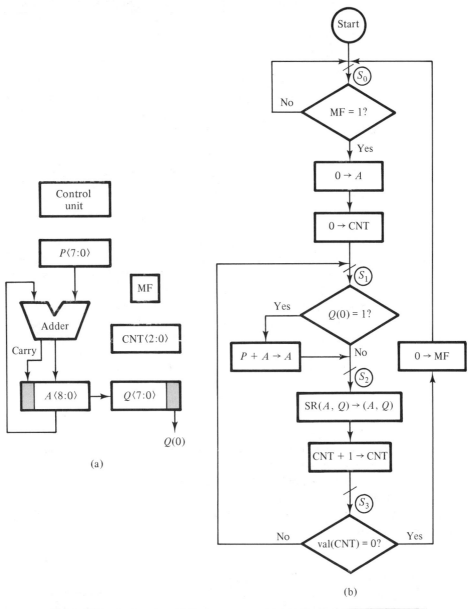

(a)

(b)

Figure 9.5.1   An 8 × 8 multiplication system. (a) Multiplier block diagram; (b) shift-and-add multiplication algorithm.

this process. This hardware consists of a register to hold the multiplier, $Q\langle 7{:}0\rangle$, a register to hold the multiplicand, $P\langle 7{:}0\rangle$, and a register to hold the sum involved in creating the partial products, $A\langle 8{:}0\rangle$. Associated with register A is an extra bit, $A\langle 8\rangle$, used to hold any carry generated when the sum is formed by the adder shown in the figure. Since the product is found after performing the add and shift process eight times, a 3-bit counter is also needed for determining when the multiplication is complete. This counter is shown as $CNT\langle 2{:}0\rangle$ in the figure. Finally, we will need some type of flag, MF in Figure 9.5.1(a), to indicate when the multiplication process is to begin. This flag flip-flop can also indicate to the ''outside world'' when the multiplication is finished. We will assume that the multiplication process is to start when the flag is set and we will indicate completion of the process by clearing the flag. Figure 9.5.1(b) shows the control flow chart describing the add-shift multiplication algorithm carried out on this hardware. In this figure, the transfer $SR(A, Q) \rightarrow (A, Q)$ is defined by the transfers

$$A\langle 8{:}1\rangle \rightarrow A\langle 7{:}0\rangle, 0 \rightarrow A(8)$$
$$Q\langle 7{:}1\rangle \rightarrow Q\langle 6{:}0\rangle, A(0) \rightarrow Q(7)$$

and the transfer $P + A \rightarrow A$ is defined as

$$P\langle 7{:}0\rangle + A\langle 7{:}0\rangle \rightarrow A\langle 7{:}0\rangle, \text{carry} \rightarrow A\langle 8\rangle.$$

Before discussing the placement of the states, we need to determine the function of each register and then specify a design for each. The function of the P register is simply to hold the multiplicand and can, therefore, be implemented by the 8-bit storage register shown in Figure 9.2.1. Although the Q register needs only to be capable of shifting right in this algorithm, it clearly must be loaded with the multiplier before the algorithm is carried out. Thus, we will implement this register with the universal shift register of Figure 9.2.6. The partial product register, A, not only must have the capability of being loaded and shifted, but it must also be capable of being cleared. These functions can all be met, once again, using the universal shift register. In this case, however, a 9-bit version is required to accommodate the carry bit. The counter, which must be capable of being cleared as well as being able to count, can be implemented using the 4-bit counter of Figure 9.2.8. Finally, the multiply flag, MF, may be implemented using the type 7474 edge-triggered D flip-flop with asynchronous Set and Clr.

The placement of states in the algorithm flow chart is based on the two-phase clock scheme presented in Section 9.4. We start with the state S_0 at the beginning of the algorithm. State S_1 must be placed prior to the test for $Q(0) = 1?$, since the resulting path taken could involve the addition of A and P, which cannot occur simultaneously with the clearing of register A. State S_2 is used to separate the modification of A, due to an addition, from the shifting of A. Finally, state S_3 separates the incrementation of CNT from a decision based on val (CNT) after this incrementation. On the basis of this state placement in the flow chart of Figure 9.5.1(b), we may construct the path-transfer table as shown in Figure 9.5.2. Note in this figure that the notation CNT(3) is used to indicate that val (CNT) = 0 and $\overline{CNT}(3)$ is used to indicate that val (CNT) $\neq 0$. This is done because we are using the 4-bit counter of Figure 9.2.8. When the fourth bit of this counter, CNT(3), goes to 1, we

Path	Path condition	Transfer	Transfer condition
$S_0\text{-}S_0$	$\overline{MF}$	–	–
$S_0\text{-}S_1$	MF	CLRA CLRCNT	– –
$S_1\text{-}S_2$	–	ADDPA	$Q(0)$
$S_2\text{-}S_3$	–	SHRAQ INCCNT	– –
$S_3\text{-}S_0$	CNT(3)	CLRMF	–
$S_3\text{-}S_1$	$\overline{\text{CNT}(3)}$	–	–

Figure 9.5.2   Path-transfer table for the serial multiplier.

have counted 8 clock pulses, and the low-order 3 bits will be 0. Thus, CNT(3) corresponds to val (CNT) = 0. The transfer and the next-state equations are easily written from this table and are given as follows:

$$(0 \rightarrow A) = \text{CLRA} = s_0\,\text{MF}$$
$$(0 \rightarrow \text{CNT}) = \text{CLRCNT} = s_0\,\text{MF}$$
$$(P + A \rightarrow A) = \text{ADDPA} = s_1 Q(0) \qquad (9.5.1)$$
$$(\text{SR}(A, Q)) \rightarrow (A, Q) = \text{SHRAQ} = s_2$$
$$(\text{CNT} + 1 \rightarrow \text{CNT}) = \text{INCCNT} = s_2$$
$$(0 \rightarrow \text{MF}) = \text{CLRMF} = s_3\,\text{CNT}(3)$$

$$S_0 = s_0\,\overline{\text{MF}} + s_3\,\text{CNT}(3)$$
$$S_1 = s_0\,\text{MF} + s_3\,\overline{\text{CNT}(3)}$$
$$S_2 = s_1 \qquad (9.5.2)$$
$$S_3 = s_2$$

Note in equation set (9.5.1) that an acronym is given to each of the transfers to simplify reference to them. These acronyms are selected so that they indicate the transfer. For example, CLRA means to *CLeaR* register A. ADDPA is used to denote the transfer *ADD* register P to register A.

From the transfer equations given in group (9.5.1) and the registers specified above, we may next derive the specific control equations required by each register:

$$\text{MF[CLR]} = \text{CLRMF} \cdot \text{SYSCLK}$$
$$A\text{[CLR]} = \text{CLRA} \cdot \text{SYSCLK}$$
$$A\text{[Load]} = \text{ADDPA}$$
$$A\text{[CLK]} = (\text{SHRAQ} + \text{ADDPA}) \cdot \text{SYSCLK} \qquad (9.5.3)$$
$$Q\text{[CLK]} = \text{SHRAQ} \cdot \text{SYSCLK}$$
$$\text{CNT}[L] = \text{CLRCNT}$$
$$\text{CNT[CLK]} = (\text{INCCNT} + \text{CLRCNT}) \cdot \text{SYSCLK}$$

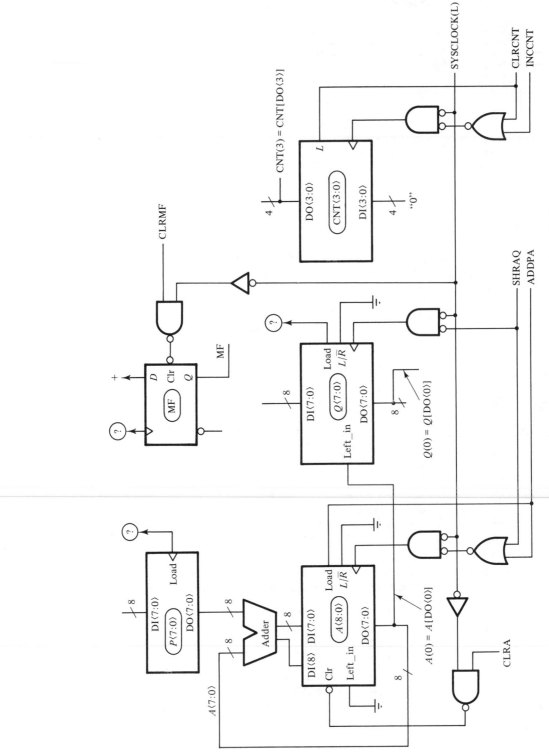

Figure 9.5.3 Schematic diagram for the processing unit of the serial multiplier.

where SYSCLK is the system clock. The resulting processing unit is shown in Figure 9.5.3. In this figure the lines labeled with a question mark are signals that must be supplied by the "outside world" to load the multiplier and multiplicand and to set the MF flag.

The equations for the state variables can be found from equation set (9.5.2) and Table 9.5.1. From this table we may obtain the equations for Y_1 and Y_0 as follows:

$$
\begin{aligned}
D_1 &= Y_1 = S_2 + S_3 = s_1 + s_2 = \bar{y}_1 y_0 + y_1 \bar{y}_0 \\
D_2 &= Y_0 = S_1 + S_3 = s_0\,\text{MF} + s_3\,\overline{\text{CNT}(3)} + s_2 \\
&= \bar{y}_1 \bar{y}_0 \text{MF} + y_1 y_0 \overline{\text{CNT}(3)} + y_1 \bar{y}_0 \\
&= \bar{y}_1\,\text{MF} + y_1 \bar{y}_0 + y_1\,\overline{\text{CNT}(3)}
\end{aligned}
\tag{9.5.4}
$$

The final realization for the control unit of the multiplier is shown in Figure 9.5.4. Before we leave this example, let us consider using a PLA device to implement this control unit. Suppose we are given a logic array IC having at least six outputs, at least five inputs, and at least seven product terms. With such a device we can implement the controller with two ICs, because a 7474 IC contains two D flip-flops in the same package. Figure 9.5.5 shows the programming diagram for the logic array, and Figure 9.5.6 shows a block diagram of the final implementation.

9.5.2 A Bicycle Speedometer

What we would like to do in this second example is design a system that can be used to keep track of the speed of a bicycle and display the speed numerically using an LED or an LCD display. Speed is a function of two factors: distance traveled and time of travel. A time scale is easily generated by using an oscillator whose output frequency can be very accurately set and maintained. The distance traveled during a specified time interval can be measured by counting the number of turns of the bicycle wheel during this time interval. Although speed is actually defined as the ratio of distance and time, we do not actually have to perform any division to obtain the bicycle's speed. This can be determined by counting the number of revolutions of the wheel during a fixed period of time and then looking up the speed corresponding to this number in a table, stored in a ROM. This, in fact, is what we did in Chapter 4 as one method for converting from one code to another.

On the basis of this simple idea, we can now begin to specify the general hardware needed to implement the speedometer. A block diagram of this hardware is shown in Figure

TABLE 9.5.1 NEXT-STATE CONDITIONS

Next state		Condition
Y_1	Y_0	
$S_0 = 0$	0	$s_0\,\overline{\text{MF}} + s_3\,\underline{\text{CNT}(3)}$
$S_1 = 0$	1	$s_0\,\text{MF} + s_3\,\text{CNT}(3)$
$S_2 = 1$	0	s_1
$S_3 = 1$	1	s_2

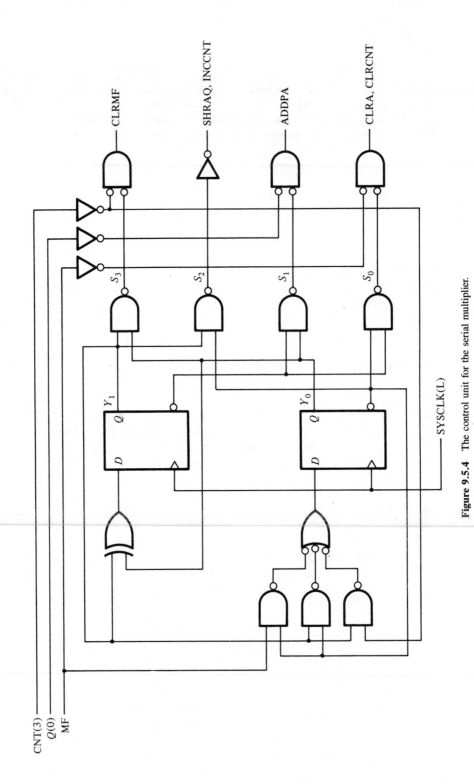

Figure 9.5.4 The control unit for the serial multiplier.

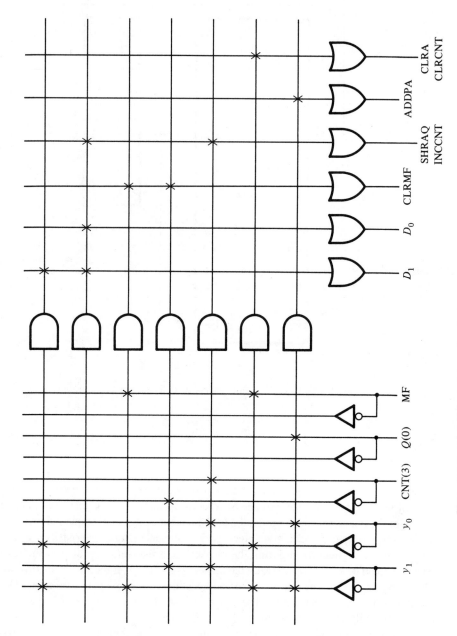

Figure 9.5.5 Programming diagram for a logic array that implements the controller for the serial multiplier system.

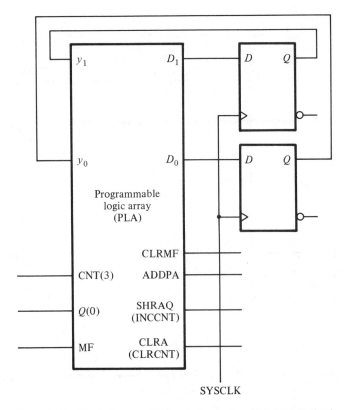

Figure 9.5.6 Block diagram of the controller for the serial multiply circuit.

9.5.7. First, we will need an oscillator to generate the time standard. This oscillator will also serve as the *system clock*, SYSCLK. The specified time interval can then be measured by counting a fixed number of clock pulses using a simple counter, which we will refer to here as the *time counter*, T. A second counter, which we will call the *revolution counter*, C, will be used to count the number of wheel revolutions during this time interval. The design of the counters T and C might be based on that of the 4-bit counter shown in Figure 9.2.8. Thus we can preset the counters as well as generate a carry out. Since the speed display should be held fixed during the measurement interval, the value of the revolution counter should be stored in a register, the *display register*, D, at the end of the time interval. The output of this register serves as the address input to the display converter ROM used for storing the conversion table. The output of the ROM would then serve to drive the display. We will describe how this is done a little later.

There are, of course, many ways in which we can determine that the wheel has gone around once. One of the simplest is to attach one or more magnets to spokes, say on the front wheel, and to attach to the fork a magnetic reed switch or other device that can detect a strong magnetic field. Each time the wheel goes around, this sensor will put out a pulse

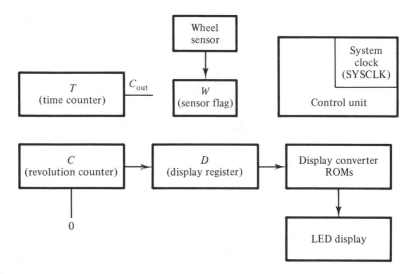

Figure 9.5.7   Block diagram of a bicycle speedometer.

which can be used to set a flip-flop, called the *sensor flag*, W, whose output is used by the system control unit to update the revolution counter C.

The control algorithm for this speedometer is easily described in words, as follows. On each clock pulse, increment the timer and check, first, to see whether the time interval is up. If it is not, then check the sensor flag W to see whether the wheel has gone around one revolution. If so, then update the revolution counter. When the time interval is up, transfer the contents of the revolution counter to the display register and clear the revolution counter for counting revolutions in the next time interval. However, make sure that if a pulse from the wheel sensor occurs during this time it will eventually be counted. Figure 9.5.8 shows the flow chart for this algorithm.

The placement of the states in Figure 9.5.8 is based on the two-phase clock scheme and the resulting guidelines given in Section 9.4. We begin by placing state S_0 at the entry point to the algorithm. It would appear that two more states are required: one between the transfer $T + 1 \rightarrow T$ and the decision "$T[C_{out}] = 1?$" and the other between the transfers $C + 1 \rightarrow C$ and $0 \rightarrow W$. The first might be required because the transfer $T + 1 \rightarrow T$ can change the decision outcome "$T[C_{out}] = 1?$"; and the second might be required because clearing W will change the decision "$W = 1?$" from yes to no. However, we may avoid putting these states in if we allow different transfers to occur on different clock edges! In particular, if we cause the transfers which cannot change decision outcomes, $C + 1 \rightarrow C$, $C \rightarrow D$, and $0 \rightarrow C$, to occur on the rising edge of the system clock (SYSCLK), and the transfers that can affect the decisions, $T + 1 \rightarrow T$ and $0 \rightarrow W$, to occur on the falling edge, then at the end of one clock cycle all of the required transfers will be carried out without conflict. *Since there will now be only one state, there will be no state variables to change and therefore there can be no problems with clock skew.* We will thus use this scheme, as shown in Figure 9.5.9, for clocking our speedometer system.

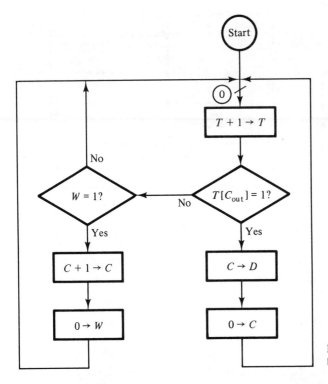

Figure 9.5.8 Control algorithm for the bicycle speedometer.

Using this two-edge clocking technique for transfers, the basic control equations can now be derived directly from the control flow chart. These equations are as follows:

$$(T + 1 \to T) = \mathrm{INCT} = 1 \text{ (which means to increment } T \text{ on each clock pulse)}$$
$$(C \to D) = \mathrm{CD} = T[C_{\mathrm{out}}]$$
$$(0 \to C) = \mathrm{CLRC} = T[C_{\mathrm{out}}] \qquad\qquad (9.5.5)$$
$$(C + 1 \to C) = \mathrm{INCC} = \overline{T}[C_{\mathrm{out}}]W$$
$$(0 \to W) = \mathrm{CLRW} = \overline{T}[C_{\mathrm{out}}]W$$

Since there is only one state, there are no state equations to deal with.

Before deriving the individual control equations for each register, we must first determine the type and size of all of the various registers. For simplicity, we will assume that the two counters, T and C, are of the same type as the presettable binary counter

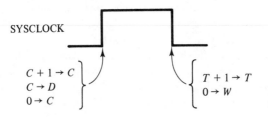

Figure 9.5.9 Transfers allowed on the rising and falling edges of SYSCLK.

shown in Figure 9.2.8, although not necessarily of the same size. We will also use the storage register of Figure 9.2.3 for the display register D. To determine the size of these counters and registers, we need to estimate time intervals, possible speeds, number of revolutions, and other quantities we will be dealing with.

To begin with, the typical, average speed for the touring cyclist is around 15 mph. Racing cyclists may run at around 30 mph, and cyclists coasting down a hill may approach 40 mph (not meant for "nervous Nellies"). Therefore, 40 mph would seem to be a reasonable upper bound on the speed. Now at 40 mph, with a bicycle having a 27-in. wheel, the cycle would be moving at 59.667 ft/s. Since the circumference of a 27-in. wheel is 7.069 ft, at 40 mph the wheel would be rotating at 9.299 rev/s. At 1 mph, the wheel would be rotating at 0.207 rev/s. This means that for a wheel having one magnet on it and moving at 1 mph, we would have to wait about 5 s for each pulse from the wheel sensor. This also implies that we would be unable to distinguish a 1-mph change in speed by counting sensor pulses over a time interval less then 5 s (why?) Updating the speed at 5-s or longer intervals would also seem to be a rather long time. To increase the update rate to, say, every 1 s, we could place five magnets around the wheel. In this case, we would have 1.037 pulses per second at 1 mph and 41.495 pulses per second at 40 mph. On the basis of these considerations, we will make the following specifications:

1. Use five magnets around the front wheel.
2. The speed range is from 0 to 40 mph.
3. Update the speed every 1 s.
4. The revolution counter C must be 6 bits to count the 41.495 pulses per second at 40 mph.

To determine the clock frequency, we observe that since we must update the revolution counter every 24.099 ms (equivalent to 1/41.495 pulses per second) at 40 mph, the clock period must be shorter than this. In fact, as we shall shortly see, this clock frequency should be greater than 3 times the rate of the revolution counter to ensure that all revolutions of the wheel are counted. Since $3 \times 41.495 = 124.485$, which is close to $2^7 = 128$, we will simply make the clock run at 128 pulses per second and, thus, make the time counter 7 bits so that $C_{out} = 1$ every 1 s. Thus,

5. The time counter T is to be 7 bits.

We have now defined all of the components for our bicycle computer except two, namely, the wheel sensor flag W and the display system. Let us first look at the sensor flag. As indicated in Figure 9.5.7, the sensor flag W is set by the occurrence of a pulse from the wheel sensor. Since this pulse occurs asynchronously with respect to the system clock and its duration is uncontrollable, we must be careful in how this flag is designed. From the control flow chart of Figure 9.5.8, we see that the sensor flag must stay set no more than one clock period, for otherwise the revolution counter will be incremented more than once for each sensor pulse. Further, since the revolution counter C is incremented on

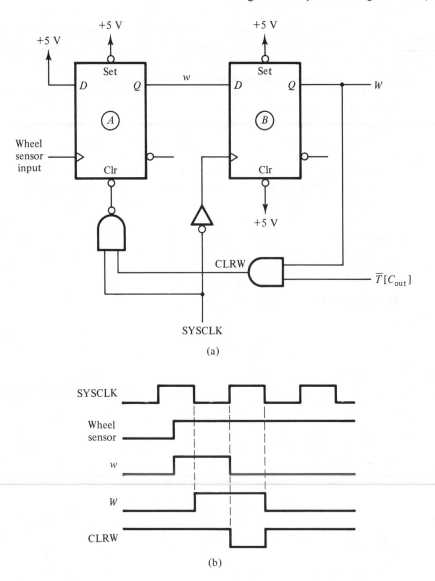

Figure 9.5.10   The wheel sensor flag circuit and timing. (a) Wheel sensor flag flip-flop. (b) Typical timing for the wheel sensor flag.

the rising edge of the system clock, we must be certain that the sensor signal is detectable on this edge. In fact, the occurrence of a sensor pulse must be detected regardless of when it happens with respect to the system clock. Figure 9.5.10(a) shows a circuit that can be used to generate the flag W under these conditions.[10] Figure 9.5.10(b) shows the corre-

---

[10] The unused asynchronous inputs in Figure 9.5.10 have been tied to a high voltage to prevent the flip-flop from being changed by external "noise."

sponding timing diagram. We see from this timing diagram that the rising edge of the wheel sensor pulse causes flip-flop A to set. Once A is set, the next falling edge of the system clock will cause flip-flop B to set. This output is just the sensor flag W. Now, if the carry out of the time counter T is 0, then on the next assertion of the system clock, flip-flop A will be cleared and will not change again until the next rising edge of the wheel sensor. Finally, when the system clock goes low, the 0 at the output of flip-flop A will be passed to the output of flip-flop B and thus W will go to 0 on the trailing edge of the system clock, as required.

However, what happens if the sensor pulse occurs coincidental with the assertion of $T[C_{out}]$? Under such circumstances, flip-flop A will not be reset, and so W will remain high for one more clock period, during which time this pulse will be recorded. Figure 9.5.11 shows a timing diagram of this situation.

As was mentioned earlier, the clock frequency should be more than 3 times the wheel sensor pulse rate to ensure proper operation. The reason for this can be seen from Figure 9.5.11. From this figure, we see that, under the worst case, a wheel sensor pulse may occur during the second half of a SYSCLK and in the cycle immediately preceding the assertion of $T[C_{out}]$. Thus, W and $T[C_{out}]$ will be asserted simultaneously, one cycle later. Now, because $T[C_{out}]$ must be low in order to reset the W flag, one more cycle will pass after $T[C_{out}]$ goes low before W is reset. Thus, a total of three cycles are required under this worst-case condition, implying that the clock rate must be at least 3 times the wheel sensor rate.

The display system for this speedometer is an LED or an LCD seven-segment display as shown in Figure 9.5.12. In this device there is one line coming in for each of the seven display segments. To turn on a segment, we need only put a 1 on the corresponding input. Thus, each of the numbers from 0 to 9 can be displayed by turning on the appropriate subset of segments as shown in Figure 9.5.12(b). Since our speedometer is to have a range

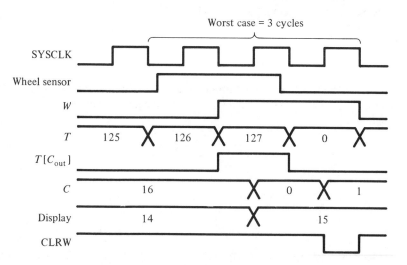

Figure 9.5.11   Worst-case timing for the wheel sensor flag.

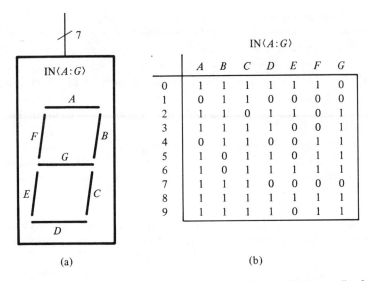

	A	B	C	D	E	F	G
0	1	1	1	1	1	1	0
1	0	1	1	0	0	0	0
2	1	1	0	1	1	0	1
3	1	1	1	1	0	0	1
4	0	1	1	0	0	1	1
5	1	0	1	1	0	1	1
6	1	0	1	1	1	1	1
7	1	1	1	0	0	0	0
8	1	1	1	1	1	1	1
9	1	1	1	1	0	1	1

(a)   (b)

Figure 9.5.12 A seven-segment display: (a) segment identification; (b) input coding for display of the ten digits.

from 0 to 40 mph, we will need two such displays. In order to cause the displays to show the speed corresponding to the count held in D, we will need to convert this count to seven lines for each display. We will do this using two ROMs, HIROM and LOROM, which will be used to convert the 6-bit count to two 7-bit codes corresponding to the appropriate segment values to display the speed. These ROMs are shown as the "display converter ROMs" in Figure 9.5.7. For example, if the count value in the D register is 21, corresponding to 20.24 mph, we would want to display 20. To do this, we would have to have stored in the two converter ROMs the following values:

$$ABCDEFG$$
$$\text{HIROM}(21) = 1101101 \quad \text{which displays a 2}$$
$$\text{LOROM}(21) = 1111110 \quad \text{which displays a 0}$$

Given the equations of group (9.5.5) and the above register specifications, we may now complete the system design by deriving the individual register control equations. These are as follows:

$$T[\text{CLK}] = \overline{\text{SYSCLOCK}}$$
$$D[\text{CLK}] = T[C_{\text{out}}]\,\text{SYSCLOCK}$$
$$C[L] = T[C_{\text{out}}]$$
$$C[\text{CLK}] = (T[C_{\text{out}}] + \overline{T}[C_{\text{out}}]W)\,\text{SYSCLOCK}$$
$$= (T[C_{\text{out}}] + W)\,\text{SYSCLOCK}$$
$$W[\text{CLK}] = \overline{T}[C_{\text{out}}]W\,\overline{\text{SYSCLOCK}}$$

(9.5.6)

Figure 9.5.13 shows the final realization for the bicycle speedometer, and Figure 9.5.14 shows the typical timing for the resulting system.

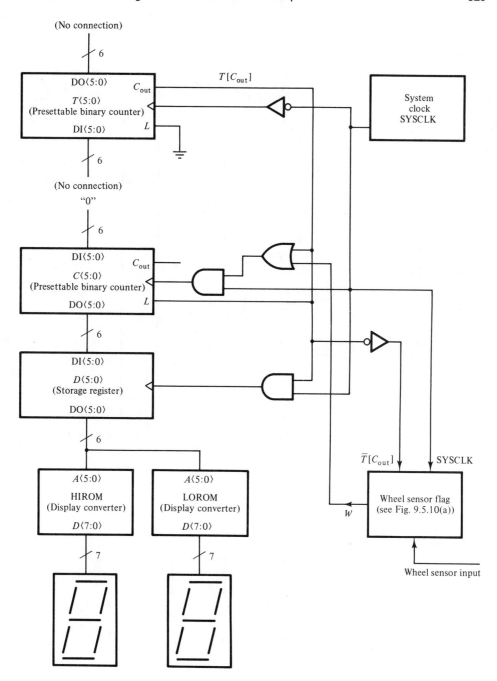

Figure 9.5.13   Final bicycle speedometer schematic diagram.

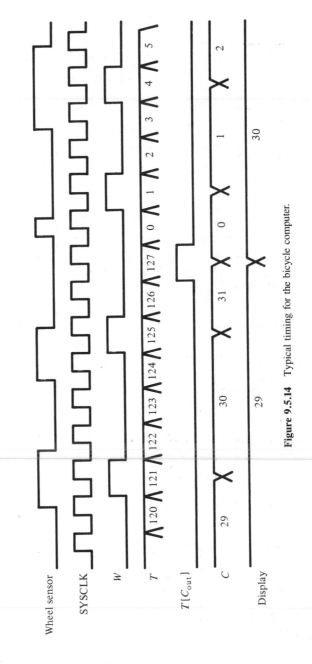

Figure 9.5.14 Typical timing for the bicycle computer.

9.6 *SOME FINAL COMMENTS AND OBSERVATIONS*

In this chapter we have demonstrated that the design of medium- to large-scale digital systems can be approached in a very systematic way. This process involves breaking the problem up into two designs: the design of the processing unit and the design of the controller. The design of the control unit is based on a careful specification of the control algorithm, in this case, by using a flow chart. The processing unit is made up of the various component parts required to implement the requisite algorithm. Each of these parts can be designed by further breaking down its function into smaller components, as was demonstrated in Section 9.2. Thus, the process of designing any large-scale system involves breaking the specification up into small components which can be easily described using methods developed in this book and then combining these elements into the larger system. As indicated in the bicycle speedometer example, however, the problem of unsynchronized inputs requires special care. In this case, the unsynchronized wheel sensor was synchronized to the system clock using the double-ranked flip-flop arrangement shown in Figure 9.5.10. Once this was done, timing in the control algorithm was based solely on this system clock. This approach is common under such circumstances and is generally a good one to take.

9.7 *AN ANNOTATED BIBLIOGRAPHY*

An excellent reference to the design procedures discussed in this chapter can be found in the classic book by Bartee, Lebow, and Reed. The more recent texts by Hayes and Mano also give a complete description of large-scale system design specifically as related to the design of computers. Hayes also discusses the use of flow charts for specifying the control algorithm. Muroga, in Chapter 9, gives a rather brief but instructive explanation of flow chart usage as well.

BARTEE, T. C., I. L. LEBOW, AND I. S. REED, *Theory and Design of Digital Machines*, McGraw-Hill, New York, 1962.

HAYS, J. P., *Computer Architecture and Organization*, McGraw-Hill, New York, 1979.

MANO, M. M., *Computer System Architecture*, Prentice-Hall, Englewood Cliffs, N.J., 1982.

MUROGA, S., *Logic Design and Switching Theory*, Wiley-Interscience, New York, 1979.

An interesting alternative to block diagrams, register transfers, and flow charts for the specification of large-scale systems and control can be found in the book by Bell and Newell. At the top level of system design, the PMS (processor-memory-switch) description system is used to specify the specific system requirements. The low-level design specification is then given by the ISP (instruction-set processor) description. The PMS system, described by Bell and Newell, has gained a good deal of acceptance for the description of computer systems. A recent book by Gorsline gives an excellent discussion of this system and uses it in a coherent computer design example.

BELL, C. G., and A. NEWELL, *Computer Structures*: *Readings and Examples*, McGraw-Hill, New York, 1971.

GORSLINE, G. W., *Computer Organization*: *Hardware/Software*, 2nd ed., Prentice-Hall, Englewood Cliffs, N.J., 1986.

9.8 PROBLEMS

9.1. Design a 4-bit counter that counts either in binary or BCD depending on a control line MODE. When MODE = 0, the counter is to count in binary, and when MODE = 1, the counter is to count in BCD.

9.2. Add circuitry to your design in Problem 9.1 to generate a carry out of the high-order bit so that two or more such counters can be cascaded to form longer binary or BCD counters.

9.3. Design a 8-bit register having three control inputs, $ROT\langle 2:0\rangle$, that rotates the register contents left as many bit positions as given by the deicmal equivalent of the number contained in ROT, namely, val $(ROT\langle 2:0\rangle)$. For example, suppose ROT = 011 = 3 (base 10) and assume that the register contains 10100110. After the clock is asserted, the register will contain 00110101. If val $(ROT\langle 2:0\rangle)$ = 0, the register is to be loaded with external information on the assertion of the clock.

9.4. Construct the register transfer equations that describe the register designed in Problem 9.3.

9.5. Write the transfers needed to implement the following statement: If A is negative, then clear register B; otherwise, make register B negative.

9.6. Let $A\langle 7:0\rangle$ and $B\langle 7:0\rangle$ be two 8-bit registers that contain two BCD digits each. Write the appropriate set of transfers to add A and B and place the result in $C\langle 8:0\rangle$, where the ninth bit contains any carry generated out of the high-order digit.

9.7. Modify the design of the programmable frequency counter so that a division factor of 1 can be obtained.

9.8. Design a system that implements the register transfer given in Equation (9.3.9).

9.9. Construct a flow chart showing the algorithm used in the programmable frequency divider.

9.10. Construct flow charts corresponding to the state diagrams shown in Figure P9.10.

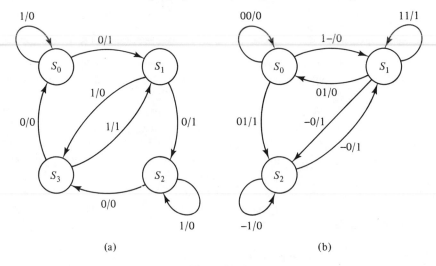

(a)   (b)

Figure P9.10

9.11. Write the control equations required to implement the algorithms of Problem 9.10.

9.12. Construct the path-transfer table corresponding to the control algorithm given in Figure P9.12.

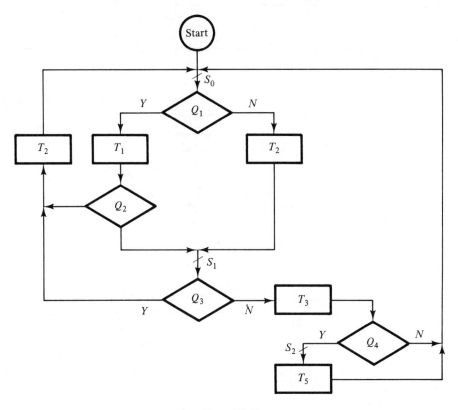

Figure P9.12

9.13. Write the appropriate next-state and transfer equations for the algorithm of Figure P9.12.

9.14. Suppose for a certain algorithm it is essential that the following three transfers each be accomplished in one clock cycle (refer to footnote 8):
   (a) $0 \rightarrow C$
   (b) $C + 1 \rightarrow C$
   (c) $0 \rightarrow C$ and $C + 1 \rightarrow C$
   Design a register capable of doing these three things in the required one clock cycle.

9.15. Based on existing small-scale TTL integrated circuits (say, the 7400 series), how many ICs would be required to implement the multiplier control unit shown in Figure 9.5.4?

9.16. Redesign the bicycle speedometer so that the output of the wheel sensor directly drives the C counter clock (i.e., eliminate the W flag). C is then to be reset by the controller using the C register's asynchronous clear line. Since the incrementation of C is now unsynchronized with respect to the system clock, some pulses of the wheel sensor may be missed. Under what conditions will this occur? Figure P9.16 shows a suggested block diagram for your design. (This situation is sometimes referred to as the unsynchronized clock problem.)

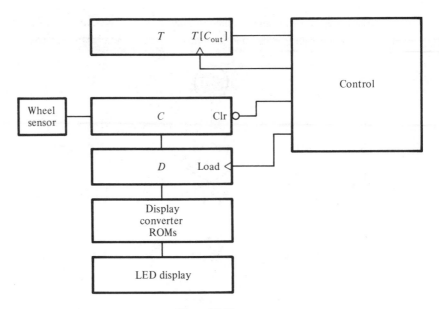

Figure P9.16

9.17. Describe how you would modify the bike speedometer design shown in Figure P9.16 so that the output value displayed is changed only if it and the last value are the same. How will this design change affect the accuracy of the speedometer output in the basic design of Problem 9.16?

9.18. How might you modify your design in Problem 9.16 so as to synchronize the incrementation of C to the system clock? (*Hint*: Refer to the original bike speedometer design given in Section 9.5.2.)

9.19. Design a system that will convert three-digit BCD numbers to binary using the algorithm discussed in Section 2.5.3. Let $B_2\langle 3:0\rangle$, $B_1\langle 3:0\rangle$, and $B_0\langle 3:0\rangle$ be the 4-bit registers used to hold the BCD numbers, and let $R\langle 9:0\rangle$ be the register used to hold the final value.
  (a) Specify the characteristics of these registers and design each.
  (b) Identify all other registers and logic necessary to carry out this transformation.
  (c) Construct a flow chart which implements the necessary control algorithm.
  (d) Write all of the necessary control equations.
  (e) Construct a schematic diagram of the completed control unit and processing unit.

9.20. Describe the modification in the design of the BCD-to-binary converter in Problem 9.19 that would be necessary to make it possible to convert binary numbers in R to a corresponding BCD equivalent in registers B_2, B_1, and B_0.

An Introduction to
IEEE Std. 91–1984

Appendix

A.1 INTRODUCTION

As we pointed out in Chapter 4, the choice of symbols used to construct logic diagrams is very important for quickly and clearly conveying information about the designer's logical intent. The symbols described in that chapter are those currently used by industry for this purpose. In effect, these symbols are the atoms or the basic building blocks for all digital system designs. However, as we pointed out in Chapter 9, when designing large-scale systems containing more complex functions such as counters, shift registers, and multiplexers, we require some type of simplification. Basically, we want to replace the detailed logic drawing for the complex function with a simple block symbol that represents the function. In Chapter 9 we did this by drawing a rectangle and labeling it with the function performed. This approach works fine as long as we have only a few different types of counter, multiplexer, or whatever. A simple perusal of any IC catalog shows, however, that there are a tremendous number of different devices available to the system designer. Thus there is a clear need for some type of simplified symbol that identifies the function performed by the device without showing the detailed logic. ANSI/IEEE Std. 91–1984 addresses this issue.

Essentially, Standard 91–1984 consists of two parts: small-scale symbols, which include the distinctive symbols introduced in Chapter 4; and large-scale symbols, which identify the function performed by a given block without small-scale detail. A third part of this symbology, and one we will mention only briefly later, shows certain physical characteristics of devices, such as drive capability, tri-state outputs, and hysteresis. The purpose of this appendix, then, is to briefly describe these symbols and show how to interpret them. We will first describe the small-scale symbols and compare them with those used in this book. We will then introduce the large-scale symbols. We will not attempt to describe all the details of this standard; this is done in the references given at the end of the Appendix. We will, however, introduce the more commonly encountered elements of the symbology and illustrate their use with examples.

One final comment before getting into some of the symbology detail. All standards are written to be as precise as possible and yet still allow room for variation in style and usage. Ultimately, it is the frequent usage of a particular form that causes it to become standardized in some loose but generally accepted manner. This is true of grammatical style in English (we rarely encounter a ''thee'' or a ''thou'' these days!) as well as programming languages. For example, the programming language Pascal allows for a tremendous variation in style. Yet the highly structured, ''properly'' indented form for Pascal programs appear to be the accepted norm. Symbology standards are no different. Since Standard 91–1984 is a relatively new standard, generally acceptable style has not yet been developed by usage. Thus, as we look through the literature, we may find a variety in the form of the IEEE standard symbol for a given device. As use of the symbology continues,[1] this style and form will become more ''standardized.'' This appendix, then, gives an indication of some of the currently used forms for the new standard symbology.

A.2 Symbols Used for Gates and Flip-Flops

There are two types of small-scale symbols: those with distinctive shape and those with uniform shape. The distinctive-shape symbols are equivalent to those used throughout this book. The uniform symbols use a rectangle to represent all gates. The type of gate represented is indicated by a *qualifying symbol* inside the rectangle. Since the uniform symbols have no apparent direction, care must be taken in identifying inputs and outputs. By convention, inputs come in on the left of the rectangle and outputs leave on the right. If, however, confusion can occur, arrows may be placed on lines to explicitly identify inputs and outputs. Figure A.2.1 shows the equivalences that exist between the uniform and the distinctive symbols. Although not part of the standard, the qualifying symbols for the OR (≥ 1) and the Exclusive OR ($= 1$) occasionally appear in the literature as $+$ for the OR and XOR for the Exclusive OR.

This standard also modifies the meaning of the ''bubble.'' The bubble, as we have used it throughout this book, means that the corresponding signal is asserted low, i.e., that the signal is interpreted as a logical 1 when its voltage is low. The bubble in Standard 91–1984 and also in the earlier IEEE Std. 91–1973 indicates a logical complementation. In this interpretation, it is assumed that *all* signals in a digital system are asserted high (positive logic) or are asserted low (negative logic) but *never* some high and some low. In other words, mixed logic is not allowed. In order to indicate a signal that is asserted low, or active low, in a mixed-logic system, a new symbol is introduced. This new asserted low

---

[1] The use of this new symbology will certainly continue, since its use is currently mandatory on logic diagrams drawn for the Department of Defense.

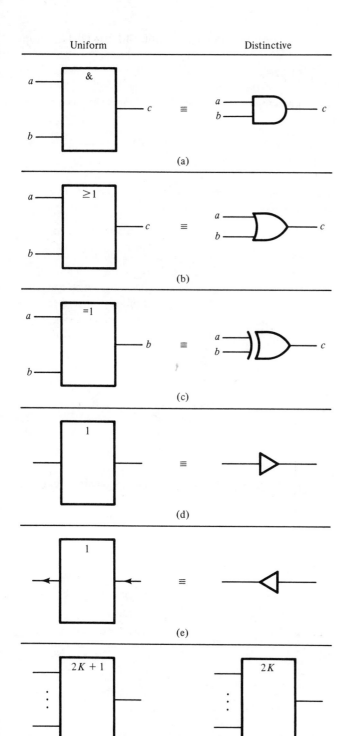

Figure A.2.1 (a–e) Equivalences between the uniform symbols and the distinctive symbols: (a) AND; (b) OR; (c) Exlusive OR (XOR); (d) buffer with left-to-right information flow; (e) buffer with right-to-left information flow. Uniform symbol labeling for odd parity (f) and even parity (g).

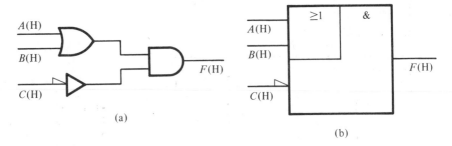

IEEE Std. 91–1984   as used in this book   **Figure A.2.2**   Assertion-level indication equivalence

indicator appears as an open ''ramp,'' or half arrow, as shown in Figure A.2.2. In what follows, we will use this symbol to indicate a signal that is asserted, or active, low.

In using the uniform symbols available in IEEE Std. 91–1984, some further economies of notation become available. For example, Figure A.2.3(a) shows a circuit, using the distinctive symbols used in this text, which realizes the function

$$f(A, B, C) = (A + B)\overline{C}$$

Figure A.2.3(b) shows the equivalent new symbol using the uniform symbols.

In Standard 91–1984, bistable devices (latches and flip-flops) are separated into four distinct categories, namely, transparent latches, edge-triggered flip-flops, pulse-triggered (master-slave) flip-flops, and data lockout flip-flops. The first three of these were discussed in Chapter 5. Figure A.2.4 shows these four flip-flop types and their interpretation in the symbology used in Chapters 4 and 5. This interpretation can be put into words as follows:

Transparent latch: The output functionally follows the input for as long as the input C is asserted.

Pulse-triggered flip-flop: The output takes on the value required of the inputs whenever the input C goes from its asserted value to its nonasserted value, or is negated. The inputs, D in this case, must not change while C is asserted or else the output may be unpredictable (refer to the discussion of the problem encountered in master-slave flip-flops caused by glitches, given in Section 5.2).

Edge-triggered flip-flops: The output takes on the value required by the inputs at the time that input C is asserted.

Flip-flops with data lockout: These flip-flops are a combination of the pulse-triggered and edge-triggered flip-flop. Basically, at the time that input C is negated, the output takes on the value that the inputs required at the time that the C input was asserted.

Figure A.2.3   Comparison of implementations of the function $(A + B)\overline{C}$: (a) distinctive symbol method; (b) uniform symbol method.

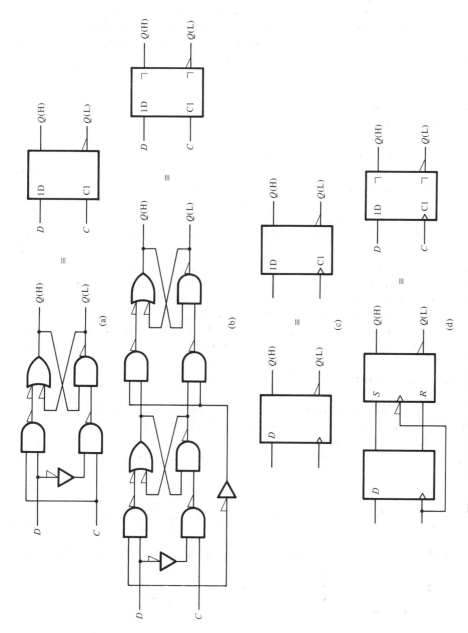

Figure A.2.4 The four basic flip-flop types, with the IEEE Std. 91-1984 symbol on the right: (a) transparent latch; (b) pulse-triggered (master-slave) flip-flop; (c) edge-triggered flip-flop; (d) flip-flop with data lockout.

Note, in these examples, that the symbol ⌐ used at the outputs of the pulse-triggered and the data lockout flip-flops simply implies that the output does not change until the input labeled C returns to its negated value after first being asserted. The labels 1D and C1, shown in the IEEE Std. 91–1984 symbols in this figure, have very specific meanings within the standard, as we shall see in the next section.

A.3 SYMBOLS FOR MEDIUM- TO LARGE-SCALE DEVICES

As mentioned above, perhaps the most important aspect of IEEE Std. 91–1984 is its ability to show the functional behavior of complex circuits with a simple symbol. The reason that the standard can accomplish this task is that it uses a special notation referred to as *dependency notation*. An example of this is the C1 and 1D labels used in the flip-flops just described. Basically, dependency notation allows the separation of the control functions from the data functions and shows, explicitly, how the control signals affect the data function. In general, identification of the controlling signal is made by a letter, indicating the dependency type, followed by a number, which is used to indicate the signal lines controlled by this controlling signal. The controlled lines are generally indicated by a number or one or more numbers followed by a letter. Thus, C1 is a controlling input of dependency type C, and 1D is the line controlled by C1, in this case, the D input of the flip-flop. In the remainder of this section, we will define some of the more commonly encountered dependency types used in the new symbology and show a number of examples of their use.

A.3.1 The G Dependency Type

The G dependency is used basically to perform a *selection function*. A very simple illustration of the use of the G dependency is found in the two-line multiplexer (MUX) shown in Figure A.3.1. The standard symbol shows a rectangle that is identified as a multiplexer by the qualifying symbol MUX. The inputs b and c are selected to appear at output d by the

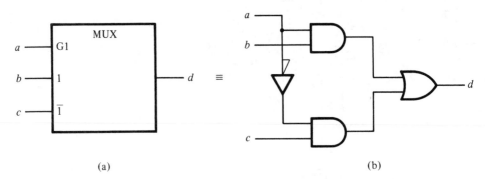

(a)   (b)

Figure A.3.1 G dependency used in a two-line multiplexer (MUX): IEEE symbol for the MUX; (b) equivalent circuit.

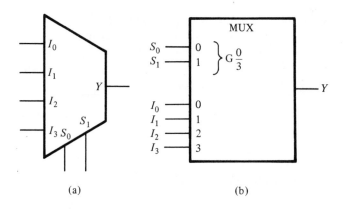

Figure A.3.2 Expanded *G* dependency in a four line MUX. (a) Block diagram symbol introduced in Chapter 4. (b) IEEE symbol.

(a)   (b)

value of the selection input, *a*. The IEEE equivalent is formed by labeling input *a* with a G followed by a number, 1 in this case, which identifies the inputs affected by this signal. Thus, if the selection input *a* is 1, then output *d* takes on the value of input *b*; and if input *a* is $\bar{1}$ (=0), then output *d* takes on the value of input *c*.

In Chapter 4 we designed a four-line MUX, as shown in Figure 4.3.8. Figure A.3.2 shows the corresponding IEEE symbol. Note that in this case, there are two selection inputs that can be used to select one of four signals. These inputs, S_0 and S_1, are labeled 0 and 1, to indicate the powers of 2 used in the coding and a "G" followed by a series of numbers, 0 through 3, which identify the signals affected by these G inputs. The selected inputs are labeled in accordance with this selection code.[2]

A demultiplexer is basically the opposite of the multiplexer. By understanding the *G* dependency just given, we should be able to interpret the operation of the demultiplexer (DMUX) shown in Figure A.3.3. This symbol states that the selection inputs, *a* and *b*, are used to choose one of the outputs, *d*, *e*, *f*, or *g*, to take on the value of input *c*. In this case, note further that the outputs are all asserted low!

Before examining the next type of dependency, let us look at one more example of the use of these symbols. A 74LS138 is an eight-line decoder-demultiplexer that has three input lines that are ANDed together to form the signal that will appear at the selected output. Figure A.3.4 shows the IEEE symbol for this device. In this case, the symbol is a compound symbol made up of a large rectangle for the DMUX that encloses a smaller rectangle showing the AND operation. As an example of the interpretation of this symbol, if the inputs (*c*, *b*, *a*) = (1, 1, 0), then output 6 (*m* in the figure), which is the decimal equivalent of 110, will take on the value of the AND of inputs *d*, *e*, and *f*, that is, *m* = *def*. All of the other outputs will be negated: in this case, they will be high. Note also that the outputs and the two inputs *d* and *e* are all asserted low.

---

[2] Although the numbers selected here correspond to the encoded values of inputs S_1 and S_0, this need not be the case. In general, these can be any four successive digits which are then associated with the encoded values of the inputs.

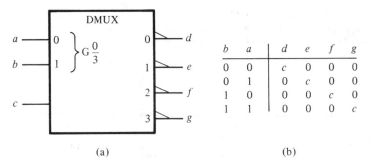

b	*a*	*d*	*e*	*f*	*g*
0	0	*c*	0	0	0
0	1	0	*c*	0	0
1	0	0	0	*c*	0
1	1	0	0	0	*c*

(a)   (b)

Figure A.3.3 The IEEE standard symbol (a) describing a DMUX. (b) DMUX truth table.

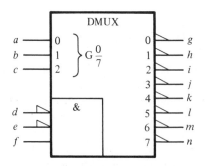

Figure A.3.4 The IEEE symbol for the 74LS138 eight-line demultiplexer.

A.3.2 The EN Dependency Type

The EN dependency is used to *enable* the functioning of a device or a set of lines, usually only outputs. For example, suppose we wish to add an additional input signal CS to the MUX shown in Figure A.3.2 that controls the output. If CS = 0, then the output will be 0 regardless of the inputs; and if CS = 1, the device will serve its normal function as a MUX. Figure A.3.5 shows the resulting symbol and a simple equivalence.

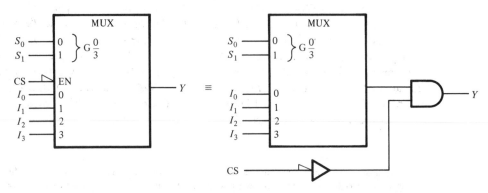

Figure A.3.5 A four-line MUX with device enable CS.

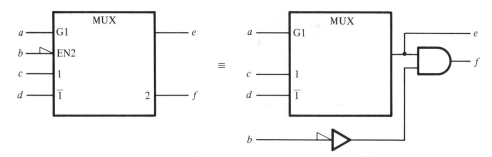

Figure A.3.6   A two-line MUX with an enable for one of the outputs.

In general, if an EN input is not followed by a number, it is assumed to affect all outputs. If the EN input is followed by a number, then it affects only those outputs which carry the same number. For example, Figure A.3.6 shows a two-line MUX having two outputs, e and f, which are identical if input b is 1 (low, in this case). On the other hand, if input b is 0, then output f will be 0 regardless of the other inputs, whereas output e will take on the required value of input c or d, depending on the value of the select input a.

A.3.3 The Common-Control Block

Before examining some of the other common dependency types, let us look at how we can put together compound symbols having common control. Suppose we wish to show a circuit having two 2-input MUXs with a common select and enable. Figure A.3.7(a) shows a symbol for this circuit that is composed of two parts. The "spade-shaped" symbol at the top of this drawing, referred to as the *common-control block*, is used to show the control signals that are common to the two multiplexers. The MUXs are indicated by the two rectangles stacked below the control block. Only the top MUX symbol shows the dependency, since the other MUX is assumed to be identical. Figure A.3.7(b) shows the physical implementation of this compound MUX.

In this figure, the qualifying symbol was placed in the common-control block. This was done because each of the blocks controlled had the same function. It can happen that there is control common to dissimilar functional blocks. For example, Figure A.3.8 shows a circuit that consists of a MUX and a DMUX both controlled by the same set of input signals. In this case, the qualifying symbols are shown in their corresponding blocks.

A.3.4 The C Dependency Type

The C dependency is used to identify a *control function*, usually associated with flip-flop operation. We have already encountered this function in the flip-flop symbols shown in Figure A.2.4. Basically, whenever a control input C is active, all of the inputs, functions, and outputs dependent on this signal perform their required function. For example, consider the edge-triggered flip-flop shown in Figure A.2.4(c). When the C1 input is asserted (goes

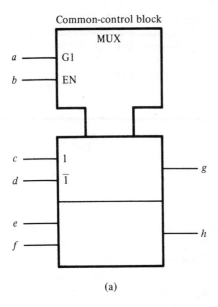

(a)

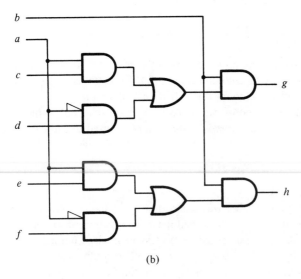

(b)

Figure A.3.7 Two 2-line MUXs with common select and enable control. (a) IEEE symbol; (b) equivalent circuit.

from a low to a high), the flip-flop functions by passing the controlled input, 1D, to the output.

In Section 9.2, we introduced the idea of a register. Figures 9.2.2 and 9.2.3 show symbols for an 8-bit storage register. The IEEE standard symbol corresponding to this storage register is shown in Figure A.3.9. This symbol is identified as a storage register by the qualifying symbol RG shown in the control block. Note the "plain English" specification for the purpose of the Load input found in square brackets in the common-

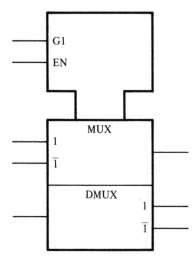

Figure A.3.8 A symbol for a MUX and a DMUX having common control.

control block. The use of these extra labels is generally a good idea, since they make the symbol's functions more quickly discernible.

Another useful register introduced in Chapter 9 was the shift register. Figure A.3.10 shows the IEEE standard symbol equivalent to the serial in–parallel out shift register discussed in Section 9.2 and shown in Figure 9.2.3. In this standard symbol, the qualifying

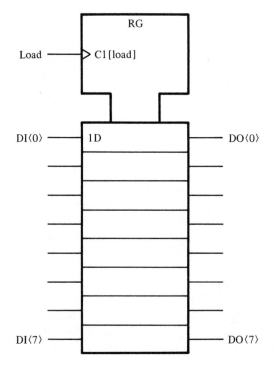

Figure A.3.9 An 8-bit storage register.

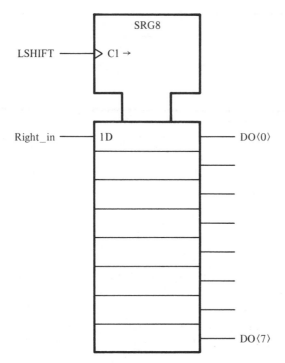

Figure A.3.10 The IEEE symbol for the 8-bit shift register of Figure 9.2.3.

symbol SRG8 is used to indicate a 8-bit shift register. The control symbol C1→ is used here to indicate that when this input goes from a low to a high (the asserted transition, in this case), the register contents are to be shifted *away* from the control block as input Right_in is loaded into the flip-flop. If the arrow had been reversed, the active transition of C1 would have caused the register to shift toward the control block. Since this device has been identified as a shift register, the internal connections from stage to stage are assumed and therefore not explicitly shown.

A.3.5 The S and R Dependency Types

In Figure 7.6.1 we showed a cmmonly encountered symbol for two popular flip-flops: the 7474 and the 74LS76. Figure A.3.11 shows the IEEE standard symbols used for these devices. In this figure, the *S* and *R* dependencies are understood to serve the asynchronous set and reset, or clear, functions described in Chapter 9. The *C* dependency was described above.

A.3.6 The M Dependency Type

Registers, counters, and other complex circuits can have several functions associated with them, as was shown by the universal shift register designed in Chapter 9. This shift register had four functions: shift left, shift right, load, and do nothing. These *modes* of operation

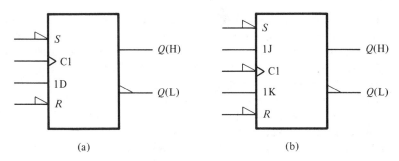

(a)   (b)

Figure A.3.11   IEEE standard symbol for (a) the 7474 and (b) the 74LS76 edge-triggered flip-flops.

were controlled by the two input control signals Load and $L/\overline{R}$. The function of the M dependency is to show the signals that control this *mode selection activity* and to show which inputs and outputs, and, perhaps, controls, are affected by these signals. Figure A.3.12 shows the IEEE symbol equivalent to the universal shift register shown in Figure 9.2.6. In this figure, we see that there are two inputs that control the mode. These are labeled 0 and 1, corresponding to inputs Load and $L/\overline{R}$, respectively. As was described for the G dependency used in Figure A.3.2, these two inputs are used to encode the modes M0 through M3. The signals which are affected by these modes are then prefixed by one of these numbers.

Most dependency is not only associated with various inputs and outputs but may be

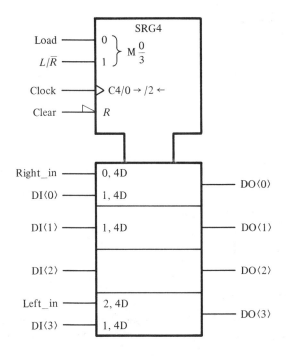

Figure A.3.12   The IEEE symbol for the universal shift register of Figure 9.2.6.

associated with other control signals as well. For example, the dependency notation that appears at the dynamic input, C4/0 →/2←, indicates some association with two of the four modes, 0 and 2 in this case. This notation consists of three parts, separated by slashes (/), and is interpreted as follows. First, the C4 indicates that this is a control input that can affect all other signals prefixed by a 4. The second part of the notation (0 →) indicates that when this control signal is active *and* mode 0 is selected, a shift *away* from the control block will be effected. Finally, the third part of the notation (2←) indicates that a shift *toward* the control block occurs on an active transition of the control input if mode 2 is selected. If neither mode 0 nor mode 2 is selected, then, although an active transition on this input will not cause a shift, some other function, such as loading, may occur. In this case, if the mode is 1, then the inputs labeled 1,4D will be active and each of the flip-flops will be loaded with the information present on its respective input. The ordering of the labels is important. In this case, mode 1 must be selected first and then control input 4 must go active. In a similar manner, the input labeled 0,4D will be active and load the top flip-flop with information on the input Right_in if mode 0 is selected (a shift away from the control block) *and* control input 4 is active. Similarly, the bottom flip-flop will be loaded with the information on input Left_in if mode 2 is selected (a shift toward the control block) *and* the control input makes an active transition. Finally, note that if mode 3 is selected, no action occurs, since no corresponding activity is shown on the control input, nor are any inputs conditioned by this mode. This is the "do nothing" mode.

Before leaving this example, we should note that the final control input *R* (Clear, in the figure) is just the asynchronous reset described earlier; it causes all four of the flip-

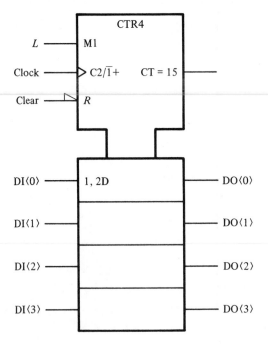

Figure A.3.13 An IEEE symbol for the presettable counter of Figure 9.2.7.

flops to reset to 0. One last comment is in order. Observe that the third flip-flop has no internal labeling. This is because it is identical to the one immediately above. In general, elements are labeled only if they differ from the elements they follow. Thus, in this example, the fourth flip-flop is labeled, since it differs from the preceding three.

Another example of the use of the M, or mode, dependency is shown in Figure A.3.13. This figure shows the standard symbol used to indicate the operation of the presettable binary counter shown in Figure 9.2.7. The 4-bit binary counter function is indicated by the qualifying symbol CTR4. If this had been a decimal counter, the qualifying symbol would have been CTRDIV10, which indicates a counter that divides by 10. In this symbol, we see that when the mode line, M1, is asserted and the dynamic input goes from a low to a high, the four flip-flops are loaded with the values appearing on their respective inputs. On the other hand, if the mode line is negated, then an active transition of the dynamic input causes the counter to increment by 1, as shown by the notation $\overline{1}+$. A counter that counts down would be indicated by the notation $\overline{1}-$. Finally, the common output shown on the control block labeled CT = 15 takes on the value 1 whenever the count reaches 15.

Figure A.3.14 shows an alternative symbol for this counter in which the mode select line has been split into two lines, labeled M1 and M2. Another difference between the symbols of Figures A.3.13 and A.3.14 is the common output element block shown at the bottom of the counter separated from the main counter body by a double line. The common

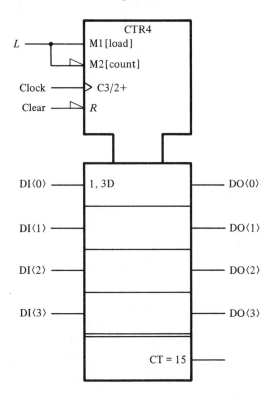

Figure A.3.14 An alternative symbol equivalent to that of Figure A.3.13.

element is used to indicate an output that is generally a function of all of the elements that appear above it.

A.3.7 The A Dependency Type

In dealing with memories such as the ROM discussed in Chapter 4, another dependency type must be introduced to indicate which character in the ROM is to be accessed. The *A* dependency is used for this function and serves to give the *address* of the required character. Figure A.3.15 shows the IEEE symbol that might be used to indicate the ROM of Figure 4.4.2. The qualifying symbol, ROM32 × 8, indicates that this is a ROM and gives its size. The five inputs *a*, *b*, *c*, *d*, and *e* form the address, which takes on values in the range 0 through 31. The *A* associated with the outputs simply indicates that the output is dependent on the value of this address.

Figure A.3.16 shows a symbol that might be used for a read-write memory (a RAM, or random-access memory). In this symbol, there are 10 address inputs, A0 through A9, which identify which character in the RAM is to be accessed. This device also has a control input, C1024, which serves the function of writing information appearing on the inputs

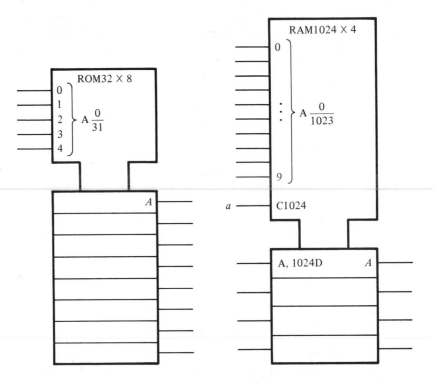

Figure A.3.15 The address dependency as used in the symbol for the 32- × 8-bit ROM of Figure 4.4.2.

Figure A.3.16 A symbol used to show a read-write memory (RAM) having 1024 four-bit characters.

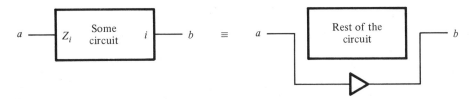

Figure A.3.17   The basic Z dependency.

into the addressed memory location. This is indicated by the notation A,1024D that appears at the inputs to the memory. *Thus, if the control input is asserted, information will be written into the addressed location. If the control input is negated, information located at the addressed location will appear at the output but will not be changed.*

A.3.8 The Z Dependency Type

The Z dependency type is used to show *interconnection*. Basically, all the Z symbol does is to identify a signal at one point in a circuit that appears at another. It simply transfers the value of the signal at the former to the latter. Figure A.3.17 shows a simple example of this usage. Although this symbol does not often appear, the reader should be aware of its presence.

A.4 SYMBOLS USED TO IDENTIFY PHYSICAL CHARACTERISTICS

We mentioned in the introduction to this appendix that a third aspect of IEEE Std. 91–1984 is a set of symbols used to identify various physical characteristics of devices. By ''physical'' characteristics, we mean electrical or electronic characteristics associated with the inputs or the outputs. Although this is not part of the subject of this text, it is useful to be aware of these symbols. It should be noted, however, that these symbols do not, in any way, change the logical interpretation of digital schematic diagrams that may be encountered; they simply add a bit more information about the electrical characteristics of the device.

   Figure A.4.1 summarizes the four most commonly encountered symbols representing physical attributes of logical devices. The first symbol, shown as Figure A.4.1(a), shows the symbol, that would appear at the output of an *open-collector* device. Open-collector outputs were introduced in Section 4.2, and an example was shown in Figure 4.2.18. Figure A.4.1(b) shows the symbol for a *tri-state* output. The output of a tri-state device has three values: a high voltage, a low voltage, and a disconnected value. The disconnected value is equivalent to a wire that is connected to nothing, at least on this end, the device output end. The third commonly encountered output characteristic is extra drive capability. An output with this capability can drive, or serve as the input to, more devices than would normally be possible. The symbol for this capability is shown in Figure A.4.1(c). Finally,

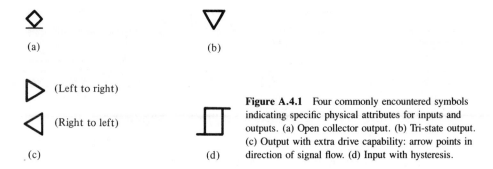

Figure A.4.1   Four commonly encountered symbols indicating specific physical attributes for inputs and outputs. (a) Open collector output. (b) Tri-state output. (c) Output with extra drive capability: arrow points in direction of signal flow. (d) Input with hysteresis.

a characteristic of inputs that is very important in many applications is that of *hysteresis*. A device with hysteresis has the capability of responding to two different threshold voltages at the input, choosing one or the other depending on whether the input is going from low to high or vice versa. This hysteresis, or bi-threshold, effect is exactly what is encountered in the household thermostat. In this case, the furnace "kicks on" when the ambient temperature falls slightly below the temperature set on the thermostat and "kicks off" when the ambient temperature rises slightly above the value set. Circuits of this type were introduced in Problem 7.1.

There are other symbols that are part of the IEEE standard to show input/output physical attributes which are encountered on occasion. These may be found in the references cited in the bibliography.

A.5 AN ANNOTATED BIBLIOGRAPHY

No attempt has been made to be comprehensive in the discussin of the IEEE Std. 91–1984 symbology presented in this appendix. However, the most commonly encountered dependency notation and usage have been introduced. A very nice booklet by Mann (1987) gives many more details and shows a large number of examples. Mann has also produced a small pamphlet (1984) which summarizes the IEEE standard.

MANN, F. A., "Overview of IEEE Std. 91–1984: Explanation of Logic Symbols," Texas Instruments, Inc., Carrollton, Tex., Publ. SDYZ001, 1984.

MANN, F. A., "Using Functional Logic Symbols: Application of IEEE Std. 91–1984," Texas Instruments, Inc., Carrollton, Tex., Publ. SZZZ003, 1987.

A copy of this standard, entitled *Standard: 091–1984 Graphics Symbols for Logic Functions*, can be obtained by writing the IEEE at the following address:

Institute of Electrical and Electronics Engineers, Inc.
345 East 47th Street
New York, New York 10017

Finally, a recent book dealing with logic design by McCluskey uses this symbology throughout. This book gives a number of practical design problems and many examples of the usage of these new IEEE symbols.

McCLUSKEY, E. J., *Logic Design Principles with Emphasis on Testable Semicustom Circuits*, Prentice-Hall, Englewood Cliffs, N.J., 1986.

Index